Mathematics of Networks

Mathematics of Networks: Modulus Theory and Convex Optimization is a book that seeks to answer the question: "What can be learned by adapting the theory of p-modulus (and related continuum analysis concepts) to discrete graphs?" As the book leads readers through the rich landscape of p-modulus on graphs, it is shown that the theory ties together many beautiful concepts from graph theory, probability, and convex optimization.

The book ought to be of interest to anyone seeking to deepen their understanding of the theoretical underpinnings of network analysis and applied graph theory. As such, it can be used as either a primary text or a useful reference for graduate courses on topics related to network analysis and applied graph theory, and even for some advanced undergraduates.

Features

- Accessible to anyone with a strong grasp of multivariable calculus and linear algebra
- Broad interdisciplinary appeal
- Numerous engaging exercises

Nathan Albin received a BA in Mathematics from the University of Hawaii and Hilo in 2001. Following a short stint as a software engineer, he attended graduate school at the University of Utah and received a PhD in Mathematics in 2006. He was then awarded an NSF Mathematical Sciences Postdoctoral Research Fellowship, hence spent one year researching at the University of Duisburg-Essen, before moving to Pasadena, CA to complete his postdoctoral research at Caltech. He was hired as an Assistant Professor of Mathematics in 2011, was promoted to Associate Professor in 2016, and to Full Professor in 2022. His primary research interests include Applied Mathematics, Network Theory, Optimization Theory, and Numerical Analysis.

Pietro Poggi-Corradini grew up in Milan, Italy, and in 1992 received a Bachelor in Mathematics from the University of Milan. He then pursued his studies at the University of Washington in Seattle (USA) and graduated in 1996 with a PhD Thesis in Complex Dynamics and Operator Theory under Professor Don Marshall. Thereafter he became a Whyburn Instructor at the University of Virginia and in 1998 joined the faculty at Kansas State University. He has been a full Professor at Kansas State University since 2005. His interests currently span Network Theory, Complex Analysis, Probability, and Applied Mathematics.

Mathematics of Networks

Modulus Theory and Convex Optimization

Nathan Albin and Pietro Poggi-Corradini

CRC Press
Taylor & Francis Group
Boca Raton London New York

CRC Press is an imprint of the
Taylor & Francis Group, an **informa** business
A CHAPMAN & HALL BOOK

Designed cover image: Getty Images

First edition published 2026
by CRC Press
2385 NW Executive Center Drive, Suite 320, Boca Raton FL 33431

and by CRC Press
4 Park Square, Milton Park, Abingdon, Oxon, OX14 4RN

CRC Press is an imprint of Taylor & Francis Group, LLC

ISBN: 978-0-367-45707-5 (hbk)
ISBN: 978-1-041-08081-7 (pbk)
ISBN: 978-1-003-02486-6 (ebk)

DOI: 10.1201/9781003024866

Typeset in LM Roman
by KnowledgeWorks Global Ltd.

Publisher's note: This book has been prepared from camera-ready copy provided by the authors.

To Courtney & Maia.
To Emmily, Anna, Giovanni, Francesca, & Carlo.

Contents

Preface

THIS BOOK grew from a collection of research notes and lecture notes written by the authors over the past decade in an attempt to answer the intriguing question: "What can be learned by adapting the theory of p-modulus (and related continuum analysis concepts) to discrete graphs?" As we explored the rich landscape of p-modulus on graphs, it became increasingly clear that the theory ties together many beautiful concepts from graph theory, probability, and convex optimization. Indeed, one of us has been overheard, on more than one occasion, declaring that "everything is modulus". Though this conjecture is yet unproven, we do believe there is sufficient evidence to support the claim that many interesting things are indeed modulus.

AUDIENCE

We hope that this book can serve as either a primary or reference textbook for graduate courses on topics related to network analysis and applied graph theory. While primarily intended for graduate students, it is likely accessible to a talented undergraduate with a strong background in multivariable calculus and linear algebra. The subject matter is drawn from the interests of the authors' research group on Network Optimization, Design, and Exploration (NODE), which was founded in 2015. This interdisciplinary group has attracted faculty, graduate students, and undergraduate students from mathematics, statistics, computer science, and electrical and computer engineering. As such, we believe this book will be of interest broadly to anyone seeking to deepen their understanding of the theoretical underpinnings of network analysis and applied graph theory.

STRUCTURE

The book is divided into several sections. Section I focuses on establishing the mathematical notation and theory for the various interconnecting subjects to be woven together, including the essentials of graph theory, electrical networks, spectral graph theory, and random walks. Section II provides a short but self-contained introduction to the theory of convex optimization. It is by no means intended to be comprehensive but contains sufficient detail to provide context and background for understanding the theory of modulus. Section III provides a comprehensive treatment of the fundamental theory of modulus on graphs, offering a basic understanding of what it means to talk about the modulus of a family of objects on a graph. Sections IV and V then discuss special features of the theory as applied to specific families. Section IV applies the theory to families of walks, which generalizes important concepts in graph metrics and electrical networks. Section V applies the theory to the family of spanning trees, yielding connections to several fascinating combinatorial concepts. Finally, Section VI provides a brief overview of the algorithms and computational techniques that can be used to experiment with modulus.

ACKNOWLEDGMENTS

The content of this book was built organically over the years. In writing it, we have drawn on lecture notes (handwritten and typeset), research manuscripts (published and unpublished), Jupyter notebooks, and recollections of discussions at various blackboards. This book has been influenced by the questions, suggestions, and feedback of many people along the way. We are truly grateful to everyone who has contributed in some fashion to this project. The list includes numerous graduate students, undergraduate students, postdocs, colleagues, collaborators, friends, and family. We are also grateful to the National Science Foundation for the research grants that have allowed us to explore the beautiful theory of modulus on graphs.

Nathan Albin
Pietro Poggi-Corradini

I

The Mathematics of Networks

CHAPTER 1

Introduction to Graph Theory

THIS IS A QUICK REVIEW of basic graph theory preliminaries needed in the rest of the book. For a more in-depth study of graph theory, we refer the reader to standard references [19, 34].

1.1 GRAPHS AND NETWORKS

A *graph* is a pair $G = (V, E)$ where V is the *vertex set* and E is the *edge set*. In general, an edge will indicate how two vertices are connected. A graph is *finite* if V and E have finite cardinality. In practice, this corresponds to labeling the vertices and edges in some way. For example, V is often taken to be $[N] := \{1, \ldots, N\}$.

We will use the terms 'graph' and 'network' interchangeably, as well as the terms 'vertex' and 'node', and 'edge' and 'link'. Also, we will mostly deal with finite undirected simple graphs. In this case, an edge can be thought of as an unordered pair of vertices, i.e., a subset of V of size 2. We denote the set of all such pairs $\binom{V}{2}$. In particular, the cardinality of the edge set E in this case is $|E| := m \leq \binom{N}{2} = N(N-1)/2$. This basic graph structure can be generalized in various ways. For example, choosing ordered pairs of vertices gives rise to *directed* graphs. If we allow *self-loops* (edges connecting a node to itself) and *multi-edges* (several edges connecting the same two nodes), then G is called a *multigraph*.

We say that two vertices x, y are *neighbors* and write $x \sim y$ if and only if $\{x, y\} \in E$. Given an integer $n \geq 1$, a string

$$\gamma := x_0 \; e_1 \; x_1 \; e_2 \; x_2 \cdots e_n \; x_n$$

DOI: 10.1201/9781003024866-2

with $x_i \in V$ for $i = 0, \ldots, n$ and $e_k = \{x_{k-1}, x_k\} \in E$, for $k = 1, \ldots, n$, is called a *walk (with n hops) from x_0 to x_n*. For simplicity, we will sometimes only list the vertices visited by the walk and write $\gamma = (x_1, \ldots, x_n)$. In the sequel, we will focus on the sequence of edges $e_1\ e_2 \cdots e_n$ that a walk visits. Indeed, to each walk γ we associate a *usage function* $\mathcal{N}(\gamma, \cdot) : E \to \mathbb{N}$ such that

$$\mathcal{N}(\gamma, e) = \text{ the number of times the walk } \gamma \text{ traverses the edge } e. \quad (1.1)$$

We think of $\mathcal{N}(\gamma, \cdot)$ as a row vector of length $|E|$ with integer non-negative entries. In particular, a family of walks Γ gives rise to a matrix $\mathcal{N}(\Gamma, \cdot)$ of size $|\Gamma| \times |E|$ and at times we will also think of Γ as a subset of $\mathbb{N}^{|E|} \subset \mathbb{R}^{|E|}$. For example, given two nodes $x, y \in V$, we let $\Gamma(x, y)$ consist of all walks that start at x and end at y. We say that $\Gamma(x, y)$ is a *connecting* family of walks. Note that this family will often be infinite. We write $\gamma : x \rightsquigarrow y$ to mean that γ is a walk connecting x to y. The graph G is *connected* if for any two vertices $a, b \in V$ there is a walk from a to b.

A walk that does not revisit any vertex is called a *(simple) path*. In particular, a path γ does not revisit edges either (the converse does not hold). So, in this case, the corresponding row vector is $\mathcal{N}(\gamma, \cdot)$ in $\{0, 1\}^E$. We let $\Gamma_{\text{path}}(x, y)$ be the family of all paths starting at x and ending at y. Note that this family is always finite on a finite graph.

Given a walk $\gamma = x_0\ e_1\ x_1\ e_2\ x_2 \cdots e_n\ x_n$, we define its *length* $\ell(\gamma)$ to be n, that is, the number of hops γ takes to go from x_0 to x_n. In other words,

$$\ell(\gamma) = \sum_{e \in E} \mathcal{N}(\gamma, e), \quad (1.2)$$

is the sum of the row corresponding to γ. Given two nodes $x, y \in V$, we let the *shortest path distance* between x and y be the length of the shortest path from x to y.

$$d(x, y) := \inf_{\gamma : x \rightsquigarrow y} \ell(\gamma). \quad (1.3)$$

Note that on a connected undirected graph G, the shortest walk from x to y is necessarily a path.

The number of edges that are incident at a vertex x is the *degree* of x and we indicate it as $\deg(x)$. The graph G is *d-regular*, if $\deg(x) \equiv d$, for $x \in V$, for some integer $d \geq 2$. Every edge is incident to two distinct vertices, contributing to two local degrees. Therefore

$$\sum_{x \in V} \deg(x) = 2|E|. \quad (1.4)$$

This identity is sometimes referred to as the *Handshake Lemma*. It says that instead of counting edges, one can add degrees, i.e., switch to $\deg(x)$ which is a function defined on V.

Given a subset of vertices $V' \subset V$, we let $E(V') \subset E$ be all the edges of G that connect pairs of vertices in V'. With this notation $G(V') = (V', E(V'))$ is a simple graph, the *subgraph induced by* V'. More generally, a *subgraph* of G is a graph $H = (V', E')$ such that $V' \subset V$ and $E' \subset E(V')$.

Remark 1.1. One can define a *curve* $\alpha = (V(\alpha), E(\alpha))$ to be a connected subgraph of G. For instance, the *trace* of a walk γ is the curve α consisting of the vertices and edges traversed by γ. Conversely, any curve α is the trace of some walk. In either case, $|E(\alpha)|$ is smaller than the number of hops γ takes. Notice also that there are finitely many curves but infinitely many walks on a finite graph.

Definition 1.1. Let $G = (V, E)$ be a finite simple graph with vertex set V and edge set E. An *orientation! of a graph* is a choice $\vec{e}$ for each edge $e = \{x, y\}$, which equals either the ordered pair (x, y) or the ordered pair (y, x). We denote an orientation as $\vec{E}$ and write $\vec{E}_R$ for the opposite orientation, where all the edges are reversed. We also define

$$\hat{E} := \{(x, y) : \{x, y\} \in E\} \quad (= \vec{E} \cup \vec{E}_R),$$

to be the set of all oriented edges given E.

1.2 TREES

Definition 1.2. A walk $\gamma = x_0\ e_1\ x_1\ e_2\ x_2 \cdots e_n\ x_n$ where $x_0\ e_1\ x_1\ e_2\ x_2 \cdots e_{n-1}\ x_{n-1}$ is a path, with $n \geq 3$, and where $x_n = x_0$, is called a *cycle*. If we drop the condition that $x_0\ e_1\ x_1\ e_2\ x_2 \cdots e_{n-1}\ x_{n-1}$ is a path, then we say that γ is a *circuit* or a *loop*.

Theorem 1.1. *Let $G = (V, E)$ be a finite connected simple graph. Then the following are equivalent:*

(1) G contains no cycles;

(2) every pair of vertices is connected by a unique path;

(3) $|E| = |V| - 1$;

(4) every edge is a cut edge, *i.e., removing it disconnects the graph.*

Definition 1.3. A graph $G = (V, E)$ is a *tree* if one of the conditions (1)–(4) in Theorem 1.1 holds.

Proof of Theorem 1.1. We show:
(1)⇒(2) It's easier to prove the contrapositive $\sim (2) \Rightarrow \sim (1)$. Hence, assume that there are a pair of nodes x and y, and two distinct paths γ_1 and γ_2 from x to y. Without loss of generality γ_1 and γ_2 share the first $k-1$ hops: $x\ e_1\ x_1 \cdots x_{k-1}$, but then γ_1 hops to x_k while γ_2 hops to $x'_k \neq x_k$. Follow γ_1 and γ_2 until they meet again, creating a cycle.
(2)⇒(3) This can be done by induction on the size N of V. When $N = 2$, then G is the path graph P_2 and $|E| = 1 = 2 - 1$. So, assume the claim holds when $|V| = N$. We want to show that it holds for $|V| = N + 1$. Pick an arbitrary node x_0 and start an exploratory walk, making sure not to revisit any node, if possible. Since G is finite, this process will terminate after n steps and will create a path $x_0\ e_1\ x_1\ e_2 \cdots x_{n-1}\ e_n\ x_n$. We claim that we must have $\deg(x_n) = 1$. Indeed, suppose that $y \sim x_n$ and $y \neq x_{n-1}$. If y is a previously visited node in $\{x_0, \ldots, x_{n-2}\}$, then x_0 and x_n can be connected by two different paths, which is impossible. If y was not previously visited, we could add the edge $\{x_n, y\}$ to extend the exploratory path, which is also impossible since we assumed the process had terminated. Remove the vertex x_n and the edge e_n leading to it. The remaining graph G_1 still satisfies property (2). By induction hypothesis, $|E(G_1)| = |V(G_1)| - 1$. Therefore $|E(G)| = |V(G)| - 1$.
(3)⇒(4) We give a proof *by smallest counterexample* based on the size N of V. Assume first that $N = 2$, then G is the path graph P_2. In this case, $|E| = 1 = |V| - 1$, and the only edge of P_2 is a cut edge. Let $N_0 \geq 3$ be the smallest integer such that the implication holds if $|V| = N_0 - 1$, but not if $|V| = N_0$. Suppose that $G = (V, E)$ is a graph with $|V| = N_0$ that satisfies property (3) but not (4). Let $e_0 \in E$ be an edge of G that is not a cut edge. Then, by the Handshake Lemma (1.4),

$$\sum_{x \in V} \deg(x) = 2|E| = 2(N_0 - 1).$$

Since G is connected, we have $\deg(x) \geq 1$ for every $x \in V$. So

$$1 \leq \frac{1}{N_0} \sum_{x \in V} \deg(x) = 2\left(1 - \frac{1}{N_0}\right) < 2.$$

This means at least one node $x \in V$ with $\deg(x) = 1$. Removing x and the edge e_1 leading to x creates a graph H on $N_0 - 1$ vertices that also satisfies (3). We claim that H also does not have property (4). Indeed,

e_1 is a cut edge of G, so we must have $e_1 \neq e_0$. Moreover, removing e_0 does not disconnect G and hence cannot disconnect H. This contradicts the minimality of N_0.
(4)⇒(1) By contradiction, if G does contain a cycle, then removing an edge along this cycle does not disconnect the graph. So, at least one edge is not a cut edge. □

Corollary 1.1. *Let $G = (V, E)$ be a simple connected graph with $|V| = N$. Then*

$$N - 1 \leq |E| \leq \frac{N(N-1)}{2}.$$

The lower bound is attained when G is a tree, while the upper bound is attained by the complete graph on N vertices K_N (the graph where any two pairs of vertices are connected).

For the proof of Corollary 1.1, we refer to Exercise 1.7.

A tree T is a *spanning tree* for G if it shares the same vertex set, $V(T) = V$, and T is a subgraph of G, i.e., $E(T) \subset E$.

Suppose that $T = (V(T), E(T))$ is a tree. Let H_2 be the graph obtained by removing an edge from T. Note that this separates the nodes into two components, V_1 and V_2 (the connected components of each endpoint of the edge that was removed). By Theorem 1.1, for $j = 1, 2$, the induced subgraph on V_j is a tree T_j. Continuing this way, after removing $k-1$ edges, we get a graph H_k on V consisting of exactly k components that are trees. This is what is known as a forest. In general, a graph $F = (V, E)$ that contains no cycles or such that every edge is a cut edge and for which $|E| = |V| - k$ for some $k = 2, 3, \ldots$, is called a *forest*.

Counting all the spanning trees of a given connected graph is difficult. A general formula in terms of the spectral properties of G exists and is known as the Matrix-Tree Theorem; see Section 17.5. There are known explicit formulas in special cases, such as the complete graph K_N.

Theorem 1.2 (Cayley's Theorem). *The complete graph K_N contains exactly N^{N-2} spanning trees.*

Proof. Label the vertices V of K_N with the numbers $\{1, \ldots, N\}$. Let $\mathcal{T}$ be the set of all spanning trees of K_N. Given a spanning tree, there are N ways to choose a distinguished vertex called a *root*. Let $\mathcal{F}_{N,1}$ be the set of rooted spanning trees. Then, $|\mathcal{F}_{N,1}| = N|\mathcal{T}|$. In a rooted tree, by property (2) of Theorem 1.1, each edge $\{x, y\}$ has an endpoint that is closer to the root in terms of hop length. Assuming that x is closer without loss of generality, we orient the edge from x to y, i.e., away from the root.

More generally, a spanning rooted forest for K_N is a forest F on the vertices V such that every connected component of F has a distinguished root. Write $\mathcal{F}_{N,k}$ for the set of all spanning rooted forests of K_N with k components. Note that $\mathcal{F}_{N,1}$ is the set of all rooted trees of K_N, while $\mathcal{F}_{N,N}$ contains only one element, namely the edge-less graph on V, thought as the forest $F_\emptyset$ with N components where every tree is an isolated root. For each forest F, we orient all the edges in a given rooted tree component to point away from the root. We say that a forest F contains a forest F' if they both have vertex set V and every oriented edge in F' is also an edge in F with the same orientation.

We will construct rooted trees one edge at a time using an algorithm that, at step k, creates a rooted forest $F_k \in \mathcal{F}_{N,N-k}$ that contains the previous rooted F_{k-1}. In particular, the number of components $N-k$ steadily decreases until, after $k = N-1$ steps, we obtain a rooted tree.

1. Start with the edgeless forest on V, i.e., Set $F_0 = F_\emptyset \in \mathcal{F}_{N,N}$.
2. For $k = 1$ to $N-1$:
   (a) Pick a node $x \in V$ uniformly.
   (b) Let $C(x)$ be the connected component of x in the forest $F_{k-1} \in \mathcal{F}_{N,N-(k-1)}$.
   (c) Since F_{k-1} has $N-k+1$ components, it must have $N-k$ other components $\{C_j\}_{j=1}^{N-k}$ aside from $C(x)$. Denote by y_j the root of component C_j. Choose j_0 uniformly among $j = 1, \ldots, N-k$ and set $y := y_{j_0}$.
   (d) Add the oriented edge (x, y) to F_{k-1} so as to merge $C(x)$ and C_{j_0} to form a new rooted tree F_k.

The algorithm ends when $k = N-1$. Then, $F_{N-1} \in \mathcal{F}_{N,1}$ is a rooted tree on V. This process can be reversed by starting with a rooted tree in $\mathcal{F}_{N,1}$ and removing a directed edge (x, y) at a time. Since each edge is a cut edge, each time an edge is removed, this creates one more component in the forest. To keep rootedness throughout, it is enough to set y to be the root of the new component containing y. With this method, the edges of a rooted tree can be removed one at a time in arbitrary order. Hence, every rooted tree can be the output of the algorithm.

Furthermore, note that the algorithm makes N arbitrary choices in (2)(a) and then $N-k$ independent choices in (2)(c), for a total of $N^{N-1}(N-1)!$ choices. This, however, is an over-count. Each output rooted tree is the result of adding one edge at a time, but the order in which these edges are added doesn't matter. So the algorithm creates $(N-1)!$ copies of the same rooted tree. Therefore, there are N^{N-1} rooted trees. And thus, there are N^{N-2} spanning trees in K_N. □

1.3 DIJKSTRA'S ALGORITHM

For this section, we follow the notes in [77]. Given a graph $G = (V, E)$, consider positive edge weights $\rho : E \to (0, \infty)$ that we interpret as determining edge lengths. This allows us to compute the length of a path or walk. Namely, if $\gamma = x_0\ e_1\ x_1 \cdots e_n\ x_n$ and $\{x_{j-1}, x_j\} = e_j \in E$ for $j = 1, \ldots, n$, then the *ρ-length* of γ is

$$\ell_\rho(\gamma) := \sum_{j=1}^{n} \rho(e_j) = \sum_{e \in E} \mathcal{N}(\gamma, e)\rho(e),$$

where $\mathcal{N}(\gamma, e)$ is the usage function introduced in (1.1).

If we fix vertices $a, x \in V$, we may define

$$d_\rho(a, x) := \inf_{\gamma \in \Gamma(a,x)} \ell_\rho(\gamma), \tag{1.5}$$

where $\Gamma(a, x)$ is the family of all walks γ starting at a and ending at x.

This is known as the *shortest path metric*. Recall the following general definition of the metric axioms.

Definition 1.4. A *metric* (or *distance*) on V is a function $d : V \times V \to [0, \infty)$ with the following properties:

1. **Non-degeneracy:** $d(a, b) = 0 \iff a = b$.
2. **Symmetry:** $d(a, b) = d(b, a)$.
3. **Triangle Inequality:** For every $a, b, c \in V$:

$$d(a, b) \leq d(a, c) + d(c, b).$$

Remark 1.2. Given $G = (V, E)$, the shortest path distance defined in (1.5) is a metric on V (see Exercise 1.9). Moreover, a walk in $\Gamma(a, x)$ that is not a path will contain a shorter subpath. So we also have

$$d_\rho(a, x) = \min_{\gamma \in \Gamma_{\text{path}}(a,x)} \ell_\rho(\gamma).$$

In particular, $d_\rho(a,x)$ is always attained by a *shortest path* $\gamma_{ax} \in \Gamma_{\text{path}}(a,x)$.

We are interested in an algorithm that, given a triple (G,a,ρ) as above, will, for every $x \in V$, compute $d_\rho(a,x)$ and also a shortest path $\gamma_{ax} \in \Gamma(a,x)$ such that $\ell_\rho(\gamma_{ax}) = d_\rho(a,x)$. Since G is finite, the infimum in (1.5) is always attained for some shortest path $\gamma \in \Gamma(a,x)$, not necessarily unique.

Definition 1.5. Given $a, x \in V$, with $x \neq a$, a vertex $y \in V$ is a *predecessor* of x if

$$d_\rho(a,x) = d_\rho(a,y) + \rho(y,x). \tag{1.6}$$

Lemma 1.1. *Assume that* $\gamma_{ax} = a\ e_1\ x_1 \cdots x_{n-1}\ e_n\ x_n$ *with* $x_n = x$ *is a shortest path in* $\Gamma(a,x)$*. Then,* x_{n-1} *is a predecessor of* x*.*

Proof. We want to show that x_{n-1} satisfies the following identity:

$$d_\rho(a,x) = d_\rho(a,x_{n-1}) + \rho(x_{n-1},x). \tag{1.7}$$

By the triangle inequality, we have:

$$d_\rho(a,x) \le d_\rho(a,x_{n-1}) + d_\rho(x_{n-1},x) \le d_\rho(a,x_{n-1}) + \rho(x_{n-1},x). \tag{1.8}$$

On the other hand, since γ_{ax} is a shortest path,

$$d_\rho(a,x) = \ell_\rho(\gamma_{ax}) = \rho(a,x_1) + \cdots + \rho(x_{n-2},x_{n-1}) + \rho(x_{n-1},x). \tag{1.9}$$

Plugging this into (1.8) and simplifying, we get

$$\rho(a,x_1) + \cdots + \rho(x_{n-2},x_{n-1}) \le d_\rho(a,x_{n-1}). \tag{1.10}$$

In other words, the path $\tilde{\gamma} = a\ e_1\ x_1 \cdots e_{n-2}\ x_{n-1}$ is a shortest path for $d_\rho(a,x_{n-1})$ and equality holds in (1.10). So, plugging this back into (1.9), we see that (1.7) holds. □

We now describe a famous algorithm known as *Dijkstra's algorithm*. During each pass in Dijkstra's algorithm, three objects are updated:

(1) A function $\delta : V \to [0,\infty]$ where $\delta(x)$ gives a better and better approximation of $d_\rho(a,x)$.

(2) A function $\text{pred}() : V \to V \cup \{\emptyset\}$, where $\text{pred}(a) = \emptyset$ and $u = \text{pred}(v)$ means that the current best guess for a shortest path from a to v reaches u immediately before reaching v.

(3) A growing set W of vertices that won't be revisited.

Throughout we will have $\delta(x) \geq d_\rho(a, x)$ for all $x \in V$, with equality if $x \in W$. Moreover, the set W will grow one vertex at a time by minimizing the function δ over the complement of W. Dijkstra's algorithm can be implemented as in Algorithm 1.

Algorithm 1 Dijkstra's algorithm

```
Require: (G, ρ, a)
W ← ∅
δ(a) ← 0
δ(u) ← ∞ and pred(u) ← ∅ for all u ∈ V \ {a}.
while There exists u ∉ W satisfying δ(u) < ∞ do
    u ← arg min{δ(u) : u ∉ W}
    W ← W ∪ {u}
    for Each v ∉ W that neighbors u do
        if δ(u) + ℓ(u, v) < δ(v) then
            δ(v) = δ(u) + ℓ(u, v)
            pred(v) ← u.
        end if
    end for
end while
```

Theorem 1.3. *After each pass of Dijkstra's algorithm:*

(i) For every $x \in V$, $\delta(x) \geq d_\rho(a, x)$.

(ii) For all $x \in W$, $\delta(x) = d_\rho(a, x)$.

(iii) For each $x \in W$, the function pred() *is determined once and for all, and it can be iterated to obtain a shortest path $\gamma(x)$ in $\Gamma(a, x)$.*

When the algorithm ends, if $W = V$, then Theorem 1.3 (iii) provides a distance and a shortest path to a for every $x \in V$. Alternatively, assume there is a point $x \in V \setminus W$. A crucial observation is that at the end of a loop, any node $v \in V \setminus W$ with a neighbor in W will necessarily satisfy $\delta(v) < \infty$. Therefore, x must belong to a different connected component than a, and we must have $\delta(x) = +\infty$.

Proof of Theorem 1.3. We first show (i). We reason by induction with respect to each loop of the algorithm. For the base case, at the end of the first pass, $\delta(a) = 0 = d_\rho(a, a)$ and $\delta(x) = +\infty > d_\rho(a, x)$ for every other vertex $x \in V \setminus \{a\}$. Assume the claim still holds after k passes. We only need to consider those vertices x for which $\delta(x)$ gets updated on the next pass. These are the neighbors of the minimizer u^* obtained in line 5 of Algorithm 1. And if the updating occurs, we set

$$\delta(x) = \delta(u^*) + \rho(u^*, x).$$

By induction hypothesis, $\delta(u^*) \geq d_\rho(a, u^*)$. And by the triangle inequality

$$d_\rho(a, u^*) + \rho(u^*, x) \geq d_\rho(a, x).$$

So we still have $\delta(x) \geq d_\rho(a, x)$.

Next, we show (ii). Again, we reason by induction on the loops of the algorithm. Initially, a is the minimizer, and therefore W is set equal to $\{a\}$. Also $\delta(a) = 0 = d_\rho(a, a)$. So, the statement holds after the first pass. Assume the claim holds after k passes. Line 5 of Algorithm 1 picks a vertex $u^* = \arg\min\{\delta(u) : u \in V \setminus W\}$, with $\delta(u^*) < \infty$. Note that if such u^* does not exist, then the algorithm ends. The key is to show that we also have

$$u^* = \arg\min\{d_\rho(a, u) : u \in V \setminus W\}.$$

To see this, suppose $x := \arg\min\{d_\rho(a, u) : u \in V \setminus W\}$. Let y be a predecessor of x. Recall that (1.6) implies $d_\rho(a, y) + \rho(y, x) = d_\rho(a, x)$. In particular, $d_\rho(a, x) > d_\rho(a, y)$. Therefore, we must have $y \in W$ because of the minimizing property of x. In particular, y was added to W before x, during some pass j with $j \in \{1, \ldots, k\}$. Line 9 of Algorithm 1, means that at the end of loop j, we must have

$$\delta(x) \leq \delta(y) + \rho(y, x),$$

and this will persist until the k-th pass is done. By induction hypothesis, since y is in W, we have $\delta(y) = d_\rho(a, y)$. So, by part (i),

$$d_\rho(a, x) \leq \delta(x) \leq d_\rho(a, y) + \rho(y, x) = d_\rho(a, x).$$

Thus, $\delta(x) = d_\rho(a, x)$. By definition, $\delta(u^*) \leq \delta(x) = d_\rho(a, x)$. And by part (i), $\delta(u^*) \geq d_\rho(a, u^*)$. Therefore

$$d_\rho(a, u^*) \leq \delta(u^*) \leq d_\rho(a, x).$$

This means that u^* also minimizes $d_\rho(a, \cdot)$ over $V \setminus W$, and we must have $d_\rho(a, u^*) = d_\rho(a, x)$. In particular, $\delta(u^*) = d_\rho(a, u^*)$. So, the claim will also hold after the $(k+1)$-st pass.

Finally, we prove (iii). Once more, we reason by induction with respect to each loop of the algorithm. Initially, $W = \{a\}$ and $\delta(a) = 0 = d_\rho(a,a)$. So, the statement holds after the first pass if we let $\gamma(a)$ be the constant walk. Assume that the claim holds after k passes and that the algorithm doesn't end on the $k+1$ pass. In particular, there is a minimizer u^* on Line 5 of Algorithm 1 with $\delta(u^*) < \infty$. This means that $\delta(u^*)$ must have been updated at least once before and that a predecessor $w = \text{pred}(u^*) \in W$ must have already been assigned (possibly several times). By the induction hypothesis, there is a walk $\gamma(w)$ such that $\ell_\rho(\gamma(w)) = d_\rho(a,w)$. Define the walk $\gamma(u^*)$ to be the walk $\gamma(w)$ followed by the edge (w,u^*). Then

$$\begin{aligned}\ell_\rho(\gamma(u^*)) &= \ell_\rho(\gamma(w)) + \rho(w,u^*) = d_\rho(a,w) + \rho(w,u^*)\\ &= \delta(v) + \rho(w,u^*) = \delta(u^*) = d_\rho(a,u^*),\end{aligned}$$

where the last equality follows from the proof of part (ii). □

1.4 FLOWS AND CUTS ON GRAPHS

Let $G = (V,E)$ be a finite simple graph with vertex set V and edge set E. To talk about flows, one must first pick an orientation for each edge as in Definition 1.1.

Definition 1.6. Let G be a finite simple graph and let $s \neq t \in V$ be two vertices. We say s is the *source* and t is the *sink*. A *flow* from s to t is a function $f : V \times V \longrightarrow \mathbb{R}$ such that

(a) $f(x,y) = 0$, if $\{x,y\} \notin E$;

(b) f is *antisymmetric*: $f(x,y) = -f(y,x)$, for every $(x,y) \in V \times V$;

(c) f satisfies the *node law* at each $x \in V \setminus \{s,t\}$:

$$\text{Div}_f(x) := \sum_{y\sim x} f(x,y) = 0. \tag{1.11}$$

The operator $\text{Div}_f(x) := \sum_{y\sim x} f(x,y)$ is called the *divergence* of the flow f at x. Also, when the conditions above are satisfied, we will refer to f as an *st- flow*.

Remark 1.3. (i) The *adjacency matrix* defined as $A(x,y) = \mathbb{1}_{x\sim y}$, is useful if one wants to extend sums over neighbors to sums over all vertices in V. For instance, we can write

$$\text{Div}_f(x) = \sum_{y\sim x} f(x,y) = \sum_{y\in V} A(x,y) f(x,y).$$

(ii) Given a flow f, when summing over all oriented edges in a given orientation, we will write $f(\vec{e})$ only if the antisymmetry matters. For instance, we will write $|f(e)|$ instead of $|f(\vec{e})|$, because $|f|$ is symmetric. Mathematically, we can think of a quotient map $\pi : \hat{E} \to E$ such that $\pi((x,y)) = \{x,y\}$ and then say that $|f|$ passes to the quotient.

Note that given a flow f on G, there is a natural orientation $\vec{E}$, which we call the *physical orientation for* f, that has the property that $f(\vec{e}) \geq 0$ for every $e \in E$.
Given an orientation $\vec{E}$ on G and a vertex $v \in V$, we let $E(v) = \{e \in E : v \in e\} = E^+(v) \cup E^-(v)$ where:

$$\begin{aligned} E^+(v) &= \{e \in E(v) : \vec{e} = (x,v) \text{ for some } x \in V\} \\ E^-(v) &= \{e \in E(v) : \vec{e} = (v,y) \text{ for some } y \in V\}. \end{aligned}$$

With these notations, the node law says that

$$\sum_{e \in E^-(v)} f(\vec{e}) = \sum_{e \in E^+(v)} f(\vec{e}) \qquad \forall v \in V \setminus \{s,t\}.$$

Also, if the orientation is the physical orientation for f, then the left-hand side represents the total flow leaving v, and the right-hand side represents the total flow entering v. Switching the orientation of one of these edges moves the corresponding flow term to the other side of the equation, and antisymmetry gives it a negative sign, so the balance between total inflow and total outflow remains unchanged.

Definition 1.7. The *value* of an st-flow is

$$\mathrm{Val}(f) := \mathrm{Div}_f(s).$$

Using the node law (1.11) and Remark 1.3 (i) we see that

$$\begin{aligned} \mathrm{Val}(f) + \mathrm{Div}_f(t) = \mathrm{Div}_f(s) + \mathrm{Div}_f(t) &= \sum_{x \in V} \mathrm{Div}_f(x) \\ &= \sum_{x \in V} \sum_{y \in V} A(x,y) f(x,y) = 0, \end{aligned}$$

where the last equality holds because $A(x,y)$ is symmetric and $f(x,y)$ is anti-symmetric, so their product is still anti-symmetric. In particular, we also have $\mathrm{Val}(f) = -\mathrm{Div}_f(t)$.

Definition 1.8. A *cut* for the flows from s to t is a set $C \subset V$ such that $s \in C$ but $t \notin C$. Let the *edge-boundary of a cut* be the set

$$\partial C := \{e = \{x, y\} \in E : x \in C \text{ and } y \notin C\}.$$

If $e = \{x, y\} \in \partial C$ is an edge with x in C and $y \notin C$, we orient it as exiting C, namely, as (x, y). We call this the *outer normal* orientation. Also, when the conditions above are satisfied, we will refer to C as an *st-cut.*

Definition 1.9. Assume that a *capacity* is given on the edges as $\sigma : E \to \mathbb{R}_{>0}$. If C is an st-cut, the *capacity* of C is

$$\sigma(C) := \sum_{e \in \partial C} \sigma(e).$$

The *minimum cut value* from s to t is the value

$$\min_{C \in \mathcal{C}} \sigma(C).$$

A cut that attains this value is called a *minimum cut*, or *min-cut* for short. An st-flow is *feasible* if

$$|f(e)| \leq \sigma(e), \qquad \text{for every } e \in E.$$

This is often called a *capacity constraint.* In this case, write $f \in \mathcal{F}$. We then define the *maximum flow value from s to t*as

$$F^* := \max_{f \in \mathcal{F}} \text{Val}(f),$$

and a feasible st-flow f^* such that $\text{Val}(f^*) = F^*$ is called a *maximum flow*, or *max-flow* for short.

Theorem 1.4 (Max-flow Min-cut). *Let $\mathcal{C}$ be the set of all st-cuts. Then*

$$F^* = \max_{f \in \mathcal{F}} \text{Val}(f) = \min_{C \in \mathcal{C}} \sigma(C).$$

Proof. To begin, assume that σ is integer-valued as follows $\sigma : E \to \{1, 2, 3, \ldots\}$. Also, since G is finite, both F^* and $\min_C \sigma(C)$ are finite and attained.

We start by showing that for every st-flow and every st-cut, we have

$$\text{Val}(f) = \sum_{e \in \partial C} f(\vec{e}), \tag{1.12}$$

where every edge $e \in \partial C$ is given the outer normal orientation. In other words, we want to show that the value of a flow, which is the net flow out of the source node, is equal to the net flow out of any cut. To see this note that:

$$\begin{aligned}
\mathrm{Val}(f) &= \sum_{x\in C} \mathrm{Div}_f(x) && \text{(Node law (1.11))}\\
&= \sum_{x\in C}\sum_{y\in V} A(x,y)f(x,y) && \text{(Remark 1.3 (i))}\\
&= \sum_{x\in C}\sum_{y\in C} A(x,y)f(x,y) + \sum_{e\in\partial C} f(\vec{e}) && \text{(outer normal orientation)}\\
&= \sum_{e\in\partial C} f(\vec{e}) && \text{(anti-symmetry.)}
\end{aligned}$$

Using (1.12), we see that for every feasible st-flow and every st-cut we must have

$$\mathrm{Val}(f) = \sum_{e\in\partial C} f(\vec{e}) \le \sum_{e\in\partial C} |f(e)| \le \sum_{e\in\partial C} \sigma(e) = \sigma(C).$$

This shows that $F^* \le \min_C \sigma(C)$.

To prove the other direction, we first discuss an algorithm. Let f be a feasible st-flow. Given a path $P = (x_0, x_1, \cdots, x_{n-1}, x_n)$ from $x_0 = s$ to $x_n = x \in V$, we say that P is *improvable* if

$$f(x_{i-1}, x_i) < \sigma(x_{i-1}, x_i) \qquad \forall i = 1, \ldots, n.$$

Suppose an improvable path P from s to t exists. Letting

$$\delta = \min_{i=1,\ldots,n} \left[\sigma(x_{i-1}, x_i) - f(x_{i-1}, x_i)\right] > 0,$$

we can define a flow f_P equal to δ on (x_{i-1}, x_i) for every $i = 1, \ldots, n$, and zero otherwise. Then $g := f + f_P$ is also feasible. To see this note that for every $i = 1, \ldots, n$,

$$\begin{aligned}
g(x_{i-1}, x_i) = f(x_{i-1}, x_i) + \delta &\le f(x_{i-1}, x_i) + \sigma(x_{i-1}, x_i) - f(x_{i-1}, x_i)\\
&= \sigma(x_{i-1}, x_i),
\end{aligned}$$

and

$$-g(x_{i-1}, x_i) = -f(x_{i-1}, x_i) - \delta \leq \sigma(x_{i-1}, x_i) - \delta \leq \sigma(x_{i-1}, x_i).$$

So, $|g(x_{i-1}, x_i)| \leq \sigma(x_{i-1}, x_i)$.

Moreover, the value of the flow increases as follows: $\text{Val}(g) = \text{Val}(f) + \delta$.

Furthermore, since σ is integer-valued and the algorithm initializes f as the zero flow, δ is also integer-valued at each iteration. Therefore, f remains integer-valued throughout, and since $\text{Val}(f)$ cannot exceed the total capacity $\sigma(E)$, the algorithm will end in finitely many steps.

When the algorithm stops, we have an integer-valued flow f with the property that no improvable path from s to t exists. In other words, if we define

$$X_f := \{x \in V : x = s \text{ or } \exists \text{ path } P : s \rightsquigarrow x \text{ that is improvable}\}$$

then $t \notin X_f$. This means that X_f is an st-cut.

Let $\vec{e} = (x, y) \in \partial X_f$ have the outer normal orientation. Then $x \in X_f$, hence there is an improvable path $P_x : s \rightsquigarrow x$. Moreover, $y \notin X_f$, so we must have $f(x, y) = \sigma(e)$; otherwise, we would be able to extend the path P_x to an improvable path from s to y, which is a contradiction. We say that the edge e is *saturated*. Thus,

$$\begin{aligned} \text{Val}(f) &= \sum_{e \in \partial X_f} f(\vec{e}) && \text{(by (1.12))} \\ &= \sum_{e \in \partial X_f} \sigma(e) = \sigma(X_f). \end{aligned}$$

This shows that $F^* \geq \min_C \sigma(C)$ and that f is a maximum flow.

The general case, when the capacities are not necessarily integer-valued, can be deduced by first considering the case of rational-valued capacities and then taking a limit, see Exercise 1.13. □

1.5 THE PATH FORMULATION FOR THE MAX-FLOW PROBLEM

Consider the problem of maximizing the flow from s to t in a graph G. Let $\Gamma = \Gamma_{\text{path}}(s, t)$ be the family of paths from s to t, and let $\mathcal{L}$ be the family of all cycles in G. If $\gamma = x_0\ x_1\ \cdots\ x_m$ is either a path in Γ or a cycle in $\mathcal{L}$ (in which case $x_m = x_0$), then we let f_γ denote the *unit path flow along* γ, by defining $f_\gamma(x_i, x_{i+1}) := 1$, for every $i = 0, \ldots, m-1$.

Recall that in the proof of Theorem 1.4 a maximum flow is constructed as a sum of weighted path flows. More generally, any st-flow f can be decomposed as a sum of (weighted) path flows and cycle flows, as follows,

$$f = \sum_{\gamma\in\Gamma} x_\gamma f_\gamma + \sum_{c\in\mathcal{L}} x_c f_c,$$

where x_γ and x_c are real numbers. If we want to maximize the flow from s to t, then cycles may be omitted because they use edge capacity while contributing a zero amount to the value of the flow. For similar reasons, negative path flows for which $x_\gamma < 0$ may also be removed from consideration. With these modifications, any maximum flow will have the property that

$$\mathrm{Val}(f) = \sum_{\gamma\in\Gamma} x_\gamma \,\mathrm{Val}(f_\gamma) = \sum_{\gamma\in\Gamma} x_\gamma.$$

This leads to the *path formulation* of the max-flow problem:

$$\begin{aligned} \underset{x\in\mathbb{R}^\Gamma_{\geq 0}}{\text{maximize}} \quad & \sum_{\gamma\in\Gamma} x_\gamma \\ \text{subject to} \quad & \sum_{\gamma\in\Gamma: e\in\gamma} x_\gamma \leq \sigma(e), \quad \forall e \in E, \\ & x_\gamma \geq 0, \quad \forall \gamma \in \Gamma. \end{aligned} \tag{1.13}$$

This form is convenient because the flow conditions in Definition 1.6 are automatically satisfied. One only needs to verify the feasibility condition in Definition 1.9, namely the capacity constraint. This requires that the sum of the flow strengths x_γ along each path using a particular edge must not exceed the capacity of that edge.

EXERCISES

1.1 *Some simple graphs.* In Figure 1.1, we will refer to graph G_1 as the square (or 4-cycle), to graph G_2 as the diamond graph, and to graph G_3 as the house graph.

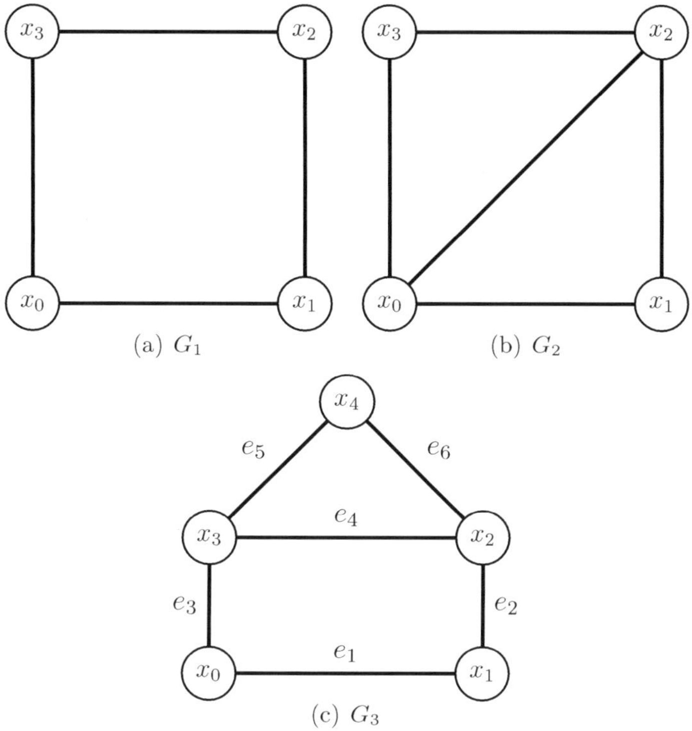

Figure 1.1 Some simple graphs: the square, the diamond, and the house.

For this exercise, familiarize yourself with the NetworkX library in Python. In particular, write code to plot these graphs with the appropriate labels on nodes and edges.

1.2 *Connecting walks on the Path graph.* Consider the path graph P_3 corresponding to the path $x_0\ e_1\ x_1\ e_2\ x_2$. Let $\Gamma = \Gamma(x_0, x_2)$ be the *connecting* family of all walks that start at x_0 and end at x_2. Note that Γ is infinite and contains walks such as

$$\gamma = (x_0, x_1, x_0, x_1, x_2, x_1, x_0, x_1, x_2).$$

(i) Compute the row vector $\mathcal{N}(\gamma, \cdot)$ corresponding to γ;

(ii) show there are other walks with the same usage function;

(iii) show that the usage matrix $\mathcal{N}$ has infinitely many rows of the form $[2k + 1, 2j + 1]$ with $k, j = 0, 1, 2, \ldots$.

1.3 *Simple paths.* For the house graph G_3 in Exercise 1.1 consider the connecting family $\Gamma = \Gamma_{\text{path}}(x_0, x_4)$.

(i) Compute the matrix $\mathcal{N}$ for the family Γ;

(ii) find the shortest path distance $d(x_0, x_4)$ and display a shortest path;

(iii) code a function that, given a graph and a pair of nodes, outputs the matrix $\mathcal{N}$, the shortest path distance between those two nodes, and a corresponding shortest path. Use the computations in parts (i) and (ii) to validate your code.

1.4 *Local degrees.* For the house graph $G = G_3$ in Exercise 1.1 consider the *adjacency* matrix $A(x, y) := \mathbb{1}_{x \sim y}$.

(i) Compute the adjacency matrix A for the graph G;

(ii) use the vector of all ones $\mathbf{1}$ of the right dimension to compute the vector of degrees $d = \deg$;

(iii) code a function that given a graph G, outputs the adjacency matrix A and the vector of degrees. Use the computations in parts (i) and (ii) to validate your code;

(iv) check that the Handshake Lemma holds for the house.

1.5 *Egalitarian orientations.* For the house graph G in Exercise 1.1 consider the orientation $\vec{E}$ induced by the order on the vertices $V = \{x_0, x_1, x_2, x_3, x_4\}$. Namely, $(x_i, x_j) \in \vec{E}$ iff $\{x_i, x_j\} \in E$ and $i < j$. Given a node $x \in V$ and an orientation $\vec{E}$, the *in-degree* at x, $\deg_{\text{in}}(x)$, is the number of oriented edges of the form (y, x) for some $y \in E$. Likewise, the *out-degree* at x, $\deg_{\text{out}}(x)$, is the number of oriented edges of the form (x, y) for some $y \in E$.

(i) Write down an oriented adjacency matrix corresponding to $\vec{E}$ and use the vector of all ones $\mathbf{1}$ to compute the in-degree and out-degree of each vertex of G;

(ii) find a reversible path, i.e., a directed path from a vertex v to a vertex u such that $\deg_{\text{in}}(v) > \deg_{\text{in}}(u) + 1$;

(iii) reverse every oriented edge along the path found in (ii). Then, recompute the in-degrees of every vertex and compare this with the in-degrees found with the orientation in (i);

(iv) code a function that takes a graph, begins with the orientation induced by the ordering of the nodes, then repeatedly finds reversible paths and reverses them, then outputs the resulting "egalitarian" orientation, which has no reversible paths (you may output a new ordering of the nodes instead). See [20, 21] for more information.

1.6 *Trees.* Reprove Theorem 1.1 by showing that

$$(1) \Rightarrow (4) \Rightarrow (3) \Rightarrow (2) \Rightarrow (1).$$

1.7 *Edge bounds.* Give a proof for Corollary 1.1.

1.8 *Cayley's Theorem.* Code the algorithm used in the proof of Theorem 1.2.

1.9 *The shortest path metric.* Check that the shortest path metric d_ρ defined in (1.5) is a distance.

1.10 *Ultrametrics.* An *ultrametric* on a set V is a function $d : V \times V \to [0, \infty)$ that satisfies the first two properties of a metric as in Definition 1.4 (non-degeneracy and symmetry), and instead of the triangle inequality, satisfies the following stronger requirement: for every $a, b, c \in V$:

$$d(a, b) \leq \max\{d(a, c), d(c, b)\}.$$

(i) Show that every ultrametric is also a metric;

(ii) Let G be the 4-cycle graph (the square). Label the vertices of G consecutively as $V = \{1, 2, 3, 4\}$. Define $d(a, b) := 1$, if a and b are neighbors, $d(a, b) = 0$ if $a = b$, and let $d(a, b) := 1/2$ otherwise. Show that d is an ultrametric.

1.11 *Dijkstra algorithm.* Code Dijkstra's algorithm. You may use a data structure called a queue.

1.12 *Breadth-First Search.* Show that when the edge weights ρ are constant, Dijkstra's algorithm becomes equivalent to the Breadth-First Search algorithm.

1.13 *Max-flow Min-cut Theorem.* Finish the proof of Theorem 1.4 by deducing the case of real-valued (positive) capacities from the case of integer-valued capacities.

1.14 *Improvable path algorithm.* Code the improvable path algorithm described in the proof of Theorem 1.4 for the case of integer-valued capacities.

1.15 *Menger's Theorem.* Given a graph $G = (V, E)$ and two vertices $a, b \in V$ with $a \neq b$, use the proof of Theorem 1.4 for integer capacities to show that if $k \in \mathbb{Z}_{\geq 0}$ is the size of the smallest ab-cut, then there are k edge-disjoint paths from a to b.

1.16 *The Min-cut metric.* Given a graph $G = (V, E)$ and two vertices $a, b \in V$ with $a \neq b$, define $\mathrm{MC}(a, b)$ to be the size of the smallest ab-cut. Show that

$$d_{\mathrm{MC}}(a, b) := \begin{cases} \frac{1}{\mathrm{MC}(a,b)} & \text{if } a \neq b \\ 0 & \text{if } a = b \end{cases}$$

is an ultrametric on V.

CHAPTER 2

Electrical Networks

WE REVIEW the basic theory of electric networks. In particular, we introduce the concept of effective resistance, which will play a crucial role later. For a more in-depth study of effective resistance, we refer the reader to standard references such as [62].

2.1 ELECTRICAL CURRENT FLOWS

Let $G = (V, E)$ be a graph with some given *edge resistances* $r(e) > 0$, for $e \in E$. We call this a *resistor network*. If f is an st-flow. as in Definition 1.6, the *energy* of f is

$$\mathcal{E}(f) := \sum_{e \in E} r(e)|f(e)|^2. \tag{2.1}$$

We also define $c(e) := 1/r(e)$ to be the *conductance* of the edge e.

Electrical current is a flow that minimizes the energy $\mathcal{E}$ among all st-flows with value 1. To formalize this, let $\mathcal{F}(s,t)$ be the set of all st-flows and set

$$\mathcal{E}^* := \min_{f \in \mathcal{F}(s,t):\mathrm{Val}(f)=1} \mathcal{E}(f). \tag{2.2}$$

Existence and uniqueness of a *minimizer* f^* such that $\mathcal{E}(f^*) = \mathcal{E}^*$ can be deduced using the theory of convex optimization and the fact that the energy functional is strictly convex. The unique minimizing flow is called the *unit*[1] *electrical current* flow from s to t. In what follows,

[1]We will sometime say that we are pushing $1A$ (Ampere) worth of current from s to t.

DOI: 10.1201/9781003024866-3

we will describe various aspects of this minimization problem in more detail.

We begin by characterizing all minimizers.

Lemma 2.1. *If f^* is a minimizer and k is an arbitrary st-flow with* $\mathrm{Val}(k) = 0$, *then*

$$\sum_{e\in E} r(e)f^*(e)k(e) = 0. \tag{2.3}$$

Remark 2.1. According to Definition 1.6, f is a function on $V \times V$. If e is an edge, then the notation $f(e)$ does not immediately make sense. The notation is used as a shorthand for special cases in which the ordering of endpoints does not matter. In particular, if $e = \{x, y\}$ then the antisymmetric property of f allows us to assign meanings to notations like

$$|f(e)| := |f(x,y)| = |f(y,x)|$$

and

$$f(e)k(e) := f(x,y)k(x,y) = f(y,x)k(y,x).$$

Proof of Lemma 2.1. Let f^* be a minimizer for the minimum energy problem (2.2) and let k be an arbitrary st-flow with $\mathrm{Val}(k) = 0$. For $\epsilon > 0$ and $\lambda \in (-\epsilon, \epsilon)$, let $g_\lambda = f^* + \lambda k$. Then, by the linearity of the divergence operator, g_λ is also a unit st-flow. Therefore, $\mathcal{E}(g_\lambda) \geq \mathcal{E}(f^*)$. Expanding the squares, we get

$$\begin{aligned}\mathcal{E}(f^*) \leq \mathcal{E}(g_\lambda) &= \sum_{e\in E} r(e)(f^*(e) + \lambda k(e))^2 \\ &= \mathcal{E}(f^*) + 2\lambda \sum_{e\in E} r(e)f^*(e)k(e) + \lambda^2\mathcal{E}(k)\end{aligned}$$

Simplifying, we see that we must have

$$\lambda\left(2\sum_{e\in E} r(e)f^*(e)k(e) + \lambda\mathcal{E}(k)\right) \geq 0.$$

Letting λ tend to zero with $\lambda > 0$, we see that $\sum_{e\in E} r(e)f^*(e)k(e) \geq 0$. On the other hand, letting λ tend to zero with $\lambda < 0$, we get that $\sum_{e\in E} r(e)f^*(e)k(e) \leq 0$. Hence, we have shown (2.3). □

Let W be a cycle, so that $W = x_0\ e_1\ x_1\ \cdots\ e_n\ x_n$ with $x_0, \ldots, x_{n-1}$ distinct nodes and $x_n = x_0$. Define a flow k_W that sends one unit around the cycle. Namely, let $k_W(x_{i-1}, x_i) := 1$ for $i = 1, \ldots, n$. One can check that k_W is a st-flow with a value of zero. Therefore, applying Lemma 2.1, we see that any minimizer f^* must satisfy the following *cycle law*:

$$\sum_{i=1}^{n} r(e_i) f^*(x_{i-1}, x_i) = 0, \tag{2.4}$$

for all cycles $x_0\ e_1\ x_1\ \cdots\ e_n\ x_n$ with $x_0 = x_n$.

Proposition 2.1. *If an st-flow f satisfies the cycle law, then it must admit a* potential, *namely, a function $h : V \to \mathbb{R}$ such that for every edge $e = \{x, y\}$:*

$$h(y) - h(x) = r(e) f(x, y). \tag{2.5}$$

The existence of a potential satisfying (2.5) is referred to as *Ohm's law*. One may recall the high school physics identity $V = RI$, indicating that the voltage drop is equal to the resistance times the intensity of the current.

Proof of Proposition 2.1. Set $h(s) = 0$. Given $x \in V \setminus \{s\}$, find a path $P_x = x_0\ e_1\ x_1\ \cdots\ e_n\ x_n$ with $x_0 = s$ and $x_n = x$. Then define

$$h(x) := \sum_{i=1}^{n} r(e_i) f(x_{i-1}, x_i).$$

Antisymmetry and the cycle law guarantee that this definition does not depend on the path $s \rightsquigarrow x$ chosen. In particular, we can choose a path P_y as P_x followed by the edge $e = \{x, y\}$, so that h satisfies (2.5). □

We begin by proving a variant of Lemma 2.1 for flows that admit a potential.

Lemma 2.2. *If f is a flow that admits a potential h, and k is an arbitrary st-flow, then*

$$\sum_{e \in E} r(e) f(e) k(e) = \mathrm{Val}(k)(h(t) - h(s)). \tag{2.6}$$

Proof. Let f be a flow that admits a potential h. This means that f and h satisfy Ohm's law. Then, if k is an arbitrary st-flow, we have:

$$\begin{aligned}
&\sum_{e\in E} r(e)f(e)k(e) \\
&= \frac{1}{2}\sum_{x\in V}\sum_{y\in V} r(x,y)f(x,y)k(x,y)A(x,y) && \text{(Remark 1.3 (i))} \\
&= \frac{1}{2}\sum_{x\in V}\sum_{y\in V} (h(y)-h(x))k(x,y)A(x,y) && \text{(Ohm's law)} \\
&= \frac{1}{2}\sum_{y\in V} h(y)\left(\sum_{x\in V} k(x,y)A(x,y)\right) \\
&\quad - \frac{1}{2}\sum_{x\in V} h(x)\left(\sum_{y\in V} k(x,y)A(x,y)\right) \\
&= -\frac{1}{2}\sum_{y\in V} h(y)\operatorname{Div}_k(y) - \frac{1}{2}\sum_{x\in V} h(x)\operatorname{Div}_k(x) && \text{(Definition 1.6)} \\
&= -\sum_{x\in V} h(x)\operatorname{Div}_k(x) \\
&= -h(s)\operatorname{Div}_k(s) - h(t)\operatorname{Div}_k(t) && \text{(Node law)} \\
&= (h(t)-h(s))\operatorname{Val}(k) && \text{(Definition 1.7)}
\end{aligned}$$

□

It turns out that Ohm's law, the cycle law, and the minimizing property are all equivalent.

Proposition 2.2. *If a unit st-flow admits a potential (therefore satisfying Ohm's law), it is an energy minimizer.*

Proof. Let f be a unit st-flow that admits a potential h. Given an arbitrary unit st-flow g, set $k := g - f$. Then, k is an st-flow with $\operatorname{Val}(k) = 0$. Hence,

$$\begin{aligned}
\mathcal{E}(g) &= \mathcal{E}(f) + \mathcal{E}(k) + 2\sum_{e\in E} r(e)f(e)k(e) && \text{(expand the squares)} \\
&= \mathcal{E}(f) + \mathcal{E}(k) + 2(h(t)-h(s))\operatorname{Val}(k) && \text{(by Lemma 2.2)} \\
&= \mathcal{E}(f) + \mathcal{E}(k) \geq \mathcal{E}(f) && \text{(since } \mathcal{E}(k) \geq 0\text{).}
\end{aligned}$$

So, f is an energy minimizer. □

We summarize this section in the following theorem.

Theorem 2.1. *Let $G = (V, E)$ be a finite connected simple graph with edge resistances $r \in \mathbb{R}^E_{>0}$. Fix a source s and a target $t \neq s$, with $s, t \in V$. Assume f is a unit st-flow. Then the following are equivalent:*

(1) f is an energy minimizer for the problem (2.2).

(2) f satisfies the cycle law (2.4).

(3) f admits a potential, so that Ohm's law (2.5) holds.

Definition 2.1. An *st*-flow f on a graph G with edge resistances r is *(electrical) current flow* if one of the conditions (1)-(3) in Theorem 2.1 holds. Also, the corresponding potential function h is called the *voltage potential.*

Remark 2.2. Note that if an *st*-flow f admits a potential h, then any translated function $h + c$, for some constant $c \in \mathbb{R}$, is also a potential for f. However, this is the only source of non-uniqueness, and often, to normalize potentials, one chooses the constant c so that $h(s) = 0$. In electrical network terms, we can think of this as "grounding" the potential at s.

In the next section, we develop the concept of harmonicity.

2.2 HARMONIC FUNCTIONS AND THE MAXIMUM PRINCIPLE

Suppose f is an *st*-flow that admits a potential, h, so that Ohm's law (2.5) holds. The node law (1.11) implies that for $x \neq s, t$:

$$\mathrm{Div}_f(x) = \sum_{y \sim x} f(x, y) = \sum_{y \sim x} c(x, y)(h(y) - h(x)) = 0, \tag{2.7}$$

where $c(x, y) = 1/r(x, y)$ is the edge conductance. If we define the generalized degree $C(x) := \sum_{y \sim x} c(x, y)$, then by (2.7), the value of h at x is a weighted average of the values of h on the neighbors of x:

$$h(x) = \frac{1}{C(x)} \sum_{y \sim x} c(x, y) h(y). \tag{2.8}$$

When (2.8) holds, we say that h satisfies the *Mean Value Property* at x, or that h is *harmonic* at x.

Theorem 2.2 (Maximum Principle). *Let $G = (V, E)$ be a finite connected simple graph with edge conductances $c \in \mathbb{R}^E_{>0}$. Fix a non-empty subset $B \subset V$, e.g., $B = \{s, t\}$, which we think of as a "boundary". Suppose that a function $h : V \to \mathbb{R}$ is harmonic on $V \setminus B$, i.e., h satisfies the mean value property (2.8), at every $x \in V \setminus B$. Then h attains its global maximum on B.*

Proof. If h is constant on V, then the global maximum is attained on B. So assume that h is not constant on V and that there is some node $x \notin B$ such that

$$h(x) = \max_{y \in V} h(y) > \max_{z \in B} h(z).$$

The mean value property implies $h(y) = h(x)$ for every neighbor $y \sim x$. Indeed, assume that there is one node $y' \sim x$ where $h(y') < h(x)$. Since for every other neighbor $y'' \sim x$ we have $h(y'') \leq h(x)$ when we average we get

$$C(x)h(x) = \sum_{y \sim x} c(x, y)h(x) > \sum_{y \sim x} c(x, y)h(y) = C(x)h(x),$$

where the last equality follows from the harmonicity at x. This, however, is a contradiction. Therefore, if the maximum is attained at x, it must also be attained at each one of its neighbors. Since G is connected, there is a walk from x to any other vertex. Therefore, repeating this argument as much as necessary will show that h must be constant. But h was assumed not to be constant; hence, we have reached a contradiction. □

Corollary 2.1. *Let $G = (V, E)$ be a finite connected simple graph with edge conductances $c \in \mathbb{R}^E_{>0}$. Suppose that a function $h : V \to \mathbb{R}$ is harmonic on V. Then, h is constant.*

Proof. This is a corollary of the proof of Theorem 2.2. Since G is finite, h attains its maximum at some node $x \in V$. By the mean value property, $h(y) = h(x)$ for every neighbor $y \sim x$. This can be repeated, and since G is connected, every node in V can be reached from x. Therefore, h is constant. □

The maximum principle can be used in an alternative proof of the uniqueness of the current flow. Suppose that f_1 and f_2 are two unit *st*-flows that minimize energy in (2.1). By Theorem (2.1), let h_1 and h_2 be the corresponding potentials normalized so that $h_1(s) = h_2(s) = 0$. By

Lemma 2.2 applied to the $f = f_j$ with $k = f_j$, $j = 1, 2$, we get that the minimum energy $\mathcal{E}^*$ satisfies

$$\mathcal{E}^* = \sum_{e \in E} r(e)|f_j(e)|^2 = \mathrm{Val}(f_j)(h_j(t) - h_j(s)) = h_j(t), \qquad \text{for } j = 1, 2.$$

Form $u := h_1 - h_2$. Then u is harmonic on $V \setminus \{s, t\}$ and $u(s) = u(t) = 0$. By the Maximum Principle (Theorem 2.2), $u(x) \leq 0$ for $x \in V$. Likewise, switching the role of h_1 and h_2, we get that $-u(x) \leq 0$, for $x \in V$. So $u(x) \equiv 0$, for all $x \in V$. Therefore, $h_1 = h_2$ and $f_1 = f_2$.

Remark 2.3. The existence of minimizers can also be established explicitly using spanning trees; see [45, Theorem 1.16].

2.3 EFFECTIVE RESISTANCE

As we have seen, if f is the unit current st-flow with potential h, then applying Lemma 2.2 with $k = f$, we get

$$\mathcal{E}^* = \sum_{e \in E} r(e)|f(e)|^2 = h(t) - h(s). \tag{2.9}$$

If we replace the whole network by a single edge $\hat{e}$ with resistance $\mathcal{R}$ between s and t, but we keep the unit flow and the potential difference as before, then $\mathcal{R}$ would have to satisfy:

$$\mathcal{R}(\hat{e})|f(\hat{e})|^2 = \mathcal{R}(\hat{e}) = h(t) - h(s). \tag{2.10}$$

We call $\mathcal{R}$ the *effective resistance* of the network from s to t and write $\mathrm{eff}\mathcal{R}(s, t)$. Effective resistance is a generalization of classical circuit theory's serial and parallel rules.

Example 2.1 (Serial Rule). Consider a path graph P_3 consisting of the walk $a\ c\ b$ as in Figure 2.1. Consider the source $s = a$ and the target $t = b$, and let f be the unit current st-flow. By the node law, $f(a, c) = f(c, b) = 1$ and by Ohm's law,

$$h(c) - h(a) = r_1 f(a, c) = r_1 \qquad \text{while} \qquad h(b) - h(c) = r_2 f(c, b) = r_2.$$

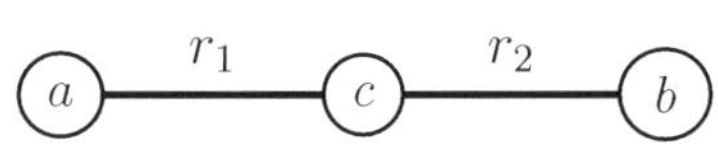

Figure 2.1 Two resistors in series.

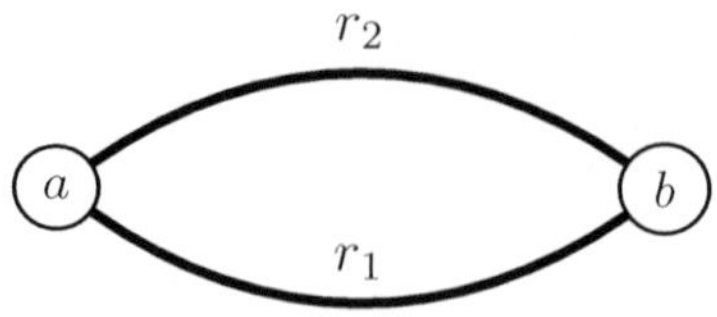

Figure 2.2 Two resistors in parallel.

In particular, $h(b) - h(a) = r_1 + r_2$. So, in this case

$$\text{eff}\mathcal{R}(a, b) = r_1 + r_2.$$

Example 2.2 (Parallel Rule). Consider a digon graph D_2, as in Figure 2.2. This is not a simple graph; it's a multigraph, but the theory extends to multigraphs. Again, consider the source $s = a$ and the target $t = b$, and let f be the unit current st-flow. Also, let $\vec{e_j}$ be the edge with resistance r_j oriented from a to b, for $j = 1, 2$. By the node law, $f(\vec{e_1}) + f(\vec{e_2}) = 1$, and by Ohm's law:

$$f(\vec{e_1})r_1 = h(b) - h(a) = f(\vec{e_2})r_2.$$

So

$$(h(b) - h(a))\left(\frac{1}{r_1} + \frac{1}{r_2}\right) = 1.$$

Thus, in this case

$$\text{eff}\mathcal{R}(a, b) = \frac{1}{\frac{1}{r_1} + \frac{1}{r_2}}.$$

Remark 2.4. Sometimes, it is more helpful to think in terms of edge conductances because they measure how well electricity can propagate through a network. The serial and parallel rules are exactly reversed in this case. For instance, k parallel edges with conductances $c_1, \ldots, c_k$ can always be replaced by a single edge with conductance $c_1 + \cdots + c_k$.

In many cases, repeated serial and parallel rule applications can reduce complex networks to much more manageable ones. However, reducing to a single edge is not always possible. The following example is the simplest example that is not reducible using the serial and parallel rule.

Example 2.3 (Wheatstone). Consider the diamond graph in Figure 2.3. The name of this example comes from the famous *Wheatstone bridge*, a circuit used in practice to measure unknown resistances in electrical circuits.

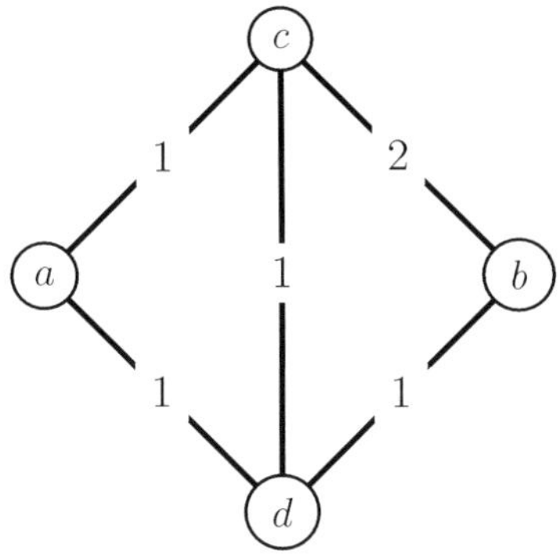

Figure 2.3 The Wheatstone example.

Assume that the edge weights in this example represent edge conductances. We want to compute the effective conductance $\text{eff}\mathcal{C}(s,t)$ for the source $s = a$ and the target $t = b$. In this example, the serial and parallel rules cannot be applied. However, a generalized version called the *star-mesh transformation* can always be performed, and that, repeatedly used in combination with the parallel rule, will reduce any graph to a single edge. Indeed, using the parallel rule, we can always assume that the graph is simple. Pick a vertex $x \in V \setminus \{s,t\}$ where $y_1, \ldots, y_k$ are the neighbors of x. Remove x and all the edges $\{x, y_j\}$ for $j = 1, \ldots, k$. Then, add a complete graph on the vertices $y_1, \ldots, y_k$. Finally, assign edge conductances

$$c(y_i, y_j) := \frac{c(x,y_i)c(x,y_j)}{\sum_{j=1}^{k} c(x,y_j)} \qquad \forall i,j = 1,\ldots,k, \quad i \neq j, \tag{2.11}$$

on these new edges. A proof that this works is given later in Exercise 6.7.

2.4 EFFECTIVE RESISTANCE IS A METRIC

One of the most celebrated properties of effective resistance on resistor graphs is that it is a metric, recall Definition 1.4, on the vertex set V.

Theorem 2.3. *Suppose $G = (V, E)$ is a finite connect simple graph with edge resistances $r \in \mathbb{R}^E_{>0}$. Then, the effective resistance is a metric on the vertex set V. Namely, the function*

$$d_{\text{ER}}(a,b) := \text{eff}\mathcal{R}(a,b), \qquad \textit{for } a,b \in V,$$

satisfies the three metric axioms in Definition 1.4.

Proof. Since G is finite simple and connected, we have $d_{\text{ER}}(a,b) < \infty$ for every $a, b \in V$. To see this, note that if $a = b$, there is no flow, and the potential is constant. So $\text{eff}\mathcal{R}(a,a) = 0$. On the other hand, if $a \neq b$, there is at least one path γ from a to b. Assume $\gamma = x_0\ x_1\ \cdots\ x_n$, with $x_0 = a$ and $x_n = b$. Let k_γ be the unit flow along the path γ, so that $k(x_{i-1}, x_i) = 1$ for $i = 1, \ldots, n$. Then, one can check that k_γ is an ab-flow. Hence, computing the energy, we get that $\text{eff}\mathcal{R}(a,b) = \mathcal{E}^* \leq \mathcal{E}(k_\gamma) < \infty$. This shows that $\text{eff}\mathcal{R}(a,b) < \infty$ by (2.9) and (2.10).

Now we show the non-degeneracy property for d_{ER}. As we already mentioned, if $a = b$, then $\text{eff}\mathcal{R}(a,a) = 0$. Conversely, if $\text{eff}\mathcal{R}(a,b) = 0$, and $a \neq b$, then the electric potential satisfies $h(b) - h(a) = 0$. So by the Maximum Principle (Theorem 2.2), $h(x) - h(a) = 0$ for all $x \in V$. However, unit current flow satisfies

$$1 = \text{Val}(f) = \text{Div}_f(a) = \sum_{y \sim a} f(a,y).$$

Since G is connected, at least one neighbor of a exists. Hence, there is $y \sim a$ such that $f(a,y) \neq 0$. But, by Ohm's law (2.5),

$$f(a,y) = c(a,y)(h(y) - h(a)) = 0.$$

This is a contradiction, so $a = b$.

Next, we show the symmetry property of d_{ER}. Suppose f is a unit current flow from a to b with potential h such that $h(a) = 0$. Then, $\text{eff}\mathcal{R}(a,b) = h(b)$. Now define $g = -f$ and $u = h(b) - h$. Then, for every edge $\{x,y\} \in E$,

$$r(x,y)g(x,y) = h(x) - h(y) = u(y) - u(x).$$

So u is a potential for g and $u(b) = 0$. In particular, $\text{eff}\mathcal{R}(b,a) = u(a) = h(b) = \text{eff}\mathcal{R}(a,b)$.

Next, we record a useful observation. Given two nodes $s \neq t$, write $f_{s,t}$ for the unit current flow from s to t and write $h_{s,t}$ for the associated potential normalized to be 0 at s. Then, $h_{s,t}(t) = \text{eff}\mathcal{R}(s,t)$ and by the Maximum Principle,

$$0 \leq h_{s,t}(x) \leq \text{eff}\mathcal{R}(s,t) \qquad \text{for every } x \in V \setminus \{s,t\}. \tag{2.12}$$

Finally, we prove the triangle inequality for d_{ER}. Without loss of generality, a, b and c are distinct. Define $g_{a,b} := f_{a,c} + f_{c,b}$. By linearity,

$$\text{Div}_{g_{a,b}}(x) = \text{Div}_{f_{a,c}}(x) + \text{Div}_{f_{c,b}}(x), \qquad \text{for all } x \in V.$$

By the node law, for $x \neq a, b, c$, $\text{Div}_{g_{a,b}}(x) = 0$. Also, for $x = c$, $\text{Div}_{f_{a,c}}(c) = -1$ and $\text{Div}_{f_{c,b}}(c) = 1$. So, $\text{Div}_{g_{a,b}}(c) = 0$ as well. In particular, $g_{a,b}$ satisfies is an ab-flow. Finally, $\text{Div}_{g_{a,b}}(a) = \text{Div}_{f_{a,c}}(a) + \text{Div}_{f_{c,b}}(a) = 1$, so $g_{a,b}$ is also a unit ab-flow. Thus,

$$\begin{aligned}
\text{eff}\mathcal{R}(a,b) &= \mathcal{E}^* && \text{(by (2.9))} \\
&\leq \mathcal{E}(g_{a,b}) && \text{(by (2.2))} \\
&= \sum_{e \in E} r(e)\left(f_{a,c}(e) + f_{c,b}(e)\right)^2 \\
&= \mathcal{E}(f_{a,c}) + \mathcal{E}(f_{c,b}) + 2\sum_{e \in E} r(e) f_{a,c}(e) f_{c,b}(e) \\
&= \text{eff}\mathcal{R}(a,c) + \text{eff}\mathcal{R}(c,b) + 2\sum_{e \in E} r(e) f_{a,c}(e) f_{c,b}(e) && \text{(by (2.9))}
\end{aligned}$$

The triangle inequality will follow if we can show that

$$\sum_{e \in E} r(e) f_{a,c}(e) f_{c,b}(e) \leq 0.$$

By mimicking the proof of Lemma 2.2, we get that

$$\begin{aligned}
&\sum_{e \in E} r(e) f_{a,c}(e) f_{c,b}(e) \\
&= \frac{1}{2}\sum_{x \in V}\sum_{y \in V} r(x,y) f_{a,c}(x,y) f_{c,b}(x,y) A(x,y) && \text{(Remark 1.3 (i))} \\
&= \frac{1}{2}\sum_{x \in V}\sum_{y \in V} (h_{a,c}(y) - h_{a,c}(x)) f_{c,b}(x,y) A(x,y) && \text{(Ohm's law)} \\
&= \frac{1}{2}\sum_{y \in V} h_{a,c}(y) \sum_{x \sim y} f_{c,b}(x,y) \\
&- \frac{1}{2}\sum_{x \in V} h_{a,c}(x) \sum_{y \sim x} f_{c,b}(x,y) \\
&= -\frac{1}{2}\sum_{y \in V} h_{a,c}(y)\,\text{Div}_{f_{c,b}}(y) - \frac{1}{2}\sum_{x \in V} h_{a,c}(x)\,\text{Div}_{f_{c,b}}(x) && \text{(Definition 1.6)} \\
&= -h_{a,c}(c) + h_{a,c}(b) && \text{(Node law)} \\
&\leq 0, && \text{(by (2.12))}
\end{aligned}$$

where the last inequality follows from (2.12) replacing $\{s, t\}$ with $\{a, c\}$. □

2.5 THE LINEAR-ALGEBRAIC APPROACH

Here, we review what we have done so far using some basic notions from linear algebra. Recall that $\hat{E}$ is the set of all oriented edges, see Definition 1.1, hence $|\hat{E}| = 2|E|$. Let $F \subset \mathbb{R}^{\hat{E}}$ be the linear subspace of anti-symmetric functions, i.e., all vectors $f \in \mathbb{R}^{\hat{E}}$ such that $f(\vec{e}) = -f(-\vec{e})$. Then, F is the set of all possible flows. We think of F as a $|E|$-dimensional vector space endowed with the weighted inner product:

$$\langle f, g\rangle_r := \sum_{e\in E} f(e)g(e)r(e), \tag{2.13}$$

where we use the abuse of notation mentioned in Remark 2.1.

With this notation, the energy of a flow f is its norm squared:

$$\mathcal{E}(f) = \|f\|_r^2 = \langle f, f\rangle_r.$$

The set of all potential functions can be thought of as the vector space $\mathbb{R}^V$, but it can also be mapped to a subspace of F. Indeed, let $p : \mathbb{R}^V \to F$ be the map that to every potential h assigns the flow $k = p(h)$, given by

$$k(x, y) := c(x, y)(h(y) - h(x)), \qquad \text{for all } x, y \in V, \text{ with } \{x, y\} \in E.$$

We will denote the flow k induced by h as cdh and call it the *gradient flow* induced by h.

Note that the energy of the gradient flow of h translates into a notion of *potential energy* (or *Dirichlet energy*) for h, since

$$\mathcal{E}(cdh) = \sum_{e=\{x,y\}\in E} c(e)^2(h(y)-h(x))^2 r(e) = \sum_{e=\{x,y\}\in E} c(e)(h(y)-h(x))^2.$$

Given a node $x \in V$, define the *star flow* at x as follows:

$$s_x(x, y) = \begin{cases} c(x, y) & \text{if } y \sim x \\ 0 & \text{otherwise.} \end{cases}$$

Then, for any flow $f \in F$, we have

$$\langle f, s_x\rangle_r = \sum_{y\sim x} c(x, y)f(x, y)r(x, y) = \sum_{y\sim x} f(x, y) = \mathrm{Div}_f(x).$$

Thus, f satisfies the node law at x if f is orthogonal to s_x in F. Consider the subspace $S := \mathrm{Span}\,(s_x : x \in V)$ of F, which is called the *star space.*

More generally, if $U \subset V$, we write $S_U := \mathrm{Span}\,(s_x : x \in U)$. Thus, if a source s and target t are given, and we let $U = V \setminus \{s, t\}$, then a flow $f \in F$ is an st-flow (i.e. satisfies the node law at each $x \in U$), if f is orthogonal to S_U.

The map $p : \mathbb{R}^V \to F$ defined above, such that $p(h) := cdh$, is a linear map. The following lemma shows that its image is in the star space.

Lemma 2.3. *Given a potential $h \in \mathbb{R}^V$, we can write its gradient flow as a linear combination of stars, as follows:*

$$cdh = -\sum_{x \in V} h(x) s_x. \tag{2.14}$$

In other words,

$$p(\mathbb{R}^V) \subset S. \tag{2.15}$$

Proof of Lemma 2.3. See Exercise 2.6. □

Furthermore, with $U = V \setminus \{s, t\}$ as above, let $\mathcal{H}_U \subset \mathbb{R}^V$ be the subspace of harmonic potential on U. Equation (2.7) can then be interpreted as saying that a potential function h is harmonic at $x \in V$ if and only if cdh is orthogonal to s_x in F. In formulas:

$$p(\mathcal{H}_U) = S_U^{\perp} \cap p(\mathbb{R}^V). \tag{2.16}$$

Recall that given a cycle W, the unit flow k_W along W was defined right before (2.4). Consider the subspace $C := \mathrm{Span}\,(k_W : \text{ cycle } W)$ of F, which is called the *cycle space*. Then, a flow $f \in F$ satisfies the cycle law (2.4) if and only if f is orthogonal to C.

Theorem 2.4. *Let S be the star space, C the cycle space, and $p : \mathbb{R}^V \to F$ the map $p(h) = cdh$. Then,*

$$S = C^{\perp} = p(\mathbb{R}^V).$$

In other words, a flow $f \in F$ is in S if and only if f satisfies the cycle law (2.4), if and only if f is a gradient flow, i.e., $f = cdh$ for some potential function $h \in \mathbb{R}^V$. Moreover, using the notion of direct sum, we see that F decomposes as the direct sum of S and C:

$$F = S \oplus C = p(\mathbb{R}^V) \oplus C.$$

Proof of Theorem 2.4. If $f = cdh$ for some potential function $h \in \mathbb{R}^V$, then for any cycle $W = x_0\ x_1\ \cdots\ x_n$ with $x_0 = x_n$, the inner product $\langle f, k_W \rangle_r$ equals

$$\sum_{i=1}^{n} r(x_{i-1}, x_i)c(x_{i-1}, x_i)(h(x_i) - h(x_{i-1})) = \sum_{i=1}^{n}(h(x_i) - h(x_{i-1})) = 0,$$

by telescoping sums. So $p(\mathbb{R}^V) \subset C^{\perp}$.

Conversely, if $f \in C^{\perp}$, fix a base node $x_0 \in V$ and define $h(x_0) = 0$. By connectedness, for every node $x \in V$, $x \neq x_0$, find a path $\gamma_x = x_0\ x_1\ \cdots\ x_k$, with $x_k = x$. Then, define

$$h(x) := \sum_{i=1}^{k} r(x_{i-1}, x_i)f(x_{i-1}, x_i).$$

This definition does not depend on the choice of path. If some other path, γ', were to yield a different sum, then concatenating γ_x and γ' into a circuit and decomposing this circuit into a sum of cycles would contradict the cycle law (2.4)..

Finally, we want to verify that, with this choice of h, we have $cdh = f$. Fix a node $x \in V$ and the corresponding path $\gamma_x = x_0\ x_1\ \cdots\ x_k$. If $y \sim x$ is a neighbor, then replace γ_y by the path $x_0\ x_1\ \cdots\ x_k\ y$. Then, since γ_x and γ_y share the first k steps, using the definition of h:

$$c(x, y)(h(y) - h(x)) = c(x, y)r(x, y)f(x, y) = f(x, y).$$

Therefore, we have shown that $p(\mathbb{R}^V) = C^{\perp}$.

Continuing, one can check directly that every star $s_x \in S$ is orthogonal to every cycle $W \in C$. Indeed, either the cycle W does not visit the node x, or it will traverse an edge (y, x) and then an edge (x, y'), where y, y' are distinct neighbors of x, and in the latter case,

$$\langle s_x, k_W \rangle_r = -c(y, x)r(y, x) + c(x, y')r(x, y') = 0.$$

This shows that $S \subset C^{\perp}$. Conversely, assume that $f \in F$ is orthogonal to C. Then, as we have seen above, $f = cdh$ for some potential function h. Thus, by (2.15), $f \in S$. □

In view of Theorem 2.4, we can rewrite (2.16) using the direct (orthogonal) sum symbol $\oplus$ and say that

$$S = S_U \oplus p(\mathcal{H}_U) = p(\mathbb{R}^V). \tag{2.17}$$

Corollary 2.2. *Let $G = (V, E)$ be a finite connected simple graph with edge resistances $r \in \mathbb{R}^E_{>0}$. Fix a source s and a target $t \neq s$, with $s, t \in V$. Set $U := V \setminus \{s, t\}$. Let f^* be the unit current st-flow. Then,*

$$F = S_U \oplus \operatorname{Span}(f^*) \oplus C.$$

Proof. Combining (2.17) and Theorem 2.4, we get that $F = S_U \oplus p(\mathcal{H}_U) \oplus C$. So it is enough to show that in this case, $p(\mathcal{H}_U) = \operatorname{Span}(f^*)$. Recall that, by Theorem 2.1, $f^* = cdh^*$, where h^* is the voltage potential, normalized to have $h^*(s) = 0$. Also, by (2.7) h^* is harmonic on U. In particular, $\operatorname{Span}(f^*) \subset p(\mathcal{H}_U)$. Conversely, assume that $h \in \mathcal{H}_U$. Then the Maximum principle (Theorem 2.2) implies that $h = h(s) + (h(t)/h^*(t))h^*$, because the function $h - h(s) - (h(t)/h^*(t))h^*$ is harmonic in U and identically equal to zero on $B = \{s, t\}$. Using the linearity of the map p and the fact that constant function $c\mathbf{1} \in \mathbb{R}^V$ is sent to the zero flow by p, we get that

$$p(h) = \frac{h(t)}{h^*(t)} p(h^*) = \frac{h(t)}{h^*(t)} f^*.$$

Therefore, $p(\mathcal{H}_U) \subset \operatorname{Span}(f^*)$, and the equality holds. □

2.6 THE DIRICHLET AND THOMSON PRINCIPLES

Let $G = (V, E)$ be a finite connected simple graph with edge conductances $c \in \mathbb{R}^E_{>0}$. Fix a source s and a target $t \neq s$, with $s, t \in V$. The *effective conductance* $\operatorname{eff}\mathcal{C}(s, t)$ is the reciprocal of the effective resistance between s and t. However, it can also be found by minimizing the total *potential energy*

$$\mathcal{E}(\phi) := \sum_{(x,y)\in E} c(x, y) \left(\phi(x) - \phi(y)\right)^2 \tag{2.18}$$

over all potential functions $\phi : V \to \mathbb{R}$ satisfying $\phi(s) = 0$ and $\phi(t) = 1$. See Corollary 2.3 below.

Note that if f is the unit current st-flow and h is the corresponding voltage potential, then $h(t)^{-1}h$ is the voltage potential for the flow $h(t)^{-1}f$ whose value is $h(t)^{-1} = \operatorname{eff}\mathcal{R}(s, t)^{-1}$. In words:

> *Effective resistance is the voltage drop needed to generate an st-flow of* 1 *Ampere, while effective conductance is the value of the st-flow generated by a voltage drop of* 1 *Volt.*

The correspondence between the problem of minimizing the power over all potential functions, also known as *Dirichlet's principle*, and the problem (2.2) of minimizing the energy of all flows, also known as *Thomson's principle*, can be established using the convex duality theory (See Appendix). Here, we will use the linear algebra approach developed in Section 2.5.

Theorem 2.5 (Dirichlet and Thomson principles). *Consider a finite connected simple graph* $G = (V, E)$ *with edge resistances* $r \in \mathbb{R}^E_{>0}$*. Fix a source* s *and a target* $t \neq s$*, with* $s, t \in V$*. Then, for any* st*-flow* f *and any potential function* $h \in \mathbb{R}^V$ *with* $h(s) \neq h(t)$*, we have*

$$\mathrm{Val}(f)|h(t) - h(s)| \leq \sqrt{\mathcal{E}(f)}\sqrt{\mathcal{E}(h)},$$

where $\mathcal{E}(f)$ *is the flow energy defined in* (2.1) *and* $\mathcal{E}(h)$ *is the potential energy defined in* (2.18).

Proof. We can assume that $h(t) > h(s)$ (if not, just consider $-h$.) Let our given flow f be equal to k in Lemma 2.2, and let the gradient flow cdh for our given potential h be equal to f in Lemma 2.2. Then

$$\langle f, cdh \rangle_r = \mathrm{Val}(f)(h(t) - h(s))$$

However, the Cauchy-Schwarz inequality in the weighted inner product space $(F, \langle \cdot, \cdot \rangle_r)$ yields

$$\langle f, cdh \rangle_r \leq \|f\|_r \|cdh\|_r.$$

□

Corollary 2.3. *Let* $G = (V, E)$ *be a finite connected simple graph with edge resistances* $r \in \mathbb{R}^E_{>0}$*. Fix a source* s *and a target* $t \neq s$*, with* $s, t \in V$*. Let* $\mathcal{F}$ *be the family of all unit* st*-flows and let* $\mathcal{H}$ *be the family of all potential functions* $h \in \mathbb{R}^V$ *with* $h(s) = 0$ *and* $h(t) = 1$*. Then,*

$$\inf_{f \in \mathcal{F}} \mathcal{E}(f) \inf_{h \in \mathcal{H}} \mathcal{E}(h) = 1.$$

Proof. Let $f \in \mathcal{F}$ and $h \in \mathcal{H}$. Then Theorem 2.5 implies that $\mathcal{E}(f)\mathcal{E}(h) \geq 1$. Since f and h were chosen arbitrarily, we get that

$$\inf_{f \in \mathcal{F}} \mathcal{E}(f) \inf_{h \in \mathcal{H}} \mathcal{E}(h) \geq 1. \tag{2.19}$$

Now let f^* be the unique minimizer with $\mathcal{E}(f^*) = \inf_{f \in \mathcal{F}} \mathcal{E}(f)$. As we saw in Theorem 2.1, f^* must admit a voltage potential h, which by (2.9) must satisfy $h(t) - h(s) = \mathcal{E}(f^*)$. Let $h^* := (h - h(s))/(h(t) - h(s))$. Then $h^* \in \mathcal{H}$ and $cdh^* = \mathcal{E}(f^*)^{-1} f^*$. In particular, $\mathcal{E}(h^*) = \mathcal{E}(f^*)^{-2}\mathcal{E}(f^*)$, i.e., $\mathcal{E}(f^*)\mathcal{E}(h^*) = 1$. Namely, equality holds in (2.19). □

EXERCISES

2.1 *A flow.* Let G be the graph in Figure 2.4. Let $s = 0$ be the source node and $t = 4$ be the target node. Consider the orientation on G is given by orienting each edge from the tail to the head of the corresponding arrow as shown in Figure 2.4. Let f be the flow determined by Figure 2.4, so that if an arrow is oriented from x to y and has label z, then $f(x, y) = z$.

(i) Compute $f(0,3)$, $f(2,4)$, $f(3,2)$, and $f(1,3) + f(3,2) + f(2,1)$;

(ii) Compute the divergence of f at each node. Is f an st-flow? If yes, what is its value $\text{Val}(f)$?

(iii) Assume all the edge resistances are equal to one: $r(e) \equiv 1$. Compute the energy of f.

2.2 *A better flow.* Staying with the graph G and flow f in Figure 2.4, consider the cycle $W = 1\ 2\ 3$ and the corresponding unit st-flow k_W, as defined above equation (2.4). Define a new flow $g := f + k_W$.

(i) Compute the divergence of g at each node. Is g an st-flow? If yes, what is its value $\text{Val}(g)$? Compare $\text{Val}(g)$ to $\text{Val}(f)$.

(ii) Assume all the edge resistances are equal to one: $r(e) \equiv 1$. Compute the energy of g. Compare $\mathcal{E}(g)$ with $\mathcal{E}(f)$.

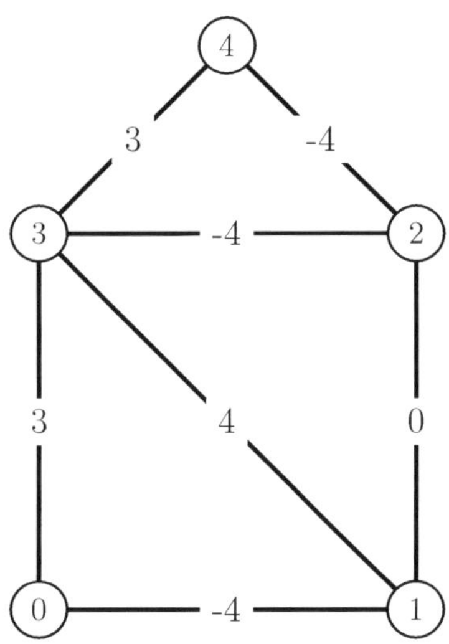

Figure 2.4 A flow.

(iii) Compute the total flow of g along the cycle W. Does g satisfy the Cycle Law (2.4)? Is g energy minimizing?

2.3 *Harmonicity.* Let G be the undirected graph underlying Figure 2.4, with unit conductances. Let $s = 0$ be the source node and $t = 4$ be the target node. Let $h : V \to \mathbb{R}$ be a function with $h(s) = 0$.

(i) Write down three linear equations, one for each node $\{1, 2, 3\} = V \setminus \{s, t\}$, imposing that h is harmonic at $x = 1, 2, 3$. Then, leave $c := h(t)$ as a parameter and solve this system of equations;

(ii) use Ohm's law to impose that the flow induced by h has value one and determine h uniquely;

(iii) deduce the flow f induced by h.

2.4 *Local maximum for harmonic functions.* Consider a square graph as in Exercise 1.1 and label its vertices A, B, C, D clockwise. Add the diagonal $\{B, D\}$ and a vertex E at the midpoint of this diagonal so that $V = \{A, B, C, D, E\}$. Now let $s = A$ be the source and $t = C$ be the target. Define a function $h \in \mathbb{R}^V$ so that $h(s) = 0$, $h(t) = 1$, and $h(B) = h(D) = 1/2$.

(i) How should h be defined at E so as to be harmonic on $V \setminus \{s, t\}$?

(ii) Deduce from (i) that a potential h as in Theorem 2.2 may attain a local maximum at a node $x \neq s, t$;

(iii) Show that a potential h as in Theorem 2.2 that attains a local maximum at a node $x \neq s, t$, may not have a neighbor $y \sim x$ with $h(y) < h(x)$.

2.5 *The Wheatstone example.* Consider Example 2.3.

(i) Compute $\text{eff}\mathcal{C}(s, t)$ by hand, using the star-mesh algorithm as described in the example.

(ii) Verify your calculations by writing code that computes the star-mesh algorithm. This function should accept as input a graph G, a source node s and a target node t. It will then find an interior node with the smallest degree and remove it with the star-mesh transformation. Finally, it should return the graph G with one less interior node but with an added complete graph on the neighbors of the removed node. Make sure to eliminate all parallel edges and update the edge conductances.

2.6 *Gradient flows and stars.* Write a proof of (2.14) by first considering the gradient flow induced by the potentials $h_x := \delta_x$ for $x \in V$, given by the indicator function of the node (one can also think of these as the standard basis vectors in $\mathbb{R}^V$).

2.7 *Potential gradient flows.* Let $G = (V, E)$ be a finite connected simple graph with edge conductances $c \in \mathbb{R}^E_{>0}$. Fix a source s and a target $t \neq s$, with $s, t \in V$. Let $h^* \in \mathbb{R}^V$ be the voltage potential with $h^*(s) = 0$ and $h^*(t) = 1$, so that h^* is harmonic at every node $x \in U := V \setminus \{s, t\}$. Also, let $f^* = cdh^*$ be the corresponding current flow. Let $h \in \mathbb{R}^V$ be an arbitrary potential function with $h(s) = 0$ and $h(t) = 1$, and let f be an unrelated arbitrary st-flow. Finally let $\pi = \mathrm{Proj}_{\mathrm{Span}(f^*)}$ be the self-map of $(F, \langle \cdot, \cdot \rangle_r)$ that carries out the orthogonal projection onto the linear subspace $\mathrm{Span}(f^*)$.

(i) Use Lemma 2.3, and the fact that $cdh - \pi(cdh) \in S_U$, to find a potential function ϕ with $\phi(s) = \phi(t) = 0$ such that $cd\phi = \pi(cdh) - cdh$.

(ii) Deduce from (i) that $\pi(cdh) = cdh^*$ and thus $\mathcal{E}(h) \geq \mathcal{E}(h^*)$.

(iii) Use orthogonality to show that:

$$\mathcal{E}(f - f^*) + \mathcal{E}(f^* - cdh) = \mathcal{E}(f - cdh).$$

See [47] for more information.

2.8 *Rayleigh's monotonicity.* Let $G = (V, E)$ be a finite connected simple graph. Use Thomson's principle to show that: if $\tilde{r}(e) \geq r(e)$ for every $e \in E$, then for every pair of nodes $a, b \in V$, $a \neq b$, we have

$$\mathrm{eff}\mathcal{R}_{\tilde{r}}(a, b) \geq \mathrm{eff}\mathcal{R}_r(a, b).$$

In particular, removing an edge is tantamount to increasing the resistance on that edge to infinity.

2.9 *Flow symmetry.* Let $G = (V, E)$ be a finite connected simple graph with edge resistances $r \in \mathbb{R}^E_{>0}$. Pick two distinct edges $\{s, t\}$ and $\{a, b\}$. Let $f_{s,t}$ be the unit current st-flow and $f_{a,b}$ be the unit current ab-flow. Compute the inner product $\langle f_{s,t}, f_{a,b} \rangle_r$, by mimicking the proof of Lemma 2.2, as done in the proof of Theorem 2.3. Then, deduce that

$$r(a,b)f_{s,t}(a,b) = r(s,t)f_{a,b}(s,t).$$

In particular, in the case of unit resistances, we get the surprising fact that flowing one Ampere of current from a to b and then measuring the positive current across $\{s,t\}$ yields the same result as flowing one Ampere from s to t and then measuring the positive current across $\{a,b\}$.

2.10 *Edge conductance monotonicity.* Let $G = (V, E)$ be a finite connected simple graph. Fix an edge $\{a,b\} \in E$. Let $c, \tilde{c} \in \mathbb{R}^E_{>0}$ be two sets of edge conductances such that $c(e') = \tilde{c}(e')$ for all $e' \neq \{a,b\}$, and $\tilde{c}(a,b) \leq c(a,b)$. Fix a source s and a target $t \neq s$, with the property that $\{s,t\} \in E$. Let $f_{s,t}$ and $\tilde{f}_{s,t}$ be the unit current *st*-flows corresponding to each set of edge conductances. Show that

$$|\tilde{f}_{s,t}(a,b)| \leq |f_{s,t}(a,b)|.$$

Hint: Use Exercise 2.9. You may also want to consider whether $\{a,b\}$ is a cut edge and then consider the parallel rule. (This problem and the previous one are from [75]; it will be generalized later in the book.)

CHAPTER 3

The Laplacian and its Spectral Theory

WE RECALL the basic spectral theory of real symmetric matrices and apply it to the Laplacian matrix of a graph. In particular, we revisit the concept of effective resistance from the matrix point of view. For a more in-depth study of the spectral theory of real symmetric matrices, we refer the reader to standard references such as [31], and for the Laplacian matrix we refer to [28,82], but especially [76].

3.1 REVIEW OF SPECTRAL THEORY

Let H be an $N \times N$ matrix. For simplicity, we will assume H is real and symmetric.

Definition 3.1. A number $\lambda \in \mathbb{R}$ is an *eigenvalue* for H if there is $u \in \mathbb{R}^N \setminus \{0\}$ such that

$$Hu = \lambda u,$$

or equivalently, if λ is a root of the *characteristic polynomial* of H:

$$p(x) := \det(xI - H).$$

Theorem 3.1 (Spectral Theorem). *If H is a real, symmetric, $N \times N$ matrix, then there exists an orthonormal basis $\{u_j\}_{j=1}^N$ of $\mathbb{R}^N$, such that each u_j is an eigenvector of H for some eigenvalue $\lambda_j \in \mathbb{R}$.*

We will prove this theorem below.

Note that for non-symmetric matrices, the situation is much more complicated. Even in the symmetric case, the eigenvalues λ_j may repeat, in which case the corresponding eigenvectors u_j are unique only

DOI: 10.1201/9781003024866-4

up to orthogonal transformations. However, for symmetric matrices, an eigenvalue's geometric multiplicity and the algebraic multiplicity always coincide.

Showing that the eigenvalues of the symmetric matrix H in Theorem 3.1 are real is an exercise that is left to the reader (argue by contradiction and then take the imaginary part). In particular, they can always be ordered, for instance, as follows:

$$\lambda_1 = \lambda_{\min}(H) \leq \lambda_2 \leq \cdots \leq \lambda_{\max}(H) = \lambda_N.$$

Before proving the Spectral Theorem 3.1, we need to establish a few important facts and definitions.

If $u \neq 0$ is a non-zero vector in $\mathbb{R}^N$, the *Rayleigh quotient* of u with respect to H is defined as:

$$Q(u) := \frac{u^T H u}{u^T u}. \tag{3.1}$$

A simple computation shows that, if u is an eigenvector for H corresponding to the eigenvalue λ, then

$$Q(u) = \frac{u^T H u}{u^T u} = \frac{u^T \lambda u}{u^T u} = \frac{\lambda u^T u}{u^T u} = \lambda. \tag{3.2}$$

Also, the Rayleigh quotient is homogeneous, in the sense that

$$Q(tu) = \frac{(tu)^T H(tu)}{(tu)^T(tu)} = \frac{u^T H u}{u^T u} = Q(u), \qquad \forall t > 0. \tag{3.3}$$

Theorem 3.2 (Maximum Stretch). *Suppose H is a real, symmetric, $N \times N$ matrix. Then,*

$$\lambda_{\max}(H) = \max_{w \neq 0} \frac{w^T H w}{w^T w}. \tag{3.4}$$

Moreover, the maximum is attained at vector $u \neq 0$ if and only if u is an eigenvector for H for the eigenvalue $\lambda_{\max}(H)$.

Remark 3.1. Recall that λ is an eigenvalue for H if and only if $-\lambda$ is an eigenvalue for $-H$. Therefore, we also have

$$\lambda_{\min}(H) = \min_{w \neq 0} \frac{w^T H w}{w^T w}. \tag{3.5}$$

as can be seen by applying the Maximum Stretch Theorem 3.2 to $-H$.

Proof. By homogeneity (3.3), the maximum in (3.4) can be written as

$$\max_{\|u\|^2=1} u^T H u.$$

Since the unit sphere $\mathcal{S} := \{u \in \mathbb{R}^N : \|u\|^2 = 1\}$ is compact, the maximum is always attained, and we are justified in using a maximum instead of a supremum.

Maximizing the quadratic form $u^T H u$ on the unit sphere $\mathcal{S}$ is a constrained optimization problem.

$$\begin{aligned} \text{maximize} \quad & u^T H u \\ \text{subject to} \quad & u^T u = 1. \end{aligned}$$

Hence, it can be solved using the method of Lagrange multipliers. First, we assign a dual variable λ to the only constraint, namely $u^T u = 1$. The Lagrangian for this problem is the function:

$$\mathcal{L}(u, \lambda) := u^T H u + \lambda(1 - u^T u) = \sum_{x,y=1}^{N} u(x)H(x,y)u(y) - \lambda \sum_{x=1}^{N} u(x)^2 + \lambda.$$

Stationarity implies that the gradient $\mathcal{L}$ with respect to the variable u vanishes for an optimal pair (u^*, λ^*):

$$\nabla_u \mathcal{L}(u^*, \lambda^*) = 2Hu^* - 2\lambda^* u^* = 0. \tag{3.6}$$

In particular, an optimal pair (u^*, λ^*) must satisfy $Hu^* = \lambda^* u^*$, i.e., u^* must be an eigenvector for H corresponding to λ^*. Taking the dot product of (3.6) with u^*, and using Lemma 3.2, we get that λ^* must equal $\lambda_{\max}(H)$. □

We are now ready to prove the Spectral Theorem 3.1.

Proof of Spectral Theorem 3.1. Use (3.5) to find a vector $u_1 \in \mathbb{R}^N$ such that

$$u_1^T u_1 = 1 \qquad \text{and} \qquad u_1^T H u_1 = \lambda_{\min}(H) = \min_{u^T u=1} u^T H u.$$

Then, recursively find u_k that solves the following constrained optimization problem:

$$\begin{aligned} \text{minimize} \quad & u^T H u \\ \text{subject to} \quad & u^T u = 1 \text{ and } u \perp \{u_1, \ldots, u_{k-1}\} \end{aligned} \tag{3.7}$$

The minimizer u^* for problem (3.7) is a unit vector, orthogonal to each u_j for $j = 1, \ldots, k-1$, that satisfies the eigenvector equation $Hu^* = \lambda_k u^*$, where λ_k is the minimum value of (3.7). To see this, note that we now have k constraints: $u^T u = 1$ and $u^T u_j = 0$ for $j = 1, \ldots, k-1$. So we need k Lagrangian dual variables $\mu_1, \ldots, \mu_k$. Then, the Lagrangian for this problem is:

$$\begin{aligned}\mathcal{L}(u, \mu_1, \ldots, \mu_k) &:= u^T Hu + \mu_k(1 - u^T u) + \sum_{j=1}^{k-1} \mu_j u_j^T u \\ &= \sum_{x,y=1}^{N} u(x)H(x,y)u(y) - \mu_k \sum_{x=1}^{N} u(x)^2 \\ &+ \sum_{j=1}^{k-1} \mu_j \sum_{x=1}^{N} u_j(x)u(x) + \mu_k.\end{aligned}$$

Stationarity implies that the gradient of $\mathcal{L}$ with respect to the variable u should vanish when $(u^*, \mu_1^*, \ldots, \mu_k^*)$ is an optimal solution. Namely:

$$\nabla_u \mathcal{L} = 2Hu^* - 2\mu_k^* u^* + \sum_{j=1}^{k-1} \mu_j^* u_j = 0. \tag{3.8}$$

Notice that by induction the unit vectors $\{u_1, \ldots, u_{k-1}, u^*\}$ are mutually orthogonal. Moreover, for $j = 1, \ldots, k-1$,

$$u_j^T Hu^* = (u^*)^T Hu_j = (u^*)^T(\lambda_j u_j) = \lambda_j u_j^T u^* = 0.$$

Taking the dot product of (3.8) with u_j, we see that $\mu_j^* = 0$ for $j = 1, \ldots, k-1$. Thus, by (3.8),

$$Hu^* = \mu_k^* u^*.$$

So u^* is an eigenvector for H corresponding to μ_k^*, and taking the dot product of (3.8) with u^*, we get that μ_k^* is the minimum value of (3.7). Finally, let $u_k := u^*$ and $\lambda_k = \mu_k^*$. □

The Spectral Theorem 3.1 can be used to give the following important *variational characterization* of the eigenvalues of a real symmetric matrix.

Theorem 3.3 (Courant-Fisher Min-Max Theorem). *Let H be an $N \times N$ real symmetric matrix. Consider the eigenvalues ordered so that $\lambda_1(H) \leq \cdots \leq \lambda_N(H)$. Then:*

$$\lambda_k(H) = \min_{S\in\mathcal{W}_k} \max_{u\in S\setminus\{0\}} \frac{u^T Hu}{u^T u} = \max_{T\in\mathcal{W}_{N-k+1}} \min_{u\in T\setminus\{0\}} \frac{u^T Hu}{u^T u} \tag{3.9}$$

where $\mathcal{W}_j := \{U \subset \mathbb{R}^N : U$ is a linear subspace of $\mathbb{R}^N$ and $\dim U = j\}$.

Moreover, if $\{u_1, \ldots, u_N\}$ is an orthonormal basis of eigenvectors such that $Hu_k = \lambda_k u_k$, then $\min_{S\in\mathcal{W}_k}$ in (3.9) is attained by $S_k = \mathrm{Span}(u_1, \ldots, u_k)$, while $\max_{T\in\mathcal{W}_{N-k+1}}$ is attained by $T_k = \mathrm{Span}(u_k, \ldots, u_N)$.

Example 3.1. Suppose $k = 1$. Then $\dim S = 1$ means the $S = \mathrm{Span}(u)$ for some $u \neq 0$. Also $\dim T = N$ means that $T = \mathbb{R}^N$. So

$$\lambda_1(H) = \min_{\|u\|=1} u^T Hu.$$

Proof of Courant-Fisher Min-Max Theorem 3.3. By the Spectral Theorem 3.1, there is an orthonormal basis $\{u_j\}_{j=1}^N$ of eigenvectors for H, with u_j corresponding to $\lambda_j := \lambda_j(H)$. As we will see, for every $k = 1, \ldots, N$, the optima in (3.9) are achieved by the subspaces $S_k = \mathrm{Span}(u_1, \ldots, u_k)$ and $T_k = \mathrm{Span}(u_k, \ldots, u_N)$. Here, we will show the min-max characterization. The max-min characterization is proved similarly and is left to the reader (see Exercise 3.5).

By orthogonality, if $u \in \mathbb{R}^N$, then we can write $u = \sum_{j=1}^N \langle u, u_j\rangle u_j$. Moreover, if $u \in S_k$, then $u = \sum_{j=1}^k \langle u, u_j\rangle u_j$. Thus,

$$Hu = \sum_{j=1}^k \langle u, u_j\rangle H(u_j) = \sum_{j=1}^k \langle u, u_j\rangle \lambda_j u_j.$$

So

$$u^T Hu = \sum_{j=1}^k \langle u, u_j\rangle^2 \lambda_j \leq \lambda_k \sum_{j=1}^k \langle u, u_j\rangle^2 = \lambda_k\, u^T u.$$

Therefore,

$$\min_{S\in\mathcal{W}_k} \max_{u\in S\setminus\{0\}} \frac{u^T Hu}{u^T u} \leq \max_{u\in S_k} \frac{u^T Hu}{u^T u} \leq \lambda_k. \tag{3.10}$$

To conclude, it will be enough to show that

$$\min_{S\in\mathcal{W}_k} \max_{u\in S\setminus\{0\}} \frac{u^T Hu}{u^T u} \geq \lambda_k. \tag{3.11}$$

To that end, fix an arbitrary subspace $S \subset \mathbb{R}^N$ with $\dim S = k$. Since $\dim(T_k) = N - k + 1$, we must have $\dim(S \cap T_k) \geq 1$ (if not, we would get $N + 1$ linearly independent vectors). In particular,

$$\max_{u \in S} \frac{u^T Hu}{u^T u} \geq \max_{u \in S \cap T_k} \frac{u^T Hu}{u^T u}. \tag{3.12}$$

Thus assume that $u \in S \cap T_k$ and $u \neq 0$. Since $u \in T_k$, we can write $u = \sum_{j=k}^N \langle u, u_j \rangle u_j$. And,

$$u^T Hu = \sum_{j=k}^{N} \langle u, u_j \rangle^2 \lambda_j \geq \lambda_k \sum_{j=k}^{N} \langle u, u_j \rangle^2 = \lambda_k \; u^T u.$$

This means that

$$\max_{u \in S \cap T_k} \frac{u^T Hu}{u^T u} \geq \lambda_k,$$

and by (3.12),

$$\max_{u \in S} \frac{u^T Hu}{u^T u} \geq \lambda_k. \tag{3.13}$$

If $\mathcal{W}_k$ is the set of all linear subspaces of $\mathbb{R}^N$ with $\dim(S) = k$, then (3.13) implies (3.11). Moreover, letting $S = S_k$ in (3.13), in combination with (3.10), we get

$$\max_{u \in S_k} \frac{u^T Hu}{u^T u} = \lambda_k.$$

Therefore, the minimum in (3.11) is attained by $S = S_k$ and equals λ_k.

□

3.2 THE LAPLACIAN MATRIX

Probably the most fundamental matrix for a finite simple graph $G = (V, E)$ with given edge conductances $\sigma : E \to (0, \infty)$ is the *Laplacian* $L : V \times V \to \mathbb{R}$:

$$L_\sigma(x, y) = \begin{cases} \deg_\sigma(x) := \sum\limits_{z \sim x} \sigma(x, z) & \text{if } x = y \\ -\sigma(x, y) & \text{if } x \sim y \\ 0 & \text{if } x \not\sim y. \end{cases} \tag{3.14}$$

Write $D_\sigma := \text{diag}(\deg_\sigma(\cdot))$ for the diagonal matrix with the generalized degrees $\deg_\sigma(x)$ for $x \in V$, and let the generalized adjacency matrix

$A_\sigma(x,y) := \sigma(x,y)\mathbb{1}_{x\sim y}$. Then

$$L_\sigma = D_\sigma - A_\sigma.$$

We often omit the subscript σ if it is clear from the context. Thinking of a function $h : V \to \mathbb{R}$ has a vector in $\mathbb{R}^N$ we get that

$$(Lh)(x) = \sum_{y\in V} L(x,y)h(y) = \deg_\sigma(x)h(x) - \sum_{y\sim x} \sigma(x,y)h(y).$$

In particular, by (2.8), $Lh = 0$ if and only if h is harmonic at every $x \in V$.

Recall that a real, symmetric $N \times N$ matrix H is said to be *positive semi-definite* if

$$v^T Hv \geq 0, \qquad \text{for all } v \in \mathbb{R}^N \setminus \{0\}. \tag{3.15}$$

It follows directly from the Courant-Fisher Theorem, Equation (3.9), that in this case the eigenvalues $\lambda(H)$ are necessarily non-negative (and in fact, this is also sufficient, see Exercise 3.4.)

Proposition 3.1. *The Laplacian matrix L is a symmetric and positive semidefinite matrix. In particular, its eigenvalues are real and non-negative, so they can be ordered as follows:*

$$0 = \lambda_1 \leq \cdots \leq \lambda_N.$$

Moreover, if k is the geometric multiplicity of $\lambda_{\min}(L) = 0$, namely $k := \max\{j : \lambda_j = 0\}$, then the corresponding graph G has exactly k connected components. In particular, if G is connected, then the eigenvalue $\lambda_1 = 0$ is simple, and the corresponding eigenspace is spanned by the constant function $u = [1\ 1\ \cdots 1]^T$.

Remark 3.2. Note that for real, symmetric $N \times N$ matrices, each eigenvalue's geometric multiplicity and algebraic multiplicity coincide; see Exercise 3.3.

Proof of Proposition 3.1. Symmetry is clear from the definition. To prove that L is positive semidefinite, we will show that L factors as the square of a rectangular matrix.

Assume that an orientation $\vec{E}$ has been chosen as in Definition 1.1.

Define the $m \times N$ *gradient matrix*

$$B(\vec{e}, x) = \begin{cases} 1 & \text{if } \vec{e} = (y, x) \text{ for some } y. \\ -1 & \text{if } \vec{e} = (x, y) \text{ for some } y. \\ 0 & \text{else} \end{cases} \tag{3.16}$$

We think of B as a "gradient" operator that turns a function h defined on V into a function $f = Bh$ defined on $\vec{E}$ by setting

$$f(\vec{e}) := h(y) - h(x) \qquad \text{if } \vec{e} = (x, y).$$

That's because every row in B is of the form $\delta_y - \delta_x$ with $\vec{e} = (x, y) \in \vec{E}$, where δ_j is the j-th standard basis vector in $\mathbb{R}^N$. The matrix B is closely related to the linear map $p : \mathbb{R}^V \to F$ and the gradient flows of the form cdh, defined in Section 2.5.

Later, we will see that the transpose B^T can be thought of as a "divergence" operator that transforms flows defined on the edges into functions defined on the vertices.

Lemma 3.1. *Let Σ be a diagonal $|E| \times |E|$ matrix listing all the edge conductance $\Sigma(e, e) := \sigma(e)$ for $e \in E$. Then*

$$L = B^T \Sigma B = \sum_{e=\{x,y\} \in E} \sigma(e)(\delta_y - \delta_x)(\delta_y - \delta_x)^T.$$

Proof of Lemma 3.1. For simplicity, let's assume that Σ is the identity matrix, namely that all the edge conductances are equal to 1. Then

$$(B^T B)(x, y) = \sum_{\vec{e} \in \vec{E}} B^T(x, \vec{e}) B(\vec{e}, y) = \sum_{e \in E} B(e, x) B(e, y),$$

where we're using the abuse of notation described in Remark 2.1. Two cases arise. If $x = y$, then the sum is over edges that are incident to x:

$$(B^T B)(x, x) = \sum_{\vec{e} \in \vec{E}} B(\vec{e}, x)^2 = \sum_{e \in E} B(e, x)^2 = \deg(x).$$

If $x \neq y$, the sum is over a single edge incident to x and y.

$$(B^T B)(x, y) = B(\vec{e}, x) B(\vec{e}, y) = -1.$$

So $(B^T B)(x, y) = D(x, y) - A(x, y)$.

Alternatively, we can use block matrix multiplication, see Exercise 3.2. Think of B as an $|E| \times 1$ column vector whose entries are $1 \times |V|$ row vectors of the form $(\delta_y - \delta_x)^T$ for each $\vec{e} = (x, y) \in \vec{E}$. Then, B^T is a $1 \times |E|$ row vector whose entries are $|V| \times 1$ column vectors of the form $(\delta_y - \delta_x)$. So, $B^T B$ can be thought of as a dot product:

$$B^T B = \sum_{e \in E} (\delta_y - \delta_x)(\delta_y - \delta_x)^T.$$

Again, the sum can be taken for $e \in E$ because the matrices $(\delta_y - \delta_x)(\delta_y - \delta_x)^T$ are symmetric in x and y.

Each matrix $(\delta_y - \delta_x)(\delta_y - \delta_x)^T$ is the Laplacian matrix of a graph on the sem vertices V, but with only one edge $\{x, y\}$. One can see that the graph Laplacian L can be built as a sum of these simple Laplacians for graphs with a single edge simply by summing them up. □

Continuing with the proof of Proposition 3.1, consider the quadratic form induced by L. Then, by Lemma 3.1,

$$u^T L u = u^T B^T \Sigma B u = \|\Sigma^{1/2} B u\|_2^2 \geq 0. \tag{3.17}$$

This shows that L is positive semidefinite.

Remark 3.3. It is useful to be able to go back and forth between matrix notation and *sigma notation* (as some people say, the expression written *in coordinates*.) For instance, in (3.17)

$$\begin{aligned} u^T L u &= \sum_{x,y \in V} u(x) L(x,y) u(y) \\ &= \|\Sigma^{1/2} B u\|_2^2 = \sum_{\vec{e} \in \vec{E}} \Sigma(e)[(Bu)(\vec{e})]^2 \\ &= \sum_{\{x,y\} \in E} \sigma(x,y)(u(x) - u(y))^2. \end{aligned}$$

In other words, the matrix identity (3.17) turns the Laplacian quadratic form from a sum over vertices $x, y \in V$, to a sum over edges $e \in E$.

Finally, we show the 'Moreover' part of Proposition 3.1. Since the row sums of L equal zero, the constant functions always belong to the kernel of L, or in other terminology, constant functions are always harmonic, independently of the choice of edge conductances. Thus,

$$\mathrm{Span}(\mathbf{1}) = \langle \mathbf{1} \rangle \subset \mathrm{Ker}(L).$$

In particular, $\lambda_1 = 0$ is always an eigenvalue of L.

Conversely, assume $Lu = 0$. By (3.17), $\mathrm{Ker}(L) = \mathrm{Ker}(B)$, so $Bu = 0$. This means that $u(x) = u(y)$ whenever $x \sim y$. When G is connected, this implies that u is a constant function. More generally, assume G has k connected components. We can order them as $G_1, \ldots, G_k$. Consider the indicator functions $u_j := \mathbb{1}_{V(G_j)}$. Note that $Lu_j = 0$ and $u_i \perp u_j$ for $i \neq j$. Thus, the geometric multiplicity of $\lambda_1 = 0$ is at least k. Conversely, if $Lu = 0$, then reasoning as above, u is constant on each component G_j, and hence $u \in \mathrm{Span}\{u_j\}_{j=1}^k$. This shows that $\dim \mathrm{Ker}(L) = k$. □

3.3 THE ALGEBRAIC CONNECTIVITY

We have seen that the Laplacian L is a real symmetric, positive semidefinite, $N \times N$ matrix, with eigenvalues $0 = \lambda_1 \leq \cdots \leq \lambda_N$. The Courant-Fisher Theorem 3.3 can be used to provide a link between the Laplacian quadratic form $u^T Lu$ (for $u \in \mathbb{R}^N$,) and the eigenvalues of L. For instance, the largest eigenvalue can be computed by maximizing the quadratic form:

$$\lambda_N = \max_{\|u\|=1} u^T Lu.$$

Note, however, that since the kernel of L is non-empty, we always have $\lambda_1 = 0 = \min_{\|u\|=1} u^T Lu$.

Moreover, we have seen that G is connected if and only if the kernel of L equals $\langle \mathbf{1} \rangle$, the space consisting of the constant functions. Namely, when G is connected, λ_1 is simple, and thus λ_2 must be positive. On the other hand, if G is disconnected, then $\lambda_2 = 0$. This implies that the magnitude of the second eigenvalue λ_2 can be thought of as a measure of connectedness for the graph G. Indeed, λ_2 is known as the *algebraic connectivity* of G.

To link λ_2 to the quadratic form, consider the orthogonal complement to the constant functions $\langle \mathbf{1} \rangle$:

$$\langle \mathbf{1} \rangle^\perp = \left\{ u \in \mathbb{R}^V : \sum_{x \in V} u(x) = 0 \right\}.$$

When G is connected, the Courant-Fisher Theorem 3.3 can be used to show that:

$$0 < \lambda_2 = \min_{\substack{u \in \langle \mathbf{1} \rangle^\perp \\ \|u\|=1}} u^T Lu. \tag{3.18}$$

To see this, note that when $k = 2$, the subspace $T_k = \langle u_2, \dots, u_N \rangle = \langle \mathbf{1} \rangle^\perp$ attains the maximum in the second inequality of Equation (3.9), see Exercise 3.5.

In particular, every choice of $u \in \langle \mathbf{1} \rangle^\perp$ yields an upper bound for λ_2. However, lower bounds are more challenging to obtain.

By the Spectral Theorem 3.1, there is an orthonormal basis of eigenvectors $\{u_j\}_{j=1}^N$ so that u_j is an eigenvector of λ_j. So if the algebraic connectivity λ_2 is simple, then the corresponding eigenvector u_2 is known as the *Fiedler vector*. Note that u_2 belongs to $\langle \mathbf{1} \rangle^\perp$ and therefore provides a way of splitting the nodes of V into two subsets, the set where $u_2 \geq 0$ and the set where $u_2 \leq 0$. This goes under the name of the *Fiedler cut*. We will say more about this later.

Example 3.2. Let $G = P_N$ be the path graph on N nodes $\{1, \dots, N\}$, so that the edges are of the form $\{i, i+1\}$, for $i = 1, \dots, N-1$. Let h be the linear function such that $h(1) = N - 1$ and $h(N) = -(N-1)$, namely set $h(j) := N + 1 - 2j$. Note that the set of nodes where $h > 0$ is $1 \leq j < \frac{N+1}{2}$. Then

$$\sum_{j=1}^N h(j) = N(N+1) - 2\sum_{j=1}^N j = 0.$$

So $h \perp \mathbf{1}$ and therefore,

$$\begin{aligned}
\lambda_2(P_N) &= \min_{u \in \langle \mathbf{1} \rangle^\perp} \frac{u^T L u}{\|u\|^2} && \text{(by (3.18))} \\
&\leq \frac{\sum\limits_{j=1}^{N-1} (h(j) - h(j+1))^2}{\sum\limits_{j=1}^{N} h(j)^2} && \text{(by Remark 3.3))} \\
&= \frac{\sum\limits_{j=1}^{N-1} 4}{\sum\limits_{j=1}^{N} (N+1-2j)^2} = \frac{12}{N(N+1)}.
\end{aligned}$$

3.4 THE GREEN OPERATOR

As we have seen above, the Laplacian matrix L admits an orthonormal basis of eigenvectors $\{u_j\}$, where $Lu_j = \lambda_j u_j$ and $0 = \lambda_1 \leq \cdots \leq \lambda_N$.

Consider the orthogonal matrix

$$U = \begin{bmatrix} | & & | \\ u_1 & \cdots & u_N \\ | & & | \end{bmatrix}$$

obtained by making the columns of U equal to these eigenvectors. Also let $\Lambda = \mathrm{Diag}(\lambda_1, \ldots, \lambda_N)$. Then, $L = U\Lambda U^T$ (see Exercise 3.3). In coordinates, for every $x, y \in V$,

$$\begin{aligned} L(x,y) &= \sum_{i,j=1}^{N} U(x,j)\Lambda(j,i)U^T(i,y) \\ &= \sum_{j=1}^{N} U(x,j)\Lambda(j,j)U(y,j) \\ &= \sum_{j=1}^{N} \lambda_j u_j(x) u_j(y) \\ &= \sum_{j=1}^{N} \lambda_j (u_j u_j^T)(x,y). \end{aligned}$$

Therefore, we can write

$$L = \sum_{j=2}^{N} \lambda_j u_j u_j^T, \tag{3.19}$$

where each matrix $u_j u_j^T$ is a rank one matrix representing the projection operator onto the line spanned by u_j. Also, since u_1 can be chosen to be the vector $\frac{1}{\sqrt{N}}[1\ 1\ \cdots 1]^T$, we have that $\mathrm{Span}\{u_2, \ldots, u_N\} = \langle \mathbf{1} \rangle^{\perp}$.

Assuming that G is connected, we get that $\lambda_2 > 0$ and that L restricted to $\langle \mathbf{1} \rangle^{\perp}$ is invertible. In particular, we can define the *Green operator* to be the pseudoinverse of L. Namely, we set

$$\mathcal{G} := \sum_{j=2}^{N} \frac{1}{\lambda_j} u_j u_j^T.$$

Then, one can verify that

$$\mathcal{G}L = L\mathcal{G} = \sum_{j=2}^{N} u_j u_j^T = \mathrm{Proj}_{\langle \mathbf{1} \rangle^{\perp}} = I - \frac{1}{N}\mathbf{1}\mathbf{1}^T. \tag{3.20}$$

A common way to denote the pseudoinverse of a matrix A is by the symbol A^+. So, the Green operator is $\mathcal{G} = L^+$.

3.5 ELECTRIC CURRENT FLOW REVISITED

Recall that, having chosen an orientation $\vec{E}$, the gradient matrix B defined in (3.16) takes a potential function h defined on V and yields a flow Bh defined on $\vec{E}$. On the other hand, the transpose B^T behaves like a divergence operator. Namely, let f be a flow. Given an orientation $\vec{E}$ on G, we can think of f as a vector $f \in \mathbb{R}^E$. Then, for $x \in V$:

$$\begin{aligned} B^T f(x) &= \sum_{\vec{e} \in \vec{E}} B^T(x, \vec{e}) f(\vec{e}) = \sum_{\vec{e} \in \vec{E}} B(\vec{e}, x) f(\vec{e}) \\ &= \sum_{\substack{y \sim x \\ (y,x) \in \vec{E}}} f(y, x) - \sum_{\substack{y \sim x \\ (x,y) \in \vec{E}}} f(x, y) \\ &= -\operatorname{Div}_f(x). \end{aligned}$$

Recall that δ_a is the $N \times 1$ column vector such that $\delta_a(a) = 1$ and $\delta_a(x) = 0$ otherwise. With that notation, we can impose that $\operatorname{Val}(f) = 1$ and that the node law holds at every $x \in V \setminus \{s, t\}$, i.e., that f is a unit s-t flow, by requiring that

$$B^T f = \delta_t - \delta_s.$$

Define an $|E| \times |E|$ diagonal matrix $R = \operatorname{diag}(r(\cdot))$ with the edge resistances on the diagonal and its inverse $\Sigma = R^{-1}$ with the conductances instead. Note that the energy of a flow as defined in (2.1) can be rewritten as

$$\mathcal{E}_r(f) = \sum_{e \in E} r(e)(f(e))^2 = f^T R f = \|R^{1/2} f\|^2.$$

The problem of finding the unit current flow from s to t is the following:

$$\begin{aligned} &\underset{f \in \mathbb{R}^E}{\text{minimize}} && \|R^{1/2} f\|^2 \\ &\text{subject to} && B^T f = \delta_t - \delta_s. \end{aligned} \tag{3.21}$$

Existence and uniqueness can then be deduced from standard results in convex optimization.

If h is the potential function of the flow f, then Ohm's law (2.5) can be rewritten as:

$$f = \Sigma B h. \tag{3.22}$$

Also, by Lemma 3.1, the Laplacian can be written as in the continuous case as the "divergence of the gradient": $L = B^T \Sigma B$. In terms of the potential, we then have

$$Lh = B^T \Sigma B h = B^T f = \delta_t - \delta_s. \tag{3.23}$$

The bad news is that L is not invertible because its row sums are zero. The good news is that when G is connected, the constant functions are the only obstruction. Namely, let $\langle \mathbf{1} \rangle$ be the set of constant functions as above, then $\left(L_{|\langle \mathbf{1} \rangle^\perp}\right)^{-1}$ makes sense and coincides with the restricted Green operator $\mathcal{G}_{|\langle \mathbf{1} \rangle^\perp}$.

In particular, since $\delta_t - \delta_s \in \langle \mathbf{1} \rangle^\perp$, we find that

$$h = \mathcal{G}(\delta_t - \delta_s) \tag{3.24}$$

is orthogonal to $\mathbf{1}$ and solves (3.23) by (3.20).

Note that $\langle \mathbf{1} \rangle$ is also the null space of B, so $\Sigma B h$ does not change if we add a constant function to h. Usually, the potential h is normalized so that $h(s) = 0$.

We can write the current flow as $f = \Sigma B \mathcal{G}(\delta_t - \delta_s)$. Then, the energy of f can be rewritten as

$$\begin{aligned}
\mathcal{E}_r(f) &= f^T R f \\
&= (\delta_t - \delta_s)^T \mathcal{G}^T B^T (\Sigma^T R) \Sigma B \mathcal{G}(\delta_t - \delta_s) \\
&= (\delta_t - \delta_s)^T \mathcal{G}^T L \mathcal{G}(\delta_t - \delta_s) \\
&= (\delta_t - \delta_s)^T \mathcal{G}(\delta_t - \delta_s),
\end{aligned}$$

where the last equality follows from (3.20) and the fact that $\mathcal{G}^T = \mathcal{G}$.

Since $h = \mathcal{G}(\delta_t - \delta_s)$ and $Lh = (\delta_t - \delta_s)$, we can also write

$$\mathcal{E}_r(f) = h^T L \mathcal{G} L h = h^T L h = (\delta_t - \delta_s)^T h = h(t) - h(s) = \operatorname{eff}\mathcal{R}(s,t). \tag{3.25}$$

Putting this together, we obtain the following expression for computing effective resistance between two nodes:

$$\operatorname{eff}\mathcal{R}(s,t) = (\delta_t - \delta_s)^T \mathcal{G}(\delta_t - \delta_s). \tag{3.26}$$

Moreover, we also have

$$\operatorname{eff}\mathcal{R}(s,t) = h^T L h = \sum_{\{x,y\} \in E} \sigma(x,y)(h(x) - h(y))^2, \tag{3.27}$$

which gives a connection between effective resistance and the voltage potential h. In particular, as seen in Section 2.5, the expression on the right is the Dirichlet energy of h.

A key observation is that Laplacian matrices are always *diagonally dominant*, and this can be shown to imply that $\mathcal{G}$ can be computed in "almost" linear time, see [78]. This makes effective resistance a computable quantity, even on large networks.

Using the fact that $\mathcal{G}$ is symmetric and positive semidefinite, we can define its square root via the singular value decomposition as follows:

$$\mathcal{G}^{1/2} := \sum_{j=2}^{N} \frac{1}{\lambda_j^{1/2}} u_j u_j^T .$$

Then (3.26) shows that effective resistance is the square of the Euclidean metric restricted to the vectors $\{\mathcal{G}^{1/2}\delta_x\}_{x\in V} \subset \mathbb{R}^V$, since we can rewrite it as

$$\operatorname{eff}\mathcal{R}(s,t) = \|\mathcal{G}^{1/2}\delta_t - \mathcal{G}^{1/2}\delta_s\|^2. \tag{3.28}$$

3.6 EFFECTIVE RESISTANCE REVISITED

Recall for (2.18) that the effective conductance $\operatorname{eff}\mathcal{C}(s,t)$ is the minimum of the energy of potential functions h such that $h(t) - h(s) \geq 1$. In particular, any choice of a potential function yields an upper bound. Lower bounds are more challenging to obtain, but we will see later that they are related to the size of s-t cuts. Here, we begin by deducing some bounds for effective conductance in terms of the algebraic connectivity.

Proposition 3.2 (Algebraic connectivity bounds). *Let $G = (V, E)$ be a graph with algebraic connectivity λ_2. Set $N = |V|$. Then,*

$$\frac{\lambda_2}{2} \leq \min_{a\neq b} \operatorname{eff}\mathcal{C}(a,b) \leq N\frac{\lambda_2}{2}, \tag{3.29}$$

where the minimum is taken over all pairs of distinct nodes $a, b \in V$.

See Exercise 3.8 (iii) for the sharpness of the lower bound, and [60, Proposition 4.2] for a generalization of the upper bound to higher eigenvalues, where $\lambda_2/2$ is replaced by λ_k/k with $k = 3, \ldots, N$.

Proof of Proposition 3.2. For the lower bound, if G is disconnected and a and b belong to different components, then $\lambda_2 = 0$ and $\operatorname{eff}\mathcal{C}(a,b) = 0$ because one can pick a potential h that is identically equal to 0 on the component containing s and identically equal to 1 everywhere else, then $\mathcal{E}(h) = 0$.

Therefore, we can assume that G is connected so that $\lambda_2 > 0$. Then, λ_2^{-1} is the largest eigenvalue of $\mathcal{G}$. By the Maximum Stretch Theorem 3.2,

$$\lambda_2^{-1} = \max_{v \neq 0} \frac{v^T \mathcal{G} v}{\|v\|^2} \geq \frac{(\delta_b - \delta_a)^T \mathcal{G} (\delta_b - \delta_a)}{\|\delta_b - \delta_a\|^2} = \frac{\operatorname{eff}\mathcal{R}(a,b)}{2}$$

The result follows by taking reciprocals.

For the upper bound, let u_2 be a unit-norm eigenvector for λ_2. Then, λ_2 is the Dirichlet energy of u_2, and hence, for any $x \neq y \in V$,

$$\begin{aligned} \lambda_2 = u_2^T L u_2 &\geq |u_2(x) - u_2(y)|^2 \operatorname{eff}\mathcal{C}(x,y) \\ &\geq |u_2(x) - u_2(y)|^2 \min_{s \neq t} \operatorname{eff}\mathcal{C}(s,t) \end{aligned}$$

Note that this estimate holds even if $x = y$. Thus, summing over all pairs of nodes, and using the fact that $\|u_2\| = 1$ and $\mathbf{1}^T u_2 = 0$, we get

$$\begin{aligned} N^2 \lambda_2 &\geq \min_{s \neq t} \operatorname{eff}\mathcal{C}(s,t) \sum_{x,y \in V} |u_2(x) - u_2(y)|^2 \\ &= \min_{s \neq t} \operatorname{eff}\mathcal{C}(s,t) \sum_{x,y \in V} \Big(|u_2(x)|^2 + |u_2(y)|^2 - 2u_2(x)u_2(y) \Big) \\ &= 2N \min_{s \neq t} \operatorname{eff}\mathcal{C}(s,t). \end{aligned}$$

Dividing by $2N$, we obtain the desired result. □

The following result is based on the fact that if $e = \{a, b\} \in E$, then $(\delta_b - \delta_a)^T$ is a row of the incidence matrix B.

Proposition 3.3. *Let $e = \{a, b\} \in E$ be an edge of G. Then*

$$\operatorname{eff}\mathcal{R}(a,b) = (B\mathcal{G}B^T)(e,e).$$

Proof. Choose an orientation $\vec{E}$. Then,

$$\begin{aligned} (B\mathcal{G}B^T)(\vec{e},\vec{e}) &= \sum_{x,y \in V} B(\vec{e},x)\mathcal{G}(x,y)B^T(y,\vec{e}) \\ &= \sum_{x,y \in V} B(\vec{e},x)B(\vec{e},y)\mathcal{G}(x,y) \\ &= \mathcal{G}(a,a) + \mathcal{G}(b,b) - \mathcal{G}(a,b) - \mathcal{G}(b,a) \\ &= (\delta_b - \delta_a)^T \mathcal{G} (\delta_b - \delta_a) \\ &= \operatorname{eff}\mathcal{R}(a,b) \end{aligned}$$

In particular, this quantity does not depend on the choice of orientation.
□

Remark 3.4. The matrix $H := B\mathcal{G}B^T$ is known as the *transfer-impedance* matrix. Having chosen an orientation $\vec{E}$, the entries of H are given by

$$H(\vec{e},\vec{f}) = (\delta_y - \delta_x)^T L^+(\delta_w - \delta_z) \qquad \forall \vec{e} = (x,y), \vec{f} = (z,w) \in \vec{E}.$$

In particular, since $\mathcal{G}$ is symmetric, H is symmetric and since we can write $H = (B\mathcal{G}^{1/2})(B\mathcal{G}^{1/2})^T$, we see that H is also positive semi-definite. Indeed,

$$H = B(B^T B)^+ B^T = \mathrm{Proj}_{\mathrm{Im} B}, \tag{3.30}$$

see Exercise 3.12.

Remark 3.5 (The triangle inequality for eff$\mathcal{R}$ revisited). In Theorem 2.3, we showed that effective resistance is a metric using the additivity of current flows. Here, we will reinterpret the proof of the triangle inequality in terms of the linearity of the Laplacian matrix.

Recall that the effective resistance eff$\mathcal{R}(a,b)$ is the potential drop between a and b necessary to force one unit of current through the network. The associated voltage potential $h: V \to \mathbb{R}$ is only defined up to an additive constant. We are going to denote by $h_{a,b}$ the voltage potential normalized so that

$$h_{a,b}(a) = 0. \tag{3.31}$$

This uniquely determines $h_{a,b}$ and implies that

$$h_{a,b}(b) = \mathrm{eff}\mathcal{R}(a,b), \tag{3.32}$$

by (2.10). Note that $h_{a,b}$ is harmonic on $V \setminus \{a,b\}$ meaning, by (3.23), that

$$Lh_{a,b} = \delta_b - \delta_a. \tag{3.33}$$

In particular, by the maximum principle (Theorem 2.2),

$$0 \leq h_{a,b}(x) \leq h_{a,b}(b) = \mathrm{eff}\mathcal{R}(a,b), \tag{3.34}$$

for every $x \in V \setminus \{a,b\}$.

Now define $h := h_{a,c} + h_{c,b}$. Then,

$$Lh = Lh_{a,c} + Lh_{c,b} = (\delta_c - \delta_a) + (\delta_b - \delta_c) = \delta_b - \delta_a.$$

In particular, $L(h - h_{a,b}) = 0$ everywhere. Since G is connected, the maximum principle (Corollary 2.1) implies that there is a constant $C \in \mathbb{R}$ so that $h - h_{a,b} = C$. Namely,

$$h_{a,c} + h_{c,b} = h_{a,b} + C. \tag{3.35}$$

Evaluating (3.35) at a and recalling that $h_{a,b}(a) = h_{a,c}(a) = 0$, we find that $C = h_{c,b}(a)$. Thus (3.34) applied to the nodes c, b, implies that $C \geq 0$. Evaluating (3.35) at b, and using (3.34) for the nodes a, c, we get

$$\begin{aligned} h_{a,b}(b) &\leq h_{a,b}(b) + C && \text{(since } C \geq 0\text{)} \\ &= h_{a,c}(b) + h_{c,b}(b) && \text{(by (3.35))} \\ &\leq h_{a,c}(c) + h_{c,b}(b) && \text{(by (3.34)).} \end{aligned}$$

Finally, using (3.32), we see that the triangle inequality holds.

EXERCISES

3.1 *Factorization of the Laplacian.* Let G be the simple house graph from Exercise 1.1.

(i) Use the enumeration of the vertices of G to write down the Laplacian matrix L for it.

(ii) Use the enumeration of its edges $e_1, \ldots, e_6$ and pick an arbitrary direction for each one. Then, use code to draw a picture where edges are made into arrows to indicate the chosen orientation.

(iii) With the chosen orientation in (ii), write down the gradient operator B and verify that $B^T B = L$ by performing the matrix multiplication.

3.2 *Block matrix multiplication.* Given two matrices $A \in \mathbb{R}^{m\times n}$ and $B \in \mathbb{R}^{p\times q}$, assume A is decomposed into $s \times t$ blocks and B into $u \times v$ blocks.

(i) Describe necessary conditions for the product

$$AB = \begin{bmatrix} [\star]_{s\times t} \end{bmatrix}_{m\times n} \begin{bmatrix} [\star]_{u\times v} \end{bmatrix}_{p\times q}$$

to make sense.

(ii) Let $a_1, \ldots, a_n$ be the columns of A and $b_1, \ldots, b_p$ the rows of B. Write the product AB in terms of the vectors a_i and b_j.

3.3 *Orthogonal diagonalization.* Recall that a real $N \times N$ matrix U is orthogonal if $U^T U = I$, where I is the *identity matrix*. Assume H is a real, symmetric, $N \times N$ matrix, and let $\lambda_1 \leq \cdots \leq \lambda_N$ be the eigenvalues H given by the Spectral Theorem 3.1.

(i) Show that there is an orthogonal matrix U such that

$$U^T H U = \Lambda,$$

where Λ is the diagonal matrix with the eigenvalues λ_j along the diagonal.

(ii) Let $\mu_1 < \cdots < \mu_k < \cdots < \mu_n$, for some $n \leq N$, be the (non-repeating) eigenvalues of H and $m_1, \ldots, m_n$ the corresponding multiplicities, so that $\lambda_1 = \cdots = \lambda_{m_1} = \mu_1$, then $\lambda_{m_1+1} = \cdots = \lambda_{m_1+m_2} = \mu_2$, etc. For $j = 1, \ldots, n$, show that m_j is equal to both the geometric and the algebraic multiplicity of μ_j.

(iii) Suppose $\mathcal{B} = \{u_j\}_{j=1}^N$ is an orthonormal basis for $\mathbb{R}^N$. Given a vector $u \in \mathbb{R}^N$, write u as a linear combination of the basis vectors: $u = \sum_{j=1}^N c_j u_j$. Show that $c_j = \langle u, u_j \rangle$. The vector of coefficients $[u]_{\mathcal{B}} = (\langle u, u_1 \rangle, \ldots, \langle u, u_N \rangle)^T$ is thought as a set of new coordinates for u with respect to the basis $\mathcal{B}$. Let U be the matrix whose columns are the basis vectors u_j. Show that U is orthogonal and use block matrix multiplication to show that $U[u]_{\mathcal{B}} = u$. Interpret the diagonalization of H in (i) as a change of variables.

3.4 *Rayleigh quotients.* Assume H is a real, symmetric, $N \times N$ matrix and let $u \in \mathbb{R}^N \setminus \{0\}$ be a non-zero vector. Recall Definition 3.1 for the Rayleigh quotient $Q(u)$.

(i) Show that $Q(u)$ is a weighted average of the eigenvalues of H;

(ii) deduce from (i) that if all the eigenvalues of H are non-negative, then H is positive semidefinite, in the sense of (3.15).

3.5 *Min-Max characterization of eigenvalues.* Complete the proof of the Courant-Fisher Min-Max Theorem 3.3 by establishing the second equality in (3.9).

3.6 *The adjacency matrix.* Let $G = (V, E)$ be a connected simple graph on $N \geq 2$ nodes. Let $A(x, y) = \mathbb{1}_{\{x \sim y\}}$ be the adjacency matrix. Since A is an $N \times N$ symmetric real matrix, the Spectral Theorem 3.1 implies that the eigenvalues of A are real and hence can be ordered as follows:

$$\mu_N \leq \cdots \leq \mu_1.$$

(i) Interpret the quantities $A\mathbf{1}, \mathbf{1}^T A\mathbf{1}, \mathbf{1}^T\mathbf{1}$, and $Q(\mathbf{1}) = \frac{\mathbf{1}^T A\mathbf{1}}{\mathbf{1}^T\mathbf{1}}$. Then, relate the average degree $\deg_{\text{avg}}$ to μ_1, and in particular deduce that $\mu_1 > 0$.

(ii) Let u_1 be an eigenvector for μ_1. Use the fact that $Au_1 = \mu_1 u_1$, evaluated at a node x_0 where $u_1(x_0) = \max_{y \in V} u_1(y)$, and show that $\mu_1 \leq \max_{x \in V} \deg(x) =: \deg_{\max}$.

(iii) Show that we always have $-\mu_1 \leq \mu_N < 0$.

3.7 *Perron-Frobenius.* Let $G = (V, E)$ be a connected simple graph on $N \geq 2$ nodes. Let $\mu_N \leq \cdots \leq \mu_1$ be the eigenvalues of the adjacency matrix A for G, and let $\{u_1, \ldots, u_N\}$ be the corresponding orthonormal basis of eigenvectors, with $Au_j = \mu_j u_j$, for $j = 1, \ldots, N$, that is guaranteed by the Spectral Theorem 3.1. The purpose of this exercise is to show that the largest eigenvalue is "positive and simple," meaning that we can choose u_1 to be positive and $\mu_2 < \mu_1$.

(i) Suppose u is an eigenvector of A, i.e., $u \in \mathbb{R}^V \setminus \{0\}$ and $Au = \mu u$ for some $\mu \in \mathbb{R}$. Show that

$$u \geq 0 \Longrightarrow u > 0.$$

(ii) Show that $|u_1|$ is also an eigenvector of A for μ_1 and deduce that u_1 never vanishes and cannot change sign. In particular, u_1 can be chosen to be positive.
Hint: Study the identity $|u_1^T A u_1| = |u_1|^T A |u_1|$.

(iii) Show that u_2 must change sign. Hence, u_2 cannot be an eigenvector for μ_1.

(iv) In Exercise 3.6, it was shown that $|\mu_N| \leq \mu_1$. Recall that the graph G is *bipartite*, if $V = U \cup W$, with $U, W \neq \emptyset$, $U \cap W = \emptyset$, and no edges connect nodes within U or within W. Show that $\mu_N = -\mu_1$ if and only if G is bipartite.
Hint: If G is bipartite, redefine u_1 to be $-u_1$ on one of the two

parts. Conversely, if $\mu_N = -\mu_1$, show that $|u_N|$ is an eigenvector for μ_1 and study the identity $|u_N^T A u_N| = |u_N|^T A |u_N|$.

3.8 *Laplacian eigenvalue.* Let $G = (V, E)$ be a graph on N nodes with eigenvalues $0 = \lambda_1 \leq \cdots \leq \lambda_N$.

(i) If G is d-regular, for some $d \geq 2$, how are the Laplacian eigenvalues λ_i related to the eigenvalues μ_j of the adjacency matrix A as defined in Exercise 3.6?

(ii) Using a similar argument as in Exercise 3.6 (ii), show that $\lambda_N \leq 2 \max_{x \in V} \deg(x)$.

(iii) If $G = K_N$ is the complete graph on N nodes, show that every $u \in \mathbb{R}^N \setminus \{0\}$ with $\mathbf{1}^T u = 0$ is an eigenvector of the Laplacian for the eigenvalue N. Hence, deduce that the lower bound (3.29) is sharp.

3.9 *1D Laplacian eigenvectors.* On the real line, the 1D Laplace operator is just the second derivative. Consider the eigenvalue problem

$$\begin{cases} u'' = \lambda u \\ u'(0) = u'(1) = 0 \end{cases}$$

If you haven't seen an eigenvalue problem for ODEs before, don't worry; it's essentially the same idea as in linear algebra. We're looking for a (generally complex) number λ and a nonzero function u such that the pair (λ, u) satisfies the above equation. Such a λ is called an eigenvalue of the 1D Laplacian with Neumann boundary conditions (since the derivative u' is specified at the boundaries), and u is called an eigenfunction associated with λ.

(i) By substituting, verify that the values

$$\lambda_k = -k^2\pi^2, \quad k = 0, 1, 2, 3, \ldots$$

are eigenvalues of the 1D Laplacian, with associated eigenfunctions

$$u_k(x) = \cos(k\pi x).$$

(Note that these eigenfunctions are not quite unique; any of the u_k could be replaced by any nonzero scalar multiple αu_k. This is analogous to the concept of eigenvector in linear algebra.)

(ii) Write code to plot the eigenfunctions of the 1D Laplacian for $k \in \{0, 1, 2, 10\}$.

(iii) Create a path graph with 50 nodes and compute its graph Laplacian L. Then compute the eigenvalues and eigenvectors of L.

(iv) Plot the eigenvectors of L in the same range as in (i). The results should look very similar to (but not identical.) Also, plot the eigenvalues and compare them to a rescaled version of the eigenvalues λ_k described above.

3.10 *Principal minors of the Laplacian.* Let $G = (V, E, \sigma)$ be a simple connected graph with edge weights $\sigma \in \mathbb{R}^E_{>0}$. Consider the corresponding Laplacian matrix

$$L = \sum_{e=\{x,y\}\in E} \sigma(x, y)(\delta_y - \delta_x)(\delta_y - \delta_x)^T$$

Fix a node $b \in V$, and let $S := V \setminus \{b\}$. Also, let L_b be the minor obtained from L by removing the row and column corresponding to node b. The goal of this exercise is to show that L_b is invertible.

(i) Show that for every $x \in S$:

$$|L_b(x, x)| \geq \sum_{y\in S\setminus\{x\}} |L_b(x, y)|, \tag{P3.1}$$

and that the inequality in (P3.1) is strict if and only if x is a neighbor of b.

(ii) Show that for every node $x' \in S$, there is a path $x_1 = x', x_2, \ldots, x_n = x$ entirely contained in S, to a node $x \in S$ that is a neighbor of b.

(iii) Let $u \in \mathbb{R}^S$ be a function with $L_b u = 0$. Assume, by contradiction, that $u \neq 0$. Pick a node $x^* \in S$ so that $|u(x^*)| = \max_{y\in S} |u(y)|$ and normalize u so that $|u(x^*)| = 1$. Pick a path $x_1 = x^*, x_2, \ldots, x_n = x$ in S from x^* to a node x, so that x is the first node along the path satisfying the strict inequality (P3.1). Then, show that $|u(x^*)| = |u(x_2)| = \cdots = |u(x)| = 1$, and reach a contradiction.

3.11 *The pseudoinverse of the Laplacian.* Assume G is a connected graph. Let L be the Laplacian for G and let $\mathcal{G}$ be the corresponding Green operator (or pseudoinverse).

(i) Verify that (3.20) holds.

(ii) Show that in this case the restriction $L|_{\langle \mathbf{1} \rangle^\perp}$ is an invertible linear map from $\langle \mathbf{1} \rangle^\perp$ to $\langle \mathbf{1} \rangle^\perp$, and identify its inverse.

(iii) Show that $L + \frac{1}{N}\mathbf{1}\mathbf{1}^T$ is invertible and write $\mathcal{G}$ in terms of its inverse.

(iv) Recall the orthonormal basis of eigenvectors $\{u_j\}$, where $Lu_j = \lambda_j u_j$ and $0 = \lambda_1 \leq \cdots \leq \lambda_N$. Use (3.28) to obtain the following formula for effective resistance in terms of the eigenpairs (u_j, λ_j) of L:

$$\operatorname{eff}\mathcal{R}(s,t) = \sum_{j=2}^{N} \frac{1}{\lambda_j}(u_j(t) - u_j(s))^2. \tag{P3.2}$$

3.12 *The transfer-impedance matrix.* Let $G = (V, E)$ be a simple connected graph on N nodes with a chosen orientation of the edges, and let B be the gradient matrix defined in (3.16).

(i) Use (3.19) to deduce the SVD decomposition for B and show that $\operatorname{rank}(B) = N - 1$.

(ii) Use (i) to show that (3.30) holds.

3.13 *The Kirchhoff index.* Let $G = (V, E)$ be a simple connected graph on N nodes, and let L denote its Laplacian matrix. Also, assume that M is an arbitrary $N \times N$ real symmetric matrix.

(i) Show that the expression $\sum_{x,y \in V} \operatorname{eff}\mathcal{R}(x,y)(LML)(x,y)$ is equal to

$$\mathbf{1}^T(LML)\operatorname{diag}(L^+) + \operatorname{diag}(L^+)^T(LML)\mathbf{1} - 2\operatorname{Tr}(LMLL^+).$$

(ii) Use (3.20) and the cyclic property of the trace to deduce from (i) that

$$\sum_{x,y \in V} \operatorname{eff}\mathcal{R}(x,y)(LML)(x,y) = -2\operatorname{Tr}(ML).$$

(iii) Choose M appropriately to establish the following two identities

$$\sum_{e \in E} \operatorname{eff}\mathcal{R}(e) = N - 1 \qquad \text{and} \qquad \frac{1}{2}\sum_{x,y \in V} \operatorname{eff}\mathcal{R}(x,y) = N \sum_{j=2}^{N} \frac{1}{\lambda_j}.$$

The latter expression is known as the *Kirchhoff index* of the graph G.

CHAPTER 4

The Language of Probability

THIS CHAPTER is a brief interlude in preparation for the next chapter on Markov chains and can be skipped by readers who already know the basic notions of probability. For a more extensive treatment of this subject, see, for instance, [46].

4.1 SAMPLE SPACES, EVENTS, AND PROBABILITY MEASURES

Unless otherwise stated, we will restrict our discussion to the discrete setting. It can be helpful to think of probability as arising from some *experiment*, such as tossing a coin, rolling a die, or drawing some objects from a bag without looking. The possible results of an experiment are called *outcomes*, and the set of all possible outcomes is called the *sample space*. So, for instance, when tossing a coin, the possible outcomes are H or T, for heads or tails, and the sample space is $\Omega = \{H, T\}$. Note that if the experiment is changed to involve tossing two distinguishable coins, then the sample space also changes to account for all the possibilities and becomes $\Omega = \{HH, HT, TH, TT\}$.

Whenever an experiment is performed, we would like to be able to ask questions. Say, if a die is rolled, in which case $\Omega = \{1, 2, 3, 4, 5, 6\}$, we would like to be able to ask whether the outcome is an even number, or is a number less than 3, etc. This gives rise to a natural algebra of questions because if we can answer whether the outcome is even, we can also answer whether the outcome is odd. The way this is codified mathematically is to equate a family of questions with a family of subsets of the sample

DOI: 10.1201/9781003024866-5

space. So, in the example of the die, the question 'Is the outcome even?' corresponds to the subset $A = \{2, 4, 6\}$ and 'Is the outcome less than 3?' corresponds to $B = \{1, 2\}$. In probability jargon, subsets of the sample space that are of interest are called *events*. There are three primary operations that can be performed on events: unions $A \cup B$, intersections $A \cap B$, and complements A^c or $\Omega \setminus A$. Thus, families of events should be closed under these three operations. We will mainly be interested in situations where one can ask questions about the individual outcomes. Namely, we often assume that every subset of Ω is an event. In this case, it makes sense to define the *power set* 2^Ω consisting of all possible subsets of Ω. For instance, when tossing a coin, the power set of the sample space is $\{\emptyset, \{H\}, \{T\}, \Omega\}$. The symbol $\emptyset$ represents the empty set and corresponds to questions that don't have an allowed answer, such as 'Is the outcome of a die roll equal to 7?'

In general, one first defines a family of events of interest, and then a *probability measure* that assigns to each event in the family a percentage representing the likelihood of that event. However, it is further assumed that these percentages must satisfy some natural properties. For instance, if the probability of an event A is 30%, then the probability of the complement A^c should be 70%. Likewise, the probabilities of disjoint events A, B (such that $A \cap B = \emptyset$) should add up, and the total probability of the whole sample space Ω should equal 1. More formally, a probability measure on a finite sample space Ω is a function $\mathbb{P} : 2^\Omega \to [0, 1]$ that satisfies

(i) *Additivity:* If $A, B \subset \Omega$ and $A \cap B = \emptyset$, then

$$\mathbb{P}(A \cup B) = \mathbb{P}(A) + \mathbb{P}(B); \tag{4.1}$$

(ii) *Probability condition:* $\mathbb{P}(\emptyset) = 0$ and $\mathbb{P}(\Omega) = 1$.

When Ω is infinite, condition (i) must be extended to countable unions. One can show that (i) and (ii) imply other properties, such as

(iii) *Complement rule:*

$$\mathbb{P}(A^c) = 1 - \mathbb{P}(A).$$

(iv) *Monotonicity:*

$$A \subset B \text{ implies } \mathbb{P}(A) \leq \mathbb{P}(B). \tag{4.2}$$

In our case, since we are assuming that every subset is an event, to define a probability measure, it is enough to define a *probability mass function* (or *pmf*) $p : \Omega \to [0, 1]$ with the following property:

$$\sum_{\omega \in \Omega} p(\omega) = 1.$$

Every pmf p yields a probability measure defined on every subset $A \subset \Omega$ as

$$\mathbb{P}(A) := \sum_{\omega \in A} p(\omega).$$

Then, one can verify that conditions (i) and (ii) for a probability measure are satisfied.

Example 4.1. The *uniform* pmf describes the case when all the outcomes are equally likely. Namely, $p(\omega) \equiv \frac{1}{|\Omega|}$. Then, for every event $A \subset \Omega$, we have

$$\mathbb{P}(A) = \frac{|A|}{|\Omega|}.$$

Example 4.2. Given $\Omega = \{\omega_1, \ldots, \omega_N\}$, it is often enough to define a set of non-negative weights $w(\omega_i) \geq 0$, for $i = 1, \ldots, N$ (not all zero). Then a pmf ensues by normalizing these weights and setting, for $i = 1, \ldots, N$,

$$p(\omega_i) := \frac{w(\omega_i)}{Z}, \qquad \text{where } Z := \sum_{j=1}^{N} w(\omega_j).$$

4.2 CONDITIONING AND INDEPENDENCE

If we fix an event $B \subset \Omega$ with $\mathbb{P}(B) > 0$ and consider it as "having happened", then that changes the probability of every other event. Namely, the *conditional probability* of an event $A \subset \Omega$ given B is defined as

$$\mathbb{P}(A \mid B) = \frac{\mathbb{P}(A \cap B)}{\mathbb{P}(B)}.$$

One can verify that the function $\mu(\cdot) := \mathbb{P}(\cdot \mid B)$ is also a probability measure on Ω, see Exercise 4.1.

Moreover, since the event B partitions the sample space as $\Omega = B \cup B^c$, we can decompose the probability of an event A as

$$\begin{aligned}\mathbb{P}(A) &= \mathbb{P}(A \cap B) + \mathbb{P}(A \cap B^c) \\ &= \mathbb{P}(A \mid B)\mathbb{P}(B) + \mathbb{P}(A \mid B^c)\mathbb{P}(B^c)\end{aligned}$$

More generally, a *partition* is a set of events $B_1, \ldots, B_k$ with $\mathbb{P}(B_i) > 0$, for $i = 1, \ldots, k$, such that the parts are mutually disjoint: $B_i \cap B_j = \emptyset$, and so that they are altogether exhaustive: $\Omega = \cup_{i=1}^k B_i$. Partitions can be used to decompose the sample space and then compute probabilities. This method is known as *conditioning.*

Theorem 4.1 (Conditioning). *Let $\{B_1, \ldots, B_n\}$ be a partition of the sample space Ω. Then, for every event A:*

$$\mathbb{P}(A) = \sum_{k=1}^{n} \mathbb{P}(A \mid B_k)\mathbb{P}(B_k).$$

Proof. $\mathbb{P}(A) = \sum_{k=1}^n \mathbb{P}(A \cap B_k) = \sum_{k=1}^n \mathbb{P}(A \mid B_k)\mathbb{P}(B_k)$. □

Given two events A and B we say that they are *independent*, if

$$\mathbb{P}(A \cap B) = \mathbb{P}(A)\mathbb{P}(B).$$

This implies both that $\mathbb{P}(A \mid B) = \mathbb{P}(A)$ and $\mathbb{P}(B \mid A) = \mathbb{P}(B)$, meaning that the probability of one event is not changed if the other event happens.

4.3 RANDOM VARIABLES

A (discrete) *random variable* is a function $X : \Omega \to \mathbb{R}$ with the property that the range $X(\Omega)$ is countable and for each value $a \in \mathbb{R}$, the set $\{\omega : X(\omega) = a\}$ is an event. From the point of view of the experiment underlying the sample space, we can think of a random variable X as a "measurement". For instance, we can think of measuring the outside temperature at regular intervals. In particular, if we plug a random variable X into a function $y = f(x)$ from Calculus, we get a new random variable $Y = f(X)$, but we do not introduce any new source of randomness. This fact can be expressed in terms of the family of events that a random variable generates. Namely, every random variable X generates a *σ-algebra* $\sigma(X)$, namely the smallest family of events that is closed under complements and (countable) unions, and that contains all the events of the form $\{X = a\}$ for some $a \in \mathbb{R}$. We think of $\sigma(X)$ as the set of all questions that it is meaningful to ask about X. From this point of view, then we can say that X and $Y = f(X)$ generate the same σ-algebra.

Also, two random variables X and Y are said to be *independent* if every event in $\sigma(X)$ is independent of every event in $\sigma(Y)$. It turns out that it is enough to check that every event of the form $\{X = a\}$ is independent of every event of the form $\{Y = b\}$.

Every (discrete) random variable X gives rise to a *probability mass distribution*, or probability mass function (pmf):

$$f_X(x) := \mathbb{P}(X = x), \qquad \text{for } x \in \mathbb{R}. \tag{4.3}$$

Two random variables that have the same pmf are said to be *equal in distribution.*

Example 4.3. We say that X has the *Poisson distribution* if $X(\Omega) = \{0, 1, 2, \dots\}$ and

$$\mathbb{P}(X = k) = \frac{1}{k!}\lambda^k e^{-\lambda}, \qquad \text{for } k = 0, 1, 2, \dots,$$

where $\lambda > 0$ is a fixed parameter.

A *continuous random variable* is a function $X : \Omega \to \mathbb{R}$ that has a *probability density function* (pdf) $f_X : \mathbb{R} \to \mathbb{R}$ so that for every $-\infty \leq a < b \leq +\infty$,

$$\mathbb{P}(a < X \leq b) = \int_a^b f_X(x)dx.$$

In particular, we must have

$$\int_{-\infty}^{+\infty} f_X(x)dx = 1. \tag{4.4}$$

4.4 EXPECTATION

The *expectation* of a discrete random variable $X : \Omega \to \mathbb{R}$ is the average of the values taken in the range $X(\Omega) \subset \mathbb{R}$ weighted by the corresponding probabilities:

$$\mathbb{E}(X) := \sum_{x \in X(\Omega)} x \cdot \mathbb{P}(X = x).$$

Note that in this definition the sample space Ω plays a lesser role, all is needed is the pdf of X. However, sometimes it is useful to think of the expectation of X as an integral, namely as an average over all possible outcomes $\omega \in \Omega$:

$$\mathbb{E}(X) = \int_\Omega X d\mathbb{P} = \int_\Omega X(\omega)d\mathbb{P}(\omega)$$

As an example of the usefulness of this point of view, consider the famous *Markov Inequality*: If $X > 0$, then for $t > 0$,

$$\mathbb{P}(X \geq t) \leq \frac{\mathbb{E}(X)}{t}. \tag{4.5}$$

To prove this, it is enough to note that

$$\mathbb{E}(X) = \int_{\Omega} X d\mathbb{P} \geq \int_{\{X \geq t\}} X d\mathbb{P} \geq t\mathbb{P}(X \geq t).$$

Another useful fact is that if we know that $X \geq 0$ *almost surely* (a.s.), meaning that $\mathbb{P}(X \geq 0) = 1$, and $\mathbb{E}(X) = 0$, then $X = 0$ a.s. This is a form of *mean value property*. Think of integrating a non-negative continuous function, then if the area under the curve is zero, the function must be zero as well.

However, probably to most important property is *linearity of expectations*:

$$\mathbb{E}(aX + b) = a\mathbb{E}(X) + b.$$

This also follows fairly immediately from the integral representation. See also Exercise 4.5.

In line with the fact that given some function $y = f(x)$, the random variable $Y = f(X)$ share the same randomness as X, the expectation of Y can also be computed in terms of the pdf of X:

$$\mathbb{E}\left[f(X)\right] = \sum_{x \in X(\Omega)} f(x)\mathbb{P}(X = x). \tag{4.6}$$

We refer to this as the *change of variables property*.

The *variance* of a random variable X is the expected square deviation:

$$\text{Var}(X) = \mathbb{E}\left[(X - \mathbb{E}(X))^2\right].$$

It measures how "spread out" the values taken on by X in $X(\Omega)$ are when their likelihood is taken into account; see Exercise 4.2.

Remark 4.1. A word about notation: suppose we have a finite set S, where s is a generic element in S. We write $\underline{s}$ to mean a random variable with values in S. Moreover, suppose that μ is a pmf on S so that

$$\mu(s) = \mathbb{P}(\underline{s} = s), \qquad \forall s \in S.$$

Then, given a function $f : S \to \mathbb{R}$, we write

$$\mathbb{E}_{\mu}\left[f\left(\underline{s}\right)\right],$$

to mean that $\underline{s}$ is a random variable with pmf μ.

4.5 COUPLING TWO RANDOM VARIABLES

Suppose X and Y are two random variables defined on two different probability triples $(\Omega_1, \mathcal{F}_1, \mathbb{P}_1)$ and $(\Omega_2, \mathcal{F}_2, \mathbb{P}_2)$. A *coupling* is a probability triple $(\Omega, \mathcal{F}, \mathbb{P})$ and two maps $\pi_j : \Omega \to \Omega_j$, $j = 1, 2$, such that $X \circ \pi_1$ is equal to X in distribution and $Y \circ \pi_2$ is equal to Y in distribution.

Example 4.4. Recall that a random variable $X : \Omega = \{0, 1\} \to \mathbb{R}$ is said to have the *Bernoulli distribution* for some value of the parameter $p \in [0, 1]$, if

$$\mathbb{P}(X = x) = \begin{cases} p & \text{if } x = 1, \\ 1 - p & \text{if } x = 0, \end{cases}$$

Assume that X and Y are both Bernoulli(p), with $\Omega_1 = \Omega_2 = \{0, 1\}$. Let $\Omega = \{0, 1\}^2$ and $\mathcal{F} = 2^{\Omega}$. Write p_{ij} for $\mathbb{P}(X = i, Y = j)$, $i, j = 0, 1$. Also, let π_1 and π_2 be the coordinate projections. Then in order for $X \circ \pi_1$ to be Bernoulli(p) we need

$$\mathbb{P}(X \circ \pi_1 = 1) = p_{10} + p_{11} = p.$$

Likewise

$$\mathbb{P}(Y \circ \pi_2 = 1) = p_{01} + p_{11} = p.$$

We have three equations in four unknowns, the third equation being $p_{00} + p_{01} + p_{10} + p_{11} = 1$. Therefore

$$p_{11} = s; \qquad p_{10} = p_{01} = p - s; \qquad \text{and} \qquad p_{00} = s + 1 - 2p.$$

In particular, $0 \le s \le p$ and $s \ge 2p - 1$, i.e., $\max\{0, 2p - 1\} \le s \le p$ (note that $2p - 1 \le p$).

If $s = p$, then $p_{01} = p_{10} = 0$, and $p_{11} = p$. Namely, in this case, X and Y are completely dependent. The *trivial coupling* is obtained when X and Y are independent. This happens when $p_{11} = s = p^2$ (note that $2p - 1 \le p^2$).

A related concept to couplings is joint mass distributions.

Definition 4.1. Suppose X and Y are discrete random variables defined on the same probability triple $(\Omega, \mathcal{F}, \mathbb{P})$, then

$$\mu_{X,Y}(x, y) := \mathbb{P}(X = x, Y = y) \qquad \forall (x, y) \in \mathbb{R}^2,$$

is called the *joint mass distribution* for X and Y.

The *marginal distributions* are defined as

$$\mathbb{P}(X = x) = \sum_y \mu_{X,Y}(x, y) \qquad \text{and} \qquad \mathbb{P}(Y = y) = \sum_x \mu_{X,Y}(x, y).$$

More generally, for a vector of random variables $(X_1, \ldots, X_N)$, the joint mass distribution is

$$\mu_{X_1,\ldots,X_N}(x_1, \ldots, x_N) := \mathbb{P}(X_1 = x_1, \ldots, X_N = x_N),$$

for all $(x_1, \ldots, x_N) \in \mathbb{R}^N$. The change of variable formula (4.6) can be extended to a situation where $Z = g(X, Y)$, then

$$\mathbb{E}(Z) = \mathbb{E}[g(X, Y)] = \sum_{(x,y)\in\mathbb{R}^2} g(x, y)\mu_{X,Y}(x, y). \tag{4.7}$$

Two coupled random variables $X, Y : \Omega \to \mathbb{R}$ are independent if and only if

$$\mu_{X,Y}(x, y) = \mu_X(x)\mu_Y(y) \qquad \forall (x, y) \in \mathbb{R}^2. \tag{4.8}$$

Theorem 4.2. *If X and Y are independent (i.e., they are coupled trivially), then*

$$\mathbb{E}(XY) = \mathbb{E}(X)\mathbb{E}(Y), \tag{4.9}$$

and in fact

$$\mathbb{E}[f(X)g(Y)] = \mathbb{E}[f(X)]\mathbb{E}[g(Y)] \tag{4.10}$$

for every pair of functions $f, g : \mathbb{R} \to \mathbb{R}$,

Conversely, if (4.10) holds for every pair of functions $f, g : \mathbb{R} \to \mathbb{R}$, then X and Y are independent.

Proof. Assuming independence, we have

$$\begin{aligned}
\mathbb{E}[f(X)g(Y)] &= \sum_{x,y} f(x)g(y)\mu_{X,Y}(x, y) \\
&= \sum_{x,y} f(x)g(y)\mu_X(x)\mu_Y(y) \\
&= \left(\sum_x f(x)\mu_X(x)\right)\left(\sum_y g(y)\mu_Y(y)\right) \\
&= \mathbb{E}[f(X)]\mathbb{E}[g(Y)]
\end{aligned}$$

Conversely, assuming (4.10), let $f = \mathbb{1}_{\{x\}}$ and $g = \mathbb{1}_{\{y\}}$. Then

$$\begin{aligned}\mathbb{P}(X = x, Y = y) &= \int_\Omega \mathbb{1}_{\{X=x,Y=y\}} d\mathbb{P} = \int_\Omega \mathbb{1}_{\{x\}}(X)\mathbb{1}_{\{y\}}(Y) d\mathbb{P} \\ &= \mathbb{E}\left[\mathbb{1}_{\{x\}}(X)\mathbb{1}_{\{y\}}(Y)\right] = \mathbb{E}\left(\mathbb{1}_{\{x\}}(X)\right)\mathbb{E}\left(\mathbb{1}_{\{x\}}(Y)\right) \\ &= \int_\Omega \mathbb{1}_{\{X=x\}} d\mathbb{P} \int_\Omega \mathbb{1}_{\{Y=y\}} d\mathbb{P} = \mathbb{P}(X = x)\mathbb{P}(Y = y)\end{aligned}$$

□

If X and Y only satisfy (4.9), they are said to be *uncorrelated.*

4.6 MOMENT GENERATING FUNCTIONS

For a general random variable $X : \Omega \to \mathbb{R}$, the *moment generating function* is

$$M_X(t) := \mathbb{E}\left(e^{tX}\right), \qquad t \in \mathbb{R}.$$

Example 4.5. Let $X \sim \text{Unif}(\{1, 2, 3, 4\})$. Then $f_X(x) = \frac{1}{4}$ for $x = 1, 2, 3, 4$, and

$$M_X(t) = \frac{1}{4}\left(e^t + e^{2t} + e^{3t} + e^{4t}\right)$$

The moment-generating function takes its name from the fact that

$$M_X(t) = \sum_{k=0}^{\infty} \frac{1}{k!}\mathbb{E}\left(X^k\right) t^k.$$

In particular, one can recover all the moments of X by computing the Taylor coefficients of $M_X(t)$:

$$\mathbb{E}\left(X^k\right) = M_X^{(k)}(0).$$

The following is a handy fact for computing the moment generating function of a sum of independent random variables.

Proposition 4.1. *If X and Y are independent, then*

$$M_{X+Y}(t) = M_X(t)M_Y(t).$$

Proof. Note that by (4.10),

$$\begin{aligned}M_{X+Y}(t) &= \mathbb{E}\left(e^{t(X+Y)}\right) \\ &= \mathbb{E}\left(e^{tX}e^{tY}\right) = \mathbb{E}\left(e^{tX}\right)\mathbb{E}\left(e^{tY}\right) = M_X(t)M_Y(t).\end{aligned}$$

□

EXERCISES

4.1 *Conditional probability measure.* Let Ω be a finite sample space. Assume that $\mathbb{P}$ is a probability measure defined on the power set 2^Ω. Fix an event $B \subset \Omega$ with $\mathbb{P}(B) > 0$.

(i) Show that the function $\mu : 2^\Omega \to [0, 1]$ defined by $\mu(A) = \mathbb{P}(A \mid B)$ is also a probability measure on 2^Ω.

(ii) Suppose that $\mathbb{P}$ is the probability measure induced by a pmf $p : \Omega \to [0, 1]$. Is μ from (i) also induced by a pmf? If so, identify the pmf and reinterpret $\mu(A)$ in terms of cardinality.

4.2 *Computing the variance.* Let $X : \Omega \to \mathbb{R}$ be a random variable.

(i) Show that

$$\mathrm{Var}(X) = \mathbb{E}(X^2) - \mathbb{E}(X)^2.$$

We say that $\mathbb{E}(X^k)$ is the k-th *moment* of X. So, the variance is the second moment minus the first moment squared.

(ii) Show that $\mathbb{E}(X)$ is the value of c that minimizes the function $f(c) := \mathbb{E}\left[(X - c)^2\right]$, and that the corresponding minimum value of this function is $\mathrm{Var}(X)$.

(iii) Compute the expectation and the variance of the Bernoulli(p) distribution, for some fixed parameter $p \in [0, 1]$.

4.3 *Binomial coefficients.* Consider strings of 0's and 1's of length N. The number of such strings that have exactly k ones is the *binomial coefficient* $\binom{N}{k}$, which we read as "N choose k".

(i) When tossing a fair coin N times, each outcome can be represented by a string of 0's and 1's, where 0 corresponds to tails, and 1 corresponds to heads (or 'success'). Use the binomial coefficients and sigma notation to express the probability of the event 'at least k heads'.

(ii) Prove Pascal's recurrence formula

$$\binom{N}{k} = \binom{N-1}{k} + \binom{N-1}{k-1}$$

in two different ways: first using a combinatorial argument and then directly from the fact that

$$\binom{N}{k} = \frac{N(N-1)\cdots(N-k+1)}{k!} = \frac{N!}{k!(N-k)!}.$$

(iii) If instead of a fair coin we toss a *biased coin* with $\mathbb{P}(\text{ heads }) = p \in [0,1]$, this gives rise to the *Binomial distribution* Namely, if we let X be the random variable that counts the number of heads in N tosses, then

$$\mathbb{P}(X = k) = \binom{N}{k} p^k (1-p)^{N-k}, \qquad \text{for } k = 0, 1, 2, \dots, N.$$

Compute $\mathbb{E}(X)$ and $\mathrm{Var}(X)$.

4.4 *The geometric distribution.* Consider the thought experiment of repeatedly tossing a biased coin such that $\mathbb{P}(\text{ heads }) = p \in [0,1]$. Let $\{X_j\}_{j=1}^{\infty}$ be a sequence of random variables so that $X_j = 1$ if the outcome of the j-th toss is heads, and $X_j = 0$ otherwise. Finally, let

$$X := \inf\{k \geq 1 : X_k = 1\}$$

be the first time the coin shows heads.

(i) Show that for $k = 1, 2, \dots,$

$$\mathbb{P}(X = k) = p(1-p)^{k-1},$$

and then verify that this defines a pmf on the set $\mathbb{Z}_{\geq 1}$.

(ii) Compute $\mathbb{E}(X)$ and $\mathrm{Var}(X)$.

(iii) Suppose X_1 and X_2 are independent random variables with the geometric distribution with parameters p_1 and p_2 respectively. Show that $Z := \min\{X_1, X_2\}$ also has the geometric distribution.

4.5 *Linearity of expectations.* Figure 4.1 shows the "skeleton" of a graph on four vertices. Each of the five dashed edges in the figure represents a potential edge in the random graph. To generate an instantiation of this random object, we flip (independently) five weighted coins (one for each edge) with probability $0 \leq p \leq 1$ of landing on

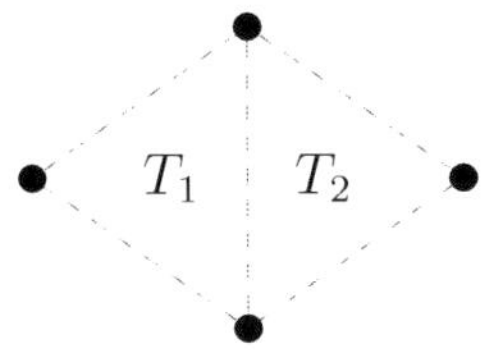

Figure 4.1 Skeleton of a random graph with two potential triangles.

heads; for each coin that comes up heads, we add the associated edge, for each coin that comes up tails, we omit the associated edge. Note that there are two potential triangles possible in the resulting random graph (labeled T_1 and T_2 in the figure).

(i) Compute the probability that each triangle occurs in the random graph.

(ii) Use linearity of expectations to compute the expected number of triangles in the random graph. It is convenient to introduce two random variables I_1 and I_2 defined such that

$$I_j = \begin{cases} 1 & \text{if } T_j \text{ is in the random graph.} \\ 0 & \text{otherwise.} \end{cases}$$

Random variables like the I_j, which signal the occurrence of some random event, are called *indicator random variables*. From these two variables define a new random variable $N_\Delta = I_1 + I_2$, so that N_Δ is the number of triangles present in the random graph, taking values in $\{0, 1, 2\}$.

4.6 *Erdős-Rényi graphs.* For a graph on N nodes, there are $M = \binom{N}{2}$ possible edges. We sample the edges using the binomial BIN (M, p) mass distribution.

(i) Fix $x \in V := \{1, \ldots, N\}$, then $\deg(x)$ is a random variable with values in $\{0, 1, 2, \ldots, N-1\}$. Compute its distribution, expectation, and variance.

(ii) Let A_j be the event that node $j \in V$ is *isolated*, meaning that $\deg(j) = 0$. Compute $\mathbb{P}(A_j)$.

(iii) Use linearity of expectations to compute the expected number of isolated nodes.

(iv) Assume that the expected degree $\mathbb{E}(\deg(x))$ of any given node is asymptotic to $2\log N$ as N tends to infinity, meaning that the ratio tends to 1 as $N \to \infty$. What happens to p in this case? What is the limit of the expected number of isolated nodes as $N \to \infty$?

4.7 *The Cavalieri principle.* Show that if X is a random variable with values in $\mathbb{Z}_{\geq 0} = \{0, 1, 2, 3, \dots\}$, then

$$\mathbb{E}(X) = \sum_{k=0}^{\infty} \mathbb{P}(X > k). \tag{P4.1}$$

To prove this, write $\mathbb{P}(X > k) = \sum_{j=k+1}^{\infty} \mathbb{P}(X = j)$, and switch the order of summation. It helps to draw a picture in the plane of the lattice

$$\{(k, j) : k = 0, 1, 2, \dots, \ j = k + 1, \dots\}$$

and then slice it horizontally. Namely, fix j in the appropriate range and figure out the corresponding range for k. Finally, recall that $\mathbb{E}(X) = \sum_{k=0}^{\infty} k\mathbb{P}(X = k)$.

4.8 *Conditioning for expectations.* Let X be a (discrete) random variable, and let B be an event with $\mathbb{P}(B) > 0$. Then the *conditional expectation* of X given B is

$$\mathbb{E}(X \mid B) = \sum_{x \in \mathbb{R}} x \cdot \mathbb{P}(X = x \mid B).$$

(i) Show that given a partition $\{B_1, \dots, B_k\}$ with $\mathbb{P}(B_j) > 0$, we can condition on these events and write

$$\mathbb{E}(X) = \sum_{j=1}^{k} \mathbb{E}(X \mid B_j)\mathbb{P}(B_j).$$

(ii) Repeatedly toss a biased coin such that $\mathbb{P}(\text{ heads }) = p \in [0, 1]$ and let L be the length of the initial run. Use Exercise 4.4 to find $\mathbb{E}(L)$ by conditioning on the outcome of the first toss.

4.9 *Dependent but uncorrelated.* Let X be uniform in $\{-1, 0, 1\}$ and let

$$Y = \begin{cases} 0 & \text{if } X = 0 \\ 1 & \text{if } X \neq 0 \end{cases}$$

Show that X and Y are dependent but uncorrelated.

4.10 *Chernoff bounds.* Here, we estimate the likelihood for a sum of Bernoulli random variables to deviate from its average. Let $\{X_k\}_{k=1}^n$ be a sequence of independent Bernoulli variables, where $X_k \in \{0,1\}$ and $\mathbb{P}(X_k = 1) = p_k$. Let $X = \sum_{k=1}^n X_k$ and $m = \mathbb{E}(X) = \sum_{k=1}^n p_k$. First, compute the moment-generating function of X_k, then compute $M_X(t)$, next use the estimate $1 + x \leq e^x$, and apply the Markov inequality 4.5. Finally, by optimizing for $t > 0$ and $t < 0$ separately, show the following, for $0 < \delta < 1$:

(i) we have

$$\mathbb{P}\left(X \geq (1+\delta)m\right) \leq \left(\frac{e^{\delta}}{(1+\delta)^{1+\delta}}\right)^m \leq \exp\left(-m\frac{\delta^2}{3}\right); \quad \text{(P4.2)}$$

(ii) moreover,

$$\mathbb{P}\left(X \leq (1-\delta)m\right) \leq \left(\frac{e^{-\delta}}{(1-\delta)^{1-\delta}}\right)^m \leq \exp\left(-m\frac{\delta^2}{2}\right); \quad \text{(P4.3)}$$

(iii) combine these two estimates and get,

$$\mathbb{P}\left(|X - m| \geq \delta m\right) \leq \exp\left(-m\frac{\delta^2}{3}\right). \quad \text{(P4.4)}$$

CHAPTER 5

Basic Theory of Markov Chains

WE REVIEW the basic theory of Markov chains with the goal of understanding random walks on graphs, which will be our running example. We cover the stable distribution, irreducible chains, reversible chains, and spectral properties. We refer the reader to [56] for a more in-depth treatment of Markov chains.

5.1 FINITE STATE SPACE DISCRETE TIME MARKOV CHAINS

Let $\mathcal{S}$ be a finite set called the *state space*. A *Markov chain* is a sequence $X_0, X_1, \dots$ of random variables, defined on a common probability space $(\Omega, \mathbb{P})$ and with values in $\mathcal{S}$, satisfying the following *Markov property*:

$$\begin{aligned} &\mathbb{P}\left(X_{k+1} = y \mid X_k = x, X_{k-1} = x_{k-1}, \dots, X_0 = x_0\right) \\ =&\mathbb{P}\left(X_{k+1} = y \mid X_k = x\right) \qquad (5.1) \\ =&P(x, y), \qquad (5.2) \end{aligned}$$

for every choice of $y, x, x_k, x_{k-1}, \dots, x_0 \in \mathcal{S}$. In words, "the future" only depends on "the present" (not the past), and in fact only depends on the current position in $\mathcal{S}$.

The matrix $P = [P(x, y)]_{x,y\in\mathcal{S}}$ is called *the transition matrix* and its rows consist of discrete probability mass functions on $\mathcal{S}$. In particular, the rows of P sum to 1:

$$\sum_{y\in\mathcal{S}} P(x, y) = \sum_{y\in\mathcal{S}} \mathbb{P}\left(X_1 = y \mid X_0 = x\right) = \mathbb{P}\left(X_1 \in \mathcal{S} \mid X_0 = x\right) = 1.$$

DOI: 10.1201/9781003024866-6

One way to think about this is that, by summing over all the possible states where X_1 might jump to, we are certain that X_1 will land somewhere in $\mathcal{S}$. A matrix whose rows are pmfs is called *row-stochastic*.

Every random variable X_k has a corresponding state space probability mass function

$$\pi_k := \{\mathbb{P}(X_k = x)\}_{x \in \mathcal{S}}.$$

The pmf π_k describes the most likely locations for the chain at time k.

The so-called method of *first-step analysis* (or simply "conditioning") yields the following relation:

$$\begin{aligned}
\pi_1(y) &= \mathbb{P}(X_1 = y) \\
&= \sum_{x \in \mathcal{S}} \mathbb{P}(X_1 = y \mid X_0 = x)\, \mathbb{P}(X_0 = x) \\
&= \sum_{x \in \mathcal{S}} \pi_0(x) P(x, y),
\end{aligned} \tag{5.3}$$

where the second equality holds by conditioning on all possible locations at time 0, and the last equality is a consequence of the time homogeneity assumed in the Markov property (5.2). Moreover, the last equation yields a matrix multiplication in sigma notation. Thus, if we interpret π_k as a row vector, we have that $\pi_1 = \pi_0 P$. More generally, we can see that

$$\pi_k = \pi_{k-1} P = \pi_{k-2} P^2 = \cdots = \pi_0 P^k.$$

Note that if π_k tends to a limit pmf π_∞ as $k \to \infty$, then the limit must satisfy $\pi_\infty = \pi_\infty P$.

Definition 5.1. A pmf π on $\mathcal{S}$ such that $\pi P = \pi$ is called a *stationary distribution* Furthermore, if π is the unique stationary distribution such for every initial distribution π_0 the sequence π_k converges to π, then we say that π is a *stable distribution*

It is common to write $\mathbb{P}_{\pi_0}$ to mean conditional probability given the initial distribution π_0, or simply $\mathbb{P}_{x_0}$ when π_0 is the delta function δ_{x_0}. Thus, by repeated applications of the Markov property, given a sequence

$x_0, \ldots, x_{k-1}$, we have

$$\begin{aligned}
&\mathbb{P}_{\pi_0}(X_k = y, X_{k-1} = x_{k-1}, \ldots, X_0 = x_0)\\
=&\mathbb{P}_{\pi_0}(X_k = y \mid X_{k-1} = x_{k-1}, \ldots, X_0 = x_0)\cdot\\
&\mathbb{P}_{\pi_0}(X_{k-1} = x_{k-1}, \ldots, X_0 = x_0)\\
=&P(x_{k-1}, y)\mathbb{P}_{\pi_0}(X_{k-1} = x_{k-1}, \ldots, X_0 = x_0)\\
=&\cdots\\
=&P(x_{k-1}, y)P(x_{k-2}, x_{k-1})\cdots P(x_1, x_2)\mathbb{P}_{\pi_0}(X_1 = x_1, X_0 = x_0)\\
=&\pi_0(x_0)P(x_0, x_1)P(x_1, x_2)\cdots P(x_{k-1}, y)
\end{aligned}$$

Hence, conditioning on the past, we obtain

$$\begin{aligned}
\pi_k(y) &= \mathbb{P}_{\pi_0}(X_k = y)\\
&= \sum_{x_0,\ldots,x_{k-1}} \mathbb{P}_{\pi_0}(X_k = y \mid X_{k-1} = x_{k-1}, \ldots, X_0 = x_0)\cdot\\
&\qquad \mathbb{P}_{\pi_0}(X_{k-1} = x_{k-1}, \ldots, X_0 = x_0)\\
&= \sum_{x_0,\ldots,x_{k-1}} \pi_0(x_0)P(x_0, x_1)P(x_1, x_2)\cdots P(x_{k-1}, y) = (\pi_0 P^k)(y).
\end{aligned}$$

If we choose $\pi_0 = \delta_x$, we get a nice interpretation of the entries of the matrix P^k. Namely,

$$P^k(x, y) = \sum_{x_1,\ldots,x_{k-1}} P(x, x_1)P(x_1, x_2)\cdots P(x_{k-1}, y). \tag{5.4}$$

In other words, to compute $P^k(x, y)$ we may look at all the walks with k steps that start at x and end at y, weigh each one with the product of the transition probabilities along the walk, and then sum all of them up.

Example 5.1 (Gambler's ruin). Perhaps the most famous example of a Markov chain is Gambler's Ruin. A gambler starts with x dollars and at every step either wins a dollar or loses a dollar with a 50/50 chance. The game ends if the gambler goes broke or if the winnings reach a pre-established goal of, say, $N \geq x$ dollars. We can think of the state space as $\mathcal{S} = \{0, \ldots, N\}$. Then, for $k = 1, \ldots, N-1$, $P(k, k \pm 1) = 1/2$, while $P(0, 0) = P(N, N) = 1$.

Define τ as the first time the gambler's fortune is either 0 or N. We are interested in the expectation $\mathbb{E}_x(\tau)$ and in the probability $\mathbb{P}_x(X_\tau = N)$ of reaching N before 0, given that the starting fortune is x.

Claim 5.1. *Assume* $0 \leq x \leq N$. *We have*

$$\mathbb{P}_x(X_\tau = N) = \frac{x}{N} \qquad \text{and} \qquad \mathbb{E}_x(\tau) = x(N - x).$$

Proof. Assume first that $0 < x < N$. The key is to define the function $h(x) := \mathbb{P}_x(X_\tau = N)$ and then condition on the first step:

$$\begin{aligned}
h(x) &= \mathbb{P}_x(X_\tau = N) \\
&= \mathbb{P}_x(X_\tau = N | X_1 = x + 1)P(x, x + 1) \\
&\quad + \mathbb{P}_x(X_\tau = N \mid X_1 = x - 1)P(x, x - 1) \\
&= \mathbb{P}_{x+1}(X_\tau = N)\frac{1}{2} + \mathbb{P}_{x-1}(X_\tau = N)\frac{1}{2} \\
&= \frac{1}{2}h(x + 1) + \frac{1}{2}h(x - 1)
\end{aligned}$$

Moreover, it can be checked directly from the definitions that $h(0) = 0$ and $h(N) = 1$. Therefore, this system of linear equations has the linear function $h(x) = x/N$ as its solution; see Exercise 5.1.

Likewise, set $g(x) := \mathbb{E}_x(\tau)$. Then $g(0) = g(N) = 0$ and the first-step analysis using conditioning for expectations, see Exercise 4.8, yields

$$\begin{aligned}
g(x) &= \mathbb{E}_x(\tau) \\
&= \mathbb{E}_x(\tau \mid X_1 = x + 1)P(x, x + 1) \\
&\quad + \mathbb{E}_x(\tau \mid X_1 = x - 1)P(x, x - 1) \\
&= (1 + \mathbb{E}_{x+1}(\tau))\frac{1}{2} + (1 + \mathbb{E}_{x-1}(\tau))\frac{1}{2} \\
&= 1 + \frac{1}{2}g(x + 1) + \frac{1}{2}g(x - 1)
\end{aligned}$$

Then, one can check that the function $g(x) = x(N - x)$ is the unique solution to this system of equations, see Exercise 5.1. □

5.2 IRREDUCIBILITY AND APERIODICITY

Definition 5.2. Given a Markov chain $X_0, X_1, \ldots$ on $\mathcal{S}$, we say that a state $y \in \mathcal{S}$ is *accessible* from state $x \in \mathcal{S}$, if

$$\mathbb{P}(\text{reaching } y \text{ from } x \text{ in } k \text{ steps}) = \mathbb{P}_x(X_k = y) > 0;$$

or, in matrix notation, if there exists a $k \geq 0$ such that

$$P^k(x, y) = \sum_{x_1, \ldots, x_{k-1}} P(x, x_1) \cdots P(x_{k-1}, y) > 0. \tag{5.5}$$

States x and y *communicate* if they are accessible from each other.

A Markov chain is *irreducible* if for every $x, y \in \mathcal{S}$, there exists $n \in \mathbb{Z}_{\geq 0}$ such that $P^n(x, y) > 0$, i.e., if every state communicates with every other state.

Example 5.2 (Random walks on graphs). The main example of a Markov chain we will consider is the special case of a Simple Random Walk on a graph. Let $G = (V, E)$ be a simple undirected finite graph. In this case, the state space is the set of nodes V, and we assume that a random walker at a node $x \in V$ chooses to jump to one of its neighbors uniformly at random. Namely

$$P(x, y) = \begin{cases} \frac{1}{\deg(x)} & \text{if } y \sim x \\ 0 & \text{else} \end{cases} \tag{5.6}$$

In particular, the rows of the transition matrix P sum to 1.

Define

$$\pi(y) := \frac{\deg(y)}{2|E|}. \tag{5.7}$$

By the Handshake Lemma, Equation (1.4), π is a pmf on V. This is an instance where the pmf π is induced by a sequence of weights, namely the degree sequence, as explained in Example 4.2. Moreover, π is stationary for the random walk:

$$\begin{aligned} (\pi P)(x) &= \sum_{y \in V} \pi(y) P(y, x) \\ &= \sum_{y \sim x} \frac{\deg(y)}{2|E|} \frac{1}{\deg(y)} \\ &= \frac{\deg(x)}{2|E|} = \pi(x). \end{aligned}$$

Remark 5.1. Note that the chain corresponding to a random walk on a graph G is irreducible if and only if G is connected.

Now we want to deal with the notion of time periodicity. The example to keep in mind is the Simple Random Walk (SRW) on the integers $\mathbb{Z}$. In this case, if the walker is at k, it either jumps to $k+1$ with probability $1/2$ or to $k-1$ with probability $1/2$. Note that if the walker is at 0 at time 0, it will visit odd integers at odd times and even integers at even times. In particular, it can only return to the origin at even times, i.e., the probability of returning to 0 in n steps is zero if n is not a multiple of 2. Therefore, we have that

$$\text{if } P^n(0,0) > 0, \text{ then } n \text{ is a multiple of } 2.$$

Moreover, this property does not hold if 2 is replaced by a larger integer $q > 2$. Indeed, this would mean that $P^2(0,0) = 0$, because 2 is not a multiple of q. However, by Equation (5.4), for the SRW we have

$$P^2(0,0) = P(0,1)P(1,0) + P(0,-1)P(-1,0) = 1/2 > 0.$$

So, in this case, we say that the origin in $\mathbb{Z}$ has period 2.

This can be generalized as follows.

Definition 5.3. The state $x \in \mathcal{S}$ has *period* $q \in \mathbb{Z}_{>0}$, if q is the largest positive integer q' with the property that

$$\text{if } P^r(x,x) > 0, \text{ then } r \text{ is a multiple of } q'.$$

Lemma 5.1. *If a Markov chain is irreducible, all states have the same period.*

In particular, for irreducible Markov chains, we can talk about the period of the chain as a whole, instead of a node.

Proof of Lemma 5.1. Consider a Markov chain on the state space $\mathcal{S}$ with transition matrix P. Fix two states x and y in $\mathcal{S}$. Let $q(x)$ and $q(y)$ be their periods. By irreducibility, there is $j, k \in \mathbb{Z}_{\geq 0}$ so that $P^j(x,y) > 0$ and $P^k(y,x) > 0$. Let $m = j + k$, then $P^m(x,x) > 0$ and $P^m(y,y) > 0$. So m must be a multiple of both $q(x)$ and $q(y)$.

Now, assume that $P^r(x,x) > 0$. Then

$$P^{r+m}(y,y) \geq P^k(y,x)P^r(x,x)P^j(x,y) > 0, \tag{5.8}$$

see Exercise 5.2. So $r + m$ must be a multiple of $q(y)$. Since m is a multiple of $q(y)$, r must also be a multiple of $q(y)$. By Definition 5.3, this implies that $q(y) \leq q(x)$. The lemma follows by reversing the roles of x and y. □

Definition 5.4. An irreducible chain is *aperiodic* if the common period of its states is 1.

If G is an undirected connected graph, then every node can travel to a neighbor and back, so the period of G is at most 2. Whether the period is 1 or 2 depends on whether a walk can return in an odd number of steps.

Definition 5.5. A graph $G = (V, E)$ is *bipartite* if one can color each node black or white in such a way that no neighbors have the same color. Mathematically, we ask for the existence of a labeling $c : V \to \{0, 1\}$ such that

$$x \sim y \Longrightarrow c(x) \neq c(y) \tag{5.9}$$

A labeling with property (5.9) is called a *coloring*.

Proposition 5.1. *Random walk on a connected undirected graph is aperiodic if and only if G is not bipartite. Moreover, if G is bipartite, its period equals* 2.

Proof. If G is bipartite, then every walk must alternate colors with each step, so a walk starting at some node x and returning to x must take an even number of steps. This means that every node has a period of at least 2.

Conversely, if G is not bipartite, then there is a walk that can return in an odd number of steps. To see this, recall the definition of hop length distance from (1.2) and consider the following natural partition.

Definition 5.6. Let $G = (V, E)$ be a connected graph and let $x_0 \in V$ be a base point. For $k = 0, 1, 2, 3, \ldots$, define the *k-th shell* $S(x_0, k)$ to be

$$S(x_0, k) = \{y \in V : d(x_0, y) = k\},$$

where $d(x_0, y)$ is the minimum number of hops for a walk from x_0 to y.

We call $\{S(x_0, k)\}_{k=0}^{\infty}$ the *shell decomposition* of G based at x_0.

Given a shell decomposition based at x_0, label the vertices in $S(x_0, k)$ with 0 or 1 depending on whether k is even or odd. If G is not bipartite, then this assignment is not a coloring, i.e., there are nodes x and y such that $x \sim y$ and both x and y belong to the same shell $S(x_0, k)$ for some k. That's because the nodes in $S(x_0, k)$ can only have neighboring nodes in $S(x_0, j)$ for $j = k - 1, k, k + 1$. If we now let γ_1 and γ_2 be the shortest paths from x_0 to x and y respectively, they both have hop length equal to k. Hence, traveling along γ_1, crossing the edge $\{x, y\}$, and traversing γ_2 in reverse order will yield a cycle of odd hop length. □

In the general case of Markov chains, an irreducible periodic chain can always be made aperiodic by setting $P(x, x) > 0$ at some state x. In terms of random walks on a graph, usually, one sets the probability that the walk stays at a node to be $1/2$. These are sometimes called *lazy* walks.

Example 5.3 (The Lazy Walk). Given a simple connected and undirected graph $G = (V, E)$, let P be the transition matrix for the random walk on G as defined in (5.6). Then, the transition matrix for the lazy version of the random walk on G is defined to be

$$\tilde{P}(x, y) = \begin{cases} \frac{1}{2\deg(x)} & \text{if } y \sim x \\ \frac{1}{2} & y = x \\ 0 & \text{else} \end{cases} \tag{5.10}$$

In other words, $\tilde{P} = \frac{1}{2}P + \frac{1}{2}I$.

5.3 HARMONICITY WITH RESPECT TO A CHAIN

As we saw in (5.3), the transition matrix P naturally acts on row vectors that are interpreted as pmfs on the state space. However, P also acts on column vectors. We will identify functions $h : \mathcal{S} \to \mathbb{R}$ as column vectors. For instance, if u is the constant function equal to 1, then row-stochasticity implies that $Pu = u$. So the following observation always holds:

> *The non-zero constant functions are eigenvectors of P for the eigenvalue* 1.

In general, P acts as an averaging operator on functions defined on $\mathcal{S}$. Indeed, let X_k be a Markov chain with $X_0 = x$, then:

$$\begin{aligned} (Ph)(x) &= \sum_{y \in \mathcal{S}} P(x, y)h(y) \\ &= \sum_{y \in \mathcal{S}} h(y)\mathbb{P}(X_1 = y \mid X_0 = x) \\ &= \mathbb{E}_x(h(X_1)). \end{aligned}$$

In other words, the value of the function Ph at x is the expected value of h over the states that can be reached from x.

Definition 5.7. The function $h : \mathcal{S} \to \mathbb{R}$ is *harmonic at x with respect to a transition matrix P* if

$$h(x) = \sum_{y \in \mathcal{S}} P(x, y)h(y) = (Ph)(x).$$

Next, we generalize the maximum principle as stated in Theorem 2.2.

Lemma 5.2 (Maximum principle for chains). *Given an irreducible Markov chain with finite state space and transition matrix P, if a function $h : \mathcal{S} \to \mathbb{R}$ is harmonic with respect to P at every $x \in \mathcal{S}$, then h is constant.*

Proof. Since $\mathcal{S}$ is finite, there is $x_0 \in \mathcal{S}$ such that $h(x_0) = \max_{x\in\mathcal{S}} h(x) = M$. Assume that there is $z \in \mathcal{S}$ such that $P(x_0, z) > 0$ and $h(z) < M$. Then, by harmonicity at x_0,

$$M = h(x_0) = \sum_{y\in\mathcal{S}} P(x_0, y)h(y) < M \sum_{y\in\mathcal{S}} P(x_0, y) = M.$$

This is a contradiction, therefore $h(z) = M$ whenever $P(x_0, z) > 0$.

Irreducibility means that for every $x \in \mathcal{S}$ there is a chain $x_0, x_1, \ldots, x_n = x$ such that $P(x_k, x_{k+1}) > 0$, for $k = 0, \ldots, n-1$. Therefore, by repeating the same argument inductively, we reach the claim. □

Example 5.4. In the case of random walks on graphs

$$(Ph)(x) = \sum_{y\sim x} \frac{1}{\deg(x)} h(y) = \frac{1}{\deg(x)} \sum_{y\sim x} h(y).$$

Recall the adjacency matrix $A = \mathbb{1}_{x\sim y}$ and the diagonal matrix $D = \operatorname{Diag}(\deg(\cdot))$, defined in Section 3.2. The transition matrix for a random walk on a graph can be written as

$$P = D^{-1}A.$$

In terms of the Laplacian

$$L = D - A = D(I - D^{-1}A) = D(I - P).$$

In particular, $(I - P)h = 0$ if and only if $Lh = 0$. So, we recover the notion of harmonicity that we encountered in (2.8).

Remark 5.2 (Uniqueness of stationary distributions). By Definition 5.1, a distribution π is stationary if and only if $\pi^T \in \operatorname{Ker}(I - P^T)$. The Rank-Nullity Theorem says that

$$\begin{aligned}\dim \operatorname{Ker}(I - P^T) &= N - \operatorname{rank}(I - P^T)\\ &= N - \operatorname{rank}(I - P) = \dim \operatorname{Ker}(I - P).\end{aligned}$$

By Definition 5.7, a function h satisfies $(I - P)h = 0$ if and only if h is harmonic at every point of $\mathcal{S}$. Also, by the Maximum Principle (Lemma 5.2), if P is irreducible, then $\dim \operatorname{Ker}(I - P) = 1$, and therefore $\dim \operatorname{Ker}(I - P^T) = 1$. This establishes the uniqueness of stationary distributions but not their existence. The problem is that we cannot immediately conclude that $\operatorname{Ker}(I - P^T)$ is spanned by a non-negative vector, which we could normalize to make it into a pmf. Here, we take the probabilistic approach, as developed, for instance, in [56]. A parallel approach is to rely exclusively on matrix theory, but this strips away the many probabilistic descriptions, which we think add richness to this topic.

5.4 STOPPING TIMES

We will prove the existence of stable distributions for irreducible chains using the concept of stopping time Green functions. Let $X_0, X_1, \dots$ be a Markov chain with values in a finite state space $\mathcal{S}$ and with transition matrix P.

A *stopping time* for the chain is a random variable τ with values in $\{0, 1, 2, \dots\}$ such that the event $\{\tau = n\}$ depends only on $X_0, \dots, X_n$, for $n \geq 0$. Technically, this means that the event $\{\tau = n\}$ is in the σ-algebra of events generated by the random variables $X_0, \dots, X_n$. In particular, the event $\{\tau > n\} = \{\tau \geq n + 1\}$ only depends on $X_0, \dots, X_n$, see Exercise 5.3.

Stopping times give rise to a generalization of the Markov property, called the *Strong Markov property*, that allows one to condition on stopping times. For instance, given a stopping time τ, we will use the following generalization of (5.1):

$$\begin{aligned}
&\mathbb{P}\left(X_{k+1} = y \mid X_k = x, X_{k-1} = x_{k-1}, \dots, X_0 = x_0, \tau > k\right) \\
=&\mathbb{P}\left(X_{k+1} = y \mid X_k = x, \tau > k\right) \\
=&\mathbb{P}\left(X_{k+1} = y \mid X_k = x\right) \\
=&P(x, y).
\end{aligned} \tag{5.11}$$

More generally, if a Markov chain $\{X_k\}_{k=0}^{\infty}$ starts at some state $X_0 = x$ at time zero, and returns to x at some time $k > 1$, then the Markov property allows one to restart the chain, in the sense that the chain $\{Y_j := X_{k+j}\}_{j=0}^{\infty}$ has the same distribution as the original chain $\{X_k\}_{k=0}^{\infty}$. Furthermore, the strong Markov property allows one to restart Markov chains that stop when a stopping time τ goes off. For more details about this we refer the reader to standard texts in probability, see for instance [37, Section 6.3].

Here are some examples of stopping times; see Exercise 5.3. Given a Markov chain $X_0, X_1, \ldots$ in a state space $\mathcal{S}$, consider the following two *hitting times* for a given state $x \in \mathcal{S}$:

$$\tau_x := \inf\{n \geq 0 : X_n = x\} \text{ and } \tau_x^+ := \inf\{n \geq 1 : X_n = x\}. \quad (5.12)$$

Namely, τ_x is the first time the chain visits x, and if $X_0 = x$, then τ_x^+ is called the *first-return time* at x.

5.5 STOPPED GREEN FUNCTIONS

Given two states $a, b \in \mathcal{S}$, and a stopping time τ, define the *(stopped) Green function*

$$G_\tau(a,b) := \textit{expected number of visits at } b\textit{, starting at } a\textit{, before } \tau.$$

Mathematically, we can count the number of visits at b using indicator functions:

$$\textit{Number of visits at } b \textit{ before } \tau = \sum_{k=0}^{\infty} \mathbb{1}_{\{X_k = b,\ k < \tau\}}$$

Also recall that if $\mathbb{P}$ is a probability measure defined on the sample space Ω, then for any measurable event $A \subset \Omega$ we have

$$\mathbb{E}(\mathbb{1}_A) = \int_\Omega \mathbb{1}_A(\omega) d\mathbb{P}(\omega) = \mathbb{P}(A).$$

Thus, by linearity of expectations,

$$G_\tau(a,b) = \mathbb{E}\left(\sum_{k=0}^{\infty} \mathbb{1}_{\{X_k = b,\ k<\tau\}}\right) = \sum_{k=0}^{\infty} \mathbb{P}_a\left(X_k = b,\ \tau > k\right). \quad (5.13)$$

In other words, one can either start the chain at a by setting $X_0 = a$ and then count the number of visits at b before τ stops the clock and then do an average; or, equivalently, one can consider the event that the chain is at b at time k and that $k < \tau$ (i.e. τ hasn't gone off yet), then compute the probability of this event conditioning on $X_0 = a$ and sum over all possible times k.

Remark 5.3. Given $a \in \mathcal{S}$, it will often be useful to have $G_\tau(a,b) < \infty$, for every $b \in \mathcal{S}$. To this end, it will be enough to require that $\mathbb{E}_a(\tau) < \infty$, since

$$\begin{aligned}
\sum_{b\in\mathcal{S}} G_\tau(a,b) &= \sum_{b\in\mathcal{S}}\sum_{k=0}^{\infty} \mathbb{P}_a\left(X_k = b,\ \tau > k\right) && \text{(by (5.13))}\\
&= \sum_{k=0}^{\infty}\sum_{b\in\mathcal{S}} \mathbb{P}_a\left(X_k = b,\ \tau > k\right) && \text{(switch sums)}\\
&= \sum_{k=0}^{\infty} \mathbb{P}_a\left(\tau > k\right) && \text{(Theorem 4.1)}\\
&= \mathbb{E}_a(\tau). && \text{(by Cavalieri (P4.1))}
\end{aligned}$$

In other words, if we add up the expected number of visits to each individual state b before τ, we get the expected number of visits before τ.

Example 5.5. Here are a couple of examples of stopping time Green functions.

(i) *Number of returns before exiting:* Consider the case when τ is the hitting time τ_b. Then

$$G_{\tau_b}(a,a) = \mathbb{E}_a\left(\#\text{ visits at } a \text{ before hitting } b\right) \tag{5.14}$$

In other words, start at a and count the number of times the chain is at a before hitting b.

(ii) *Number of hits before returning:* Now let τ be the first-return time τ_a^+. Then

$$G_{\tau_a^+}(a,b) = \mathbb{E}_a\left(\#\text{ visits at } b \text{ before returning to } a\right). \tag{5.15}$$

The following lemma establishes a formula for the *expected number of traversals* from state x to state y.

Lemma 5.3. *Let $X_0, X_1, \ldots$ be a Markov chain with values in a finite state space $\mathcal{S}$ and with transition matrix P. Let τ be a stopping time and $G_\tau(a,b)$ the corresponding Green function. Then, given a starting state $a \in \mathcal{S}$, for every $x, y \in \mathcal{S}$:*

$$\mathbb{E}_a\left(\#\text{ times chain jumps from } x \text{ to } y \text{ before } \tau\right) = G_\tau(a,x)P(x,y). \tag{5.16}$$

Proof. Indeed,

$$
\begin{aligned}
& G_\tau(a,x)P(x,y) \\
&= \sum_{k=0}^{\infty} \mathbb{P}_a\left(X_k = x,\ \tau > k\right) P(x,y) && \text{(by (5.13))} \\
&= \sum_{k=0}^{\infty} \mathbb{P}_a\left(X_k = x,\ \tau > k\right) \mathbb{P}_a\left(X_{k+1} = y \mid X_k = x, \tau > k\right) && \text{(by (5.11))} \\
&= \sum_{k=0}^{\infty} \mathbb{P}_a\left(X_{k+1} = y,\ X_k = x,\ \tau > k\right) \\
&= \mathbb{E}_a\left(\#\text{ times } x \text{ is followed by } y \text{ before } \tau\right),
\end{aligned}
$$

where the last equality follows as in (5.13) by the linearity of expectations. □

Definition 5.8. Given a state $z \in \mathcal{S}$, we say that τ is a *time stopping at* z if τ is a stopping time such that:

$$\mathbb{P}(X_\tau = z) = 1.$$

Note that a time stopping at z is not necessarily a hitting time; namely, it is not necessarily the first time z has been visited.

Proposition 5.2. *Fix $a, z \in \mathcal{S}$. Suppose τ is a time stopping at z for the Markov chain $\{X_k\}_{k=0}^{\infty}$ such that $\mathbb{E}_a(\tau) < \infty$. For every $y \in \mathcal{S}$, consider the Green function $G_\tau(a,y)$, which counts the expected number of visits at y, when the chain starts at a and before τ goes off. Then,*

$$G_\tau(a,y) = \sum_{x \in \mathcal{S}} G_\tau(a,x)P(x,y) + \mathbb{P}_a(\tau > 0)(\delta_z - \delta_a)(y), \tag{5.17}$$

where $\delta_x(\cdot)$ is the indicator function at x.

Proof. First note that, by Remark 5.3, $G_\tau(a,y) < \infty$, for every $y \in \mathcal{S}$. Next, recall from the proof of Lemma 5.3, that

$$G_\tau(a,x)P(x,y) = \sum_{k=0}^{\infty} \mathbb{P}_a\left(X_{k+1} = y,\ X_k = x,\ \tau > k\right).$$

Therefore,

$$
\begin{aligned}
&\sum_{x\in\mathcal{S}} G_\tau(a,x)P(x,y)\\
=&\sum_{k=0}^{\infty}\sum_{x\in\mathcal{S}} \mathbb{P}_a\left(X_k=x, X_{k+1}=y, \tau>k\right) && \text{(switch sums)}\\
=&\sum_{k=0}^{\infty} \mathbb{P}_a\left(X_{k+1}=y, \tau\geq k+1\right) && \text{(Theorem 4.1)}\\
=&\sum_{n=1}^{\infty} \mathbb{P}_a\left(X_n=y, \tau\geq n\right) && \text{(relabel } n=k+1\text{)}\\
=&\sum_{n=1}^{\infty} \left[\mathbb{P}_a\left(X_n=y, \tau> n\right)+\mathbb{P}_a\left(X_n=y, \tau= n\right)\right] && \text{(by (4.1))}\\
=&G_\tau(a,y)-\mathbb{P}_a\left(X_0=y,\tau>0\right)\\
&+\sum_{n=1}^{\infty}\mathbb{P}_a\left(X_n=y,\tau=n\right) && \text{(by (5.13))}
\end{aligned}
$$

Now note that

$$\mathbb{P}_a\left(X_0=y,\tau>0\right)=\mathbb{P}_a(\tau>0)\delta_a(y),$$

and

$$
\begin{aligned}
\sum_{n=1}^{\infty}\mathbb{P}_a\left(X_n=y,\tau=n\right)&=\left(\sum_{n=1}^{\infty}\mathbb{P}_a\left(\tau=n\right)\right)\delta_z(y)\\
&=\mathbb{P}_a(\tau>0)\delta_z(y).
\end{aligned}
$$

□

5.6 THE STATIONARY DISTRIBUTION

For irreducible chains, if a stationary distribution exists, it must be unique; see Remark 5.2. This can be used to write a formula for the stationary distribution in terms of return times.

Theorem 5.1. *If a Markov chain on $\mathcal{S}$ is irreducible, then there exists a (unique) stationary distribution π on $\mathcal{S}$, satisfying*

$$\pi(x)=\frac{1}{\mathbb{E}_x(\tau_x^+)}>0 \qquad \forall x\in\mathcal{S}. \tag{5.18}$$

This means that the larger the expected first-return time at a given state, the lower the probability of finding the chain at that state in the long run.

Before we can prove Theorem 5.1, we need to show that, for any $x, y \in \mathcal{S}$, $\mathbb{E}_x(\tau_y^+) < \infty$, so that we can apply Proposition 5.2.

Claim 5.2 (Lemma 1.13 in [56]). *When the chain is irreducible, for any $x, y \in \mathcal{S}$,*

$$\mathbb{E}_x(\tau_y^+) < \infty. \tag{5.19}$$

Proof. Fix $y \in \mathcal{S}$. Since the chain is irreducible, for any $x \in \mathcal{S}$, there is a smallest integer $r(x) \in \mathbb{Z}_{\geq 0}$ such that $P^{r(x)}(x, y) > 0$. Namely, this is the smallest number of steps for which there is a chance of walking from x to y. In particular, as soon as $x \neq y$, we have $r(x) \geq 1$. Set $n := \max_{x \in \mathcal{S}} r(x)$. Then, if $\mathcal{S}$ has more than one state, we must have $n \geq 1$. Further, set $\delta := \min_{x \in \mathcal{S}} P^{r(x)}(x, y)$. Note that we have $\delta > 0$ since it is a minimum of finitely many positive numbers. With these definitions, we see that, for any $x \in \mathcal{S}$, there is at least a chance δ to reach y in at most n steps from x. We can reinterpret this in terms of $\tau_y^+ = \inf\{k \geq 1 : X_k = y\}$. Namely, we have $\mathbb{P}_x\left(\tau_y^+ \leq n\right) \geq \delta$. In other words,

$$\mathbb{P}_x\left(\tau_y^+ > n\right) \leq 1 - \delta. \tag{5.20}$$

The goal is to show that the tail probabilities $\mathbb{P}_x\left(\tau_y^+ > kn\right)$ decay geometrically. This will follow from induction, because for $k > 1$,

$$\mathbb{P}_x\left(\tau_y^+ > kn\right) \leq (1 - \delta)\mathbb{P}_x\left(\tau_y^+ > (k-1)n\right). \tag{5.21}$$

The proof of (5.21) is deduced in Exercise 5.4. Iterating this we get that, for $k > 1$,

$$\mathbb{P}_x\left(\tau_y^+ > kn\right) \leq (1 - \delta)^k. \tag{5.22}$$

To conclude, using Cavalieri's principle (P4.1), we see that

$$\begin{aligned}\mathbb{E}_x(\tau_y^+) &= \sum_{j=0}^{\infty} \mathbb{P}_x(\tau_y^+ > j) = \sum_{k=0}^{\infty}\sum_{r=0}^{n-1} \mathbb{P}_x(\tau_y^+ > kn + r) \\ &\leq \sum_{k=0}^{\infty} n\mathbb{P}_x(\tau_y^+ > kn) \leq \sum_{k=0}^{\infty} n(1-\delta)^k < \infty.\end{aligned}$$

□

Proof of Theorem 5.1. Assume that $X_0 = a$ for some $a \in \mathcal{S}$. Consider the first return time τ_a^+. By (5.19), $\mathbb{E}_a(\tau_a^+) < \infty$ and also $\mathbb{P}_a\left(X_{\tau_a^+} = a\right) = 1$. Therefore, by Proposition 5.2 with $z = a$, the function

$$\pi(x) := \frac{G_{\tau_a^+}(a,x)}{\mathbb{E}_a(\tau_a^+)} \qquad \forall x \in \mathcal{S},$$

satisfies

$$\pi(y) = \sum_{x \in \mathcal{S}} \pi(x) P(x,y) \qquad \forall y \in \mathcal{S}.$$

Also, we have $\pi(x) \geq 0$ and $\sum_{x \in \mathcal{S}} \pi(x) = 1$ follows from Remark 5.3. So, π is a stationary distribution for the Markov chain.

Evaluating at $x = a$, we get that

$$\pi(a) = \frac{1}{\mathbb{E}_a(\tau_a^+)}.$$

However, the stationary distribution must be unique by Remark 5.2 and a was arbitrary, so (5.18) is verified. □

5.7 REVERSIBLE CHAINS

Definition 5.9. A chain is *reversible* if there exists a distribution π on $\mathcal{S}$ such that

$$\pi(x)P(x,y) = \pi(y)P(y,x) \qquad \forall x,y \in \mathcal{S}. \tag{5.23}$$

This is known as the *detailed balance equation.*

Proposition 5.3. *If π satisfies the detailed balance equation* (5.23), *then π is stationary.*

Proof. Note that for $x \in \mathcal{S}$:

$$\begin{aligned}(\pi P)(x) &= \sum_{y \in \mathcal{S}} \pi(y)P(y,x) = \sum_{y \in \mathcal{S}} \pi(x)P(x,y) \\ &= \pi(x) \sum_{y \in \mathcal{S}} P(x,y) = \pi(x).\end{aligned}$$

□

The term 'reversible' comes from observing that the law of every sequence $(X_0, \ldots, X_n)$ is the same as that of the reversed sequence $(X_n, \ldots, X_0)$. To see this, we use conditional probability and note that

$$\begin{aligned}
&\mathbb{P}_\pi(X_0 = x_0, \ldots, X_n = x_n) \\
&= \mathbb{P}_\pi(X_n = x_n \mid X_{n-1} = x_{n-1}, \ldots, X_0 = x_0)\cdot \\
&\qquad \mathbb{P}_\pi(X_0 = x_0, \ldots, X_{n-1} = x_{n-1}) && \text{(cond. probability)} \\
&= P(x_{n-1}, x_n)\mathbb{P}_\pi(X_0 = x_0, \ldots, X_{n-1} = x_{n-1}) && \text{(Markov property)} \\
&= \ldots && \text{(repeat with } x_{n-1}\text{...)} \\
&= \pi(x_0)P(x_0, x_1)P(x_1, x_2)\cdots P(x_{n-1}, x_n) && \text{(this is the end result)} \\
&= P(x_1, x_0)\pi(x_1)P(x_1, x_2)\cdots P(x_{n-1}, x_n) && \text{(balance equation)} \\
&= \ldots && \text{(repeat with } x_1,\ x_2\text{...)} \\
&= \pi(x_n)P(x_n, x_{n-1})\cdots P(x_1, x_0) && \text{(this is the end result)} \\
&= \mathbb{P}_\pi(X_0 = x_n, \ldots, X_n = x_0) && \text{(retrace steps above).}
\end{aligned}$$

Example 5.6. Random walk on a simple undirected graph is a reversible chain. To see this, recall the transition matrix P, equation (5.6), and the stationary distribution π, equation (5.7), in this case. Then check that

$$\pi(x)P(x, y) = \pi(y)P(y, x) = \begin{cases} \frac{1}{2|E|} & \text{If } y \sim x, \\ 0 & \text{else.} \end{cases}$$

For an example of a chain that is not reversible, see Exercise 5.5.

5.8 THE SPECTRUM OF REVERSIBLE CHAINS

Recall that $\mathbf{1}$ denotes the constant function equal to 1, i.e., the column vector of all ones.

Lemma 5.4. *Consider a Markov chain with transition matrix P. Then,*

1. *$\lambda = 1$ is always an eigenvalue for P and $\mathbf{1}$ is always an eigenfunction for $\lambda = 1$.*

2. *If the chain is irreducible, then $\lambda = 1$ is simple, meaning that the corresponding eigenspace is spanned by $\mathbf{1}$, i.e., it's the space of all constant functions.*

3. *Since P is not generally symmetric, it might have complex eigenvalues. However, if λ is an eigenvalue for P, then $|\lambda| \le 1$.*

Proof. For 1., note that P is row-stochastic, which is equivalent to the equation $Pu = u$, whenever $u = c\mathbf{1}$. So $\lambda = 1$ is always an eigenvalue, and the constant functions $u = c\mathbf{1}$ are always eigenvectors.

For 2., when the chain is irreducible, as we have already seen, if h satisfies $Ph = h$, then h is harmonic everywhere and hence must be constant by the maximum principle (Lemma 5.2).

For 3., suppose that $Ph = \lambda h$ for some eigenvector $h \neq 0$. We can always assume that $\max_{x\in\mathcal{S}} |h(x)| = 1$ by dividing h by $\|h\|_\infty$. Then

$$\begin{aligned}
|\lambda||h(x)| &= |(\lambda h)(x)| = |(Ph)(x)| && (h \text{ is an eigenvector})\\
&= \left|\sum_{y\in\mathcal{S}} P(x,y)h(y)\right| \le \sum_{y\in\mathcal{S}} P(x,y)|h(y)| && (\text{triangle inequality})\\
&\le \sum_{y\in\mathcal{S}} P(x,y) = 1. && (\max_{y\in\mathcal{S}} |h(y)| = 1)
\end{aligned}$$

So, maximizing over x yields, $|\lambda| \le 1$. □

Reversible chains have the symmetry needed to make the spectral theory more approachable. Suppose a chain is reversible with transition P and stationary distribution π. Define

$$S(x,y) := \pi(x)^{1/2}\pi(y)^{-1/2}P(x,y). \tag{5.24}$$

Then, S is a symmetric matrix. Thus, by the Spectral Theorem 3.1, the eigenvalues of S are real, and S admits an orthonormal basis of eigenvectors. Namely, there are numbers $\lambda_1, \ldots, \lambda_N \in \mathbb{R}$ and an orthonormal basis of eigenfunctions $\{\phi_j\}_{j=1}^N$ so that $S\phi_j = \lambda_j\phi_j$, for $j = 1, \ldots, N$.

Proposition 5.4. *Suppose a chain is reversible with transition P and stationary distribution π. Consider the symmetric matrix S as defined in (5.24), and its corresponding eigenpairs (λ_j, ϕ_j), for $j = 1, \ldots, N$. Then, P shares the same eigenvalues as S, and the functions*

$$f_j(x) := \pi(x)^{-1/2}\phi_j(x) \qquad \forall x \in \mathcal{S}$$

form a basis of eigenfunctions for P.

Proof. Compute:

$$\begin{aligned}
(Pf_j)(x) &= \sum_{y\in\mathcal{S}} P(x,y)f_j(y) = \sum_{y\in\mathcal{S}} P(x,y)\pi(y)^{-1/2}\phi_j(y)\\
&= \pi(x)^{-1/2}\sum_{y\in\mathcal{S}} P(x,y)\phi_j(y) = \pi(x)^{-1/2}\lambda_j\phi_j(x)\\
&= \lambda_j f_j(x).
\end{aligned}$$

The linear independence of the f_j's follows from the fact that they are obtained from the ϕ_j's by multiplication by the diagonal matrix $D_\pi^{-1/2}$, where

$$D_\pi = \operatorname{diag}(\pi(\cdot)). \tag{5.25}$$

□

EXERCISES

5.1 *Gambler's ruin.* With the notations of Example 5.1, recall that $h(x) = \mathbb{P}_x(X_\tau = N)$ and $g(x) = \mathbb{E}_x(\tau)$.

(i) Verify that $h(x) := x/N$ satisfy the system of linear equations

$$\begin{cases} h(x) = \frac{1}{2}h(x-1) + \frac{1}{2}h(x+1), & \text{for } x = 1, \ldots, N-1, \\ h(0) = 0, \\ h(N) = 1. \end{cases}$$

(ii) Use the maximum principle to argue that h is the unique such solution.

(iii) Study the differences $f(x) := g(x) - g(x-1)$, for $x = 1, \ldots, N$ and show that $g(x) = x(N-x)$ is the unique solution for the system

$$\begin{cases} g(x) = 1 + \frac{1}{2}g(x-1) + \frac{1}{2}g(x+1), & \text{for } x = 1, \ldots, N-1, \\ g(0) = 0, \\ g(N) = 0. \end{cases}$$

5.2 *Markov property.* The purpose of this problem is to give two proofs of (5.8).

(i) The first approach is to use the formula for matrix multiplication in (5.5).

(ii) The second approach is to make use of the probabilistic interpretation

$$P^n(x,y) = \mathbb{P}(X_n = y \mid X_0 = x) = \mathbb{P}_x(X_n = y),$$

and then using conditioning and the Markov property.

5.3 *The stopping time property.* Show that the stopping time property implies that one can determine whether the stopping time τ hasn't occurred yet by observing the past and the present. More specifically, show that the event $\{\tau > n\}$ belongs to the σ-algebra generated by the random variables $X_0, \ldots, X_n$, i.e., by events of the form $\{X_k = x\}$ for $0 \le k \le n$ and $x \in \mathcal{S}$.

5.4 *Tail probabilities.* Here, we prove the induction step (5.21) leading to the tail probability estimate in (5.22).

(i) First show that it is enough to estimate

$$\mathbb{P}_x\left(\tau_y^+ > kn \mid \tau_y^+ > (k-1)n\right). \tag{P5.1}$$

(ii) Next, condition the probability in (P5.1) on the position of the chain at time $(k-1)n$.

(iii) Now, use the strong Markov property to show that, for every $z \in \mathcal{S}$,

$$\mathbb{P}_x\left(\tau_y^+ > kn \mid X_{(k-1)n} = z, \tau_y^+ > (k-1)n\right) \le \mathbb{P}_z\left(\tau_y^+ > n\right).$$

(iv) Finally, invoke (5.20).

5.5 *Reversibility.* Consider the cycle graph C_N on N nodes labeled consecutively as $x_0, x_1, \ldots, x_N$, where $x_N = x_0$. Consider a biased random walk on C_N with transition

$$P(x_k, x_{k+1}) = p,$$

for some $0 < p < 1/2$. Consider the event that such a random walk starts at x_0 and goes around C_N once in one direction, namely

$$X_0 = x_0, X_1 = x_1, \ldots, X_{N-1} = x_{N-1}, \quad \text{and} \quad X_N = x_0.$$

Compute the probability of this event, then compare it to the event that the walk goes around C_N in the reverse order.

5.6 *Spectrum of reversible chains.* Suppose a chain is reversible with transition P and stationary distribution π.

(i) Deduce that the spectrum of the matrix P lies in the interval $[-1, 1]$.

(ii) Deduce a range for the spectrum of the lazy version $\tilde{P} := (P + I)/2$.

5.7 *Substochastic matrices.* This exercise is a rephrasing of Exercise 3.10 in the language of transition matrices. Let $G = (V, E, \sigma)$ be a simple connected graph with edge weights $\sigma \in \mathbb{R}^E_{>0}$. Consider the corresponding transition matrix

$$P(x, y) = \frac{\sigma(x, y)}{\sum\limits_{z \sim x} \sigma(x, z)} \mathbb{1}_{x \sim y}$$

Fix a node $b \in V$, and let $A = V \setminus \{b\}$. Also, let Q be the minor obtained from P by removing the row and column corresponding to node b.

(i) Show that Q is sub-stochastic, meaning that the entries are non-negative and the rows sum to at most 1. Also, show that for all $k \geq 2$ and for all $x \in A$:

$$\sum_{y \in A} Q^k(x, y) \leq \sum_{y \in A} Q(x, y)$$

(ii) Show that for every node $x' \in A$, there is a path $x_1 = x', x_2, \ldots, x_n = x$ in A, to a node $x \in A$ with the property that

$$\sum_{y \in A} Q(x, y) < 1.$$

(iii) Deduce that Q^N, with $N = |V|$, has the property that all of its rows sum to a number strictly less than 1, and therefore that there is a number $0 < \delta < 1$ and a constant $C < \infty$, such that

$$Q^k(x, y) < C\delta^k$$

for all $x, y \in A$, and for all $k \geq 2$.

(iv) Show that any eigenvalue λ of Q must satisfy $|\lambda| < 1$.

(v) Show that the partial sums of the geometric series $\sum_{k=0}^{n} Q^k$ converge entry-wise to the inverse of $I - Q$.

CHAPTER 6

Connections between Electrical Networks and Random Walks

THROUGHOUT this section, we will assume that $G = (V, E, \sigma)$ is a finite simple connected graph with weights $\sigma \in \mathbb{R}^E_{>0}$ on the edges, representing electrical conductances.

6.1 TRANSITION MATRICES AND LAPLACIAN MATRICES

Write $\deg_\sigma(x) := \sum_{y \sim x} \sigma(x,y)$ for the generalized degree at x, and set

$$\sigma(E) := \sum_{e \in E} \sigma(e) = \frac{1}{2} \sum_{x \in V} \deg_\sigma(x).$$

The transition matrix P and stationary distribution π for the random walk on G take the form

$$P(x,y) = \frac{\sigma(x,y)}{\deg_\sigma(x)} \mathbb{1}_{y \sim x} \qquad \text{and} \qquad \pi(x) = \frac{\deg_\sigma(x)}{2\sigma(E)}. \tag{6.1}$$

In particular, if $A(x,y) = \sigma(x,y)\mathbb{1}_{y \sim x}$ is the weighted adjacency matrix and $D = \mathrm{diag}(\deg_\sigma(\cdot))$ the diagonal matrix of generalized degrees, then the weighted Laplacian is again $L = D - A$, and we have the following relations:

$$P = D^{-1}A \qquad \text{and} \qquad L = D(I - P).$$

DOI: 10.1201/9781003024866-7

Moreover, $\mathcal{L} = D^{-1/2}LD^{-1/2}$ is called *the normalized Laplacian*, and $D^{-1}L = I - P$ is *the random walk Laplacian.*

In particular, the mean value property (2.8) for a harmonic function h can be written as $Lh = 0$ or $(I - P)h = 0$ interchangeably. So, the notion of harmonicity with respect to the random walk viewed as a Markov chain is the same as the one we saw in the context of electrical networks. Also, the random walk is a reversible Markov chain, as we saw in Example 5.6.

Recall the diagonal matrix D_π defined in (5.25) and the symmetric matrix $S = D_\pi^{1/2}PD_\pi^{-1/2}$. Notice that $2\sigma(E)D_\pi = D$. So $S = D^{1/2}PD^{-1/2}$ as well, and

$$I - S = I - D^{-1/2}PD^{1/2} = D^{1/2}(I - P)D^{-1/2} = D^{-1/2}LD^{-1/2} = \mathcal{L}. \tag{6.2}$$

By Proposition 5.4, the spectrum of P coincides with the spectrum of S, and lies in $[-1, 1]$, see Exercise 5.6.

Since $\det(\lambda I - (I - S)) = (-1)^N \det((1 - \lambda)I - S)$, we see that the spectrum of the normalized Laplacian $\mathcal{L}$ lies in $[0, 2]$ and the eigenvalue 1 for P corresponds to the eigenvalue 0 for $\mathcal{L}$ (under the transformation $z \mapsto 1 - z$). Moreover, the orthonormal basis of eigenvectors $\{\phi_j\}$ introduced above Proposition 5.4, is also an orthonormal basis of eigenvectors for $\mathcal{L}$.

6.2 THE DIRICHLET PROBLEM

Let $B \subset V$ be a set of nodes that will play the role of the *boundary.* We are given some *boundary data* in the form of a function $h_B : B \to \mathbb{R}$. *The Dirichlet Problem* is the following:

$$\begin{cases} \text{Find } h : V \to \mathbb{R} \text{ such that} \\ h \text{ is harmonic in } V \setminus B \\ \text{and } h = h_B \text{ on } B \end{cases} \tag{6.3}$$

Theorem 6.1 (Solution of the Dirichlet Problem). *Let $\emptyset \neq B \subset V$ and let h_B be some given boundary data. Consider the random walk $\{X_k\}_{k=0}^{\infty}$ on G and let*

$$\tau_B := \inf\{k \geq 0 : X_k \in B\}$$

be the hitting time for B. Then, the unique solution to the Dirichlet Problem is

$$h(x) := \mathbb{E}_x \left[h_B(X_{\tau_B}) \right], \qquad \forall x \in V. \tag{6.4}$$

Proof. Uniqueness follows from the Maximum Principle (Theorem 2.2) since the difference of two candidate solutions has to equal 0 on B.

To establish existence, it is enough to check that (6.4) satisfies the boundary values and is harmonic on $V \setminus B$. The latter follows from a "first-step analysis", using conditioning for expectations as described in Exercise 4.8. Given $x \in V \setminus B$, assume $X_0 = x$. Then, conditioning on the location of X_1, we get:

$$\begin{aligned} h(x) &= \mathbb{E}_x \left[h_B(X_{\tau_B}) \right] \\ &= \sum_{y \in V} \mathbb{E}_x \left[h_B(X_{\tau_B}) \mid X_1 = y \right] \mathbb{P}_x(X_1 = y) && \text{(conditioning)} \\ &= \sum_{y \in V} \mathbb{E}_x \left[h_B(X_{\tau_B}) \mid X_1 = y \right] P(x, y). && \text{(by (5.2))} \end{aligned}$$

Using the Strong Markov property as explained after (5.11), and the fact that $\tau_B \geq 1$, since $x \notin B$, we can restart the chain and write:

$$\mathbb{E}_x \left[h_B(X_{\tau_B}) \mid X_1 = y \right] = \mathbb{E}_y \left[h_B(X_{\tau_B}) \right].$$

So, we have shown that $h(x) = \sum_{y \in V} P(x, y) h(y)$. This means that h is harmonic at x. □

6.3 RANDOM WALKS AND EFFECTIVE RESISTANCE

As we saw in Section 2.2 if we hook a battery at nodes $a, b \in V$ so as to create a unit current flow $f_{a,b}$ from a to b, the corresponding voltage potential $h_{a,b}$ is harmonic on $V \setminus \{a, b\}$. Moreover, since potentials are defined up to an additive constant, we can normalize the voltage so that

$$h_{a,b}(a) = 0 \qquad \text{and} \qquad h_{a,b}(b) = \operatorname{eff}\mathcal{R}(a, b).$$

In particular, the function

$$u_{a,b} := \frac{h_{a,b}}{\operatorname{eff}\mathcal{R}(a, b)}$$

solves the Dirichlet Problem with boundary data $h_B = \mathbb{1}_{\{b\}}$, where $B = \{a, b\}$.

By Theorem 6.1, we must have

$$u_{a,b}(x) = \mathbb{E}_x \left[h_B(X_{\tau_B}) \right].$$

So, this gives a probabilistic interpretation of the voltage potential.

Moreover, X_{τ_B} is a random variable that takes values in $\{a, b\}$ and $h_B(X_{\tau_B}) = \mathbb{1}_{\{X_{\tau_B}=b\}}$. In particular, recalling the hitting time τ_x defined in (5.12), we have

$$X_{\tau_B} = b \text{ if and only if } \tau_b < \tau_a.$$

We get the following connection between probability and the electrical potential: for $x \neq a, b$,

$$\begin{aligned} \frac{h_{a,b}(x)}{\text{eff}\mathcal{R}(a,b)} &= \mathbb{E}_x \left[h_B(X_{\tau_B}) \right] \\ &= \mathbb{E}_x \left[\mathbb{1}_{\{X_{\tau_B} = b\}} \right] \\ &= \mathbb{P}_x \left(X_{\tau_B} = b \right) \\ &= \mathbb{P}_x \left(\tau_b < \tau_a \right). \end{aligned} \tag{6.5}$$

The probability that a random walk starting at a hits b before returning to a

$$\mathbb{P}_a \left(\tau_b < \tau_a^+ \right)$$

is called *the hitting probability* from a to b.

We can relate hitting probability to effective conductance.

Proposition 6.1. *Given two nodes $a, b \in V$:*

$$\mathbb{P}_a \left(\tau_b < \tau_a^+ \right) = \frac{\text{eff}\mathcal{C}(a,b)}{\deg_c(a)}.$$

Proof. We do a first-step analysis, conditioning on the location after one step is taken:

$$
\begin{aligned}
&\mathbb{P}_a\left(\tau_b < \tau_a^+\right) \\
&= \sum_{x \in V} \mathbb{P}_a\left(\tau_b < \tau_a^+ \mid X_1 = x\right) P(a,x) && \text{(Conditioning)} \\
&= \sum_{x \in V} \mathbb{P}\left(\tau_b < \tau_a^+ \mid X_1 = x\right) P(a,x) && \text{(Markov property)} \\
&= \sum_{x \in V} \mathbb{P}_x\left(\tau_b < \tau_a\right) P(a,x) && \text{(Restart and } \tau_a^+ = \tau_a \text{ at } x\text{)} \\
&= \sum_{x \in V} \mathrm{eff}\mathcal{C}(a,b) h_{a,b}(x) P(a,x) && \text{(By (6.5))} \\
&= \frac{\mathrm{eff}\mathcal{C}(a,b)}{\deg_\sigma(a)} \sum_{x \sim a} h_{a,b}(x)\sigma(a,x) && \text{(By (6.1))} \\
&= \frac{\mathrm{eff}\mathcal{C}(a,b)}{\deg_\sigma(a)} \sum_{x \sim a} f_{a,b}(a,x) && \text{(By Ohm's Law (2.5))} \\
&= \frac{\mathrm{eff}\mathcal{C}(a,b)}{\deg_\sigma(a)} && (f \text{ is a unit flow})
\end{aligned}
$$

□

Corollary 6.1. *Consider the Green's function $G_{\tau_b}(a,a)$ defined in (5.14), namely, the expected number of times a random walk, starting at node a, visits a before hitting b. We have:*

$$G_{\tau_b}(a,a) = \deg_\sigma(a)\ \mathrm{eff}\mathcal{R}(a,b).$$

Proof. By the strong Markov property, every time the walker returns to a having failed to hit b, it will start a new walk and try again to exit at b. This is analogous to tossing a biased coin with a probability equal to the hitting probability from a to b:

$$p := \mathbb{P}_a\left(\tau_b < \tau_a^+\right),$$

and then asking for the expected number of tosses needed before winning. This corresponds to defining a random variable with the geometric distribution $X \sim \mathrm{Geom}(p)$ where $\mathbb{P}(X = k) = p(1-p)^{k-1}$ for $k = 1, 2, \ldots$, and noticing that the mean of X is $1/p$, see Exercise 4.4.

The result follows from Proposition 6.1. □

6.4 PROBABILISTIC INTERPRETATION OF CURRENT

Assume that a unit current flow $f_{a,b}$ flows between a and b, and let $x \in V$. Define

$$u(x) := G_{\mathcal{T}_b}(a,x). \tag{6.6}$$

Namely, $u(x)$ is the expected number of times a random walk, started at a, visits x before hitting b. By Proposition 5.2, we have

$$u(x) = \sum_{y\sim x} u(y)P(y,x) + (\delta_b - \delta_a)(x). \tag{6.7}$$

The transition matrix and stationary distributions, in this case, are given in (6.1). In particular, random walk on an undirected graph is reversible, so P and π satisfy the detailed balance equation (5.23). Therefore, $P(y,x) = \frac{\pi(x)}{\pi(y)} P(x,y)$, and (6.7) becomes:

$$u(x) = \sum_{y\sim x} \frac{\pi(x)}{\pi(y)} P(x,y)u(y) + (\delta_b - \delta_a)(x).$$

Therefore,

$$\frac{u(x)}{\pi(x)} = \sum_{y\sim x} P(x,y)\frac{u(y)}{\pi(y)} + \frac{1}{\pi(x)}(\delta_b - \delta_a)(x).$$

This can be interpreted as saying that the function $v(\cdot) := u(\cdot)/\pi(\cdot)$ is harmonic at $x \notin \{a,b\}$. Moreover,

$$v(b) = \frac{G_{\mathcal{T}_b}(a,b)}{\pi(b)} = 0,$$

and, by Corollary 6.1 and the formula for π in (6.1),

$$v(a) = \frac{G_{\mathcal{T}_b}(a,a)}{\pi(a)} = 2\sigma(E)\ \mathrm{eff}\mathcal{R}(a,b).$$

By the maximum principle, uniqueness implies that the voltage potential $h_{a,b}$ satisfies

$$h_{a,b} = \mathrm{eff}\mathcal{R}(a,b) - \frac{v}{2\sigma(E)}. \tag{6.8}$$

In other words, we can generalize Corollary 6.1 as follows:

$$G_{\mathcal{T}_b}(a,x) = \deg_\sigma(x)\left[h_{a,b}(b) - h_{a,b}(x)\right], \qquad \forall x \in V. \tag{6.9}$$

Now, we look at the current:

$$\begin{aligned}
f_{a,b}(x,y) &= \sigma(x,y)(h_{a,b}(y) - h_{a,b}(x)) && \text{(by Ohm's Law (2.5))} \\
&= \frac{\sigma(x,y)}{2\sigma(E)}(v(x) - v(y)) && \text{(by (6.8))} \\
&= \frac{\sigma(x,y)}{2\sigma(E)}\left(\frac{u(x)}{\pi(x)} - \frac{u(y)}{\pi(y)}\right) \\
&= \sigma(x,y)\left(\frac{u(x)}{\deg_\sigma(x)} - \frac{u(y)}{\deg_\sigma(y)}\right) \\
&= u(x)P(x,y) - u(y)P(y,x). && \text{(by (6.1))}
\end{aligned}$$

Thus,

$$f_{a,b}(x,y) = u(x)P(x,y) - u(y)P(y,x). \tag{6.10}$$

Recall that Equation (5.16) for the expected number of traversals from x to y and (6.6) give:

$$u(x)P(x,y) = \mathbb{E}_a\left(\#\text{ times } x \text{ is followed by } y \text{ before hitting } b\right). \tag{6.11}$$

Therefore, when a unit current flow is established between a and b, the current flowing through an edge $f_{a,b}(x,y)$ is

> *the expected net amount of times the random walk starting at a crosses from x to y before hitting b.*

6.5 EFFECTIVE RESISTANCE IN TERMS OF COMMUTE TIMES

Definition 6.1. Given $a, b \in V$, let the *commute time* between a and b be

$$\tau_{a,b} := \inf\{k \geq \tau_b : X_k = a\}. \tag{6.12}$$

This is the first time the random walk hits a after hitting b for the first time.

Proposition 6.2. *Consider the random walk on a weighted graph $G = (V, E, \sigma)$. Given $a, b \in V$, let $\tau_{a,b}$ be the commute time defined in (6.12) and recall the hitting times τ_a and τ_b defined in (5.12). Then*

$$2\sigma(E)\mathrm{eff}\mathcal{R}(a,b) = \mathbb{E}_a\left(\tau_{a,b}\right) = \mathbb{E}_a\left(\tau_b\right) + \mathbb{E}_b\left(\tau_a\right).$$

Proof. By the strong Markov property, we can start a walk at a, wait until it first visits b, and then restart a new walk from b and wait until it first hits a. Thus, $\mathbb{E}_a(\tau_{a,b}) = \mathbb{E}_a(\tau_b) + \mathbb{E}_b(\tau_a)$. In particular, by (5.19), $\mathbb{E}_a(\tau_{a,b}) < \infty$. By Proposition 5.2, with $\tau = \tau_{a,b}$ and $z = a$, and the uniqueness of the stationary distribution, we find that

$$\frac{G_{\tau_{a,b}}(a,a)}{\mathbb{E}_a\left(\tau_{a,b}\right)} = \pi(a) = \frac{\deg_\sigma(a)}{2\sigma(E)}.$$

Note that $G_{\tau_{a,b}}(a,a) = G_{\tau_b}(a,a)$, because a walk starting at a will revisit a a certain amount of times before visiting b. But after τ_b, the walk will not visit a again before $\tau_{a,b}$. Thus, $G_{\tau_{a,b}}(a,a) = \deg_\sigma(a)\mathrm{eff}\mathcal{R}(a,b)$, by Corollary 6.1, and the proposition follows. □

EXERCISES

6.1 *Normalized Laplacian.* Let $G = (V,E)$ be a simple connected graph with unit edge conductances. Recall the Laplacian matrix L and the normalized Laplacian $\mathcal{L} = D^{-1/2}LD^{-1/2}$, where $D = \mathrm{diag}(\deg(\cdot))$ is the diagonal matrix of degrees. Also, let $\{\phi_j\}_{j=1}^N$ be the orthogonal basis of eigenvectors for $\mathcal{L}$ given by the relation with the matrix S in (6.2). Also, let μ_j be the corresponding eigenvalues for $\mathcal{L}$, so that $\mu_1 = 0 < \mu_2 \leq \cdots \leq \mu_N \leq 2$.

(i) Show that $\phi_1 = \frac{1}{2|E|}D^{1/2}\mathbf{1}$, then deduce that $\operatorname{Ker}\mathcal{L} = \operatorname{Span}(\phi_1)$ and

$$\mathcal{L}\mathcal{L}^+ = \mathcal{L}^+\mathcal{L} = \operatorname{Proj}_{(\operatorname{Ker}\mathcal{L})^\perp} = I - \phi_1\phi_1^T.$$

(ii) Set $u := \frac{1}{\sqrt{\deg(t)}}\delta_t - \frac{1}{\sqrt{\deg(s)}}\delta_s$, and show that

$$\mathrm{eff}\mathcal{R}(s,t) = u^T\mathcal{L}^+u.$$

(iii) Verify that

$$\mathcal{L}^+ = \left(I - \phi_1\phi_1^T\right) D^{1/2}L^+D^{1/2}\left(I - \phi_1\phi_1^T\right)$$

6.2 *Green symmetry.* Let $G = (V,E,\sigma)$ be a simple connected graph with edge conductances $c \in \mathbb{R}^E_{>0}$. If $a, b \in V$ are a source-target pair of distinct nodes, let $h_{a,b}$ be the voltage potential needed to generate a unit current flow from a to b, normalized so that $h_{a,b}(a) = 0$, and thus, so that $h_{a,b}(b) = \mathrm{eff}\mathcal{R}(a,b)$.

(i) Use the fact that $h_{a,b} = \mathcal{G}(\delta_b - \delta_a) + (\mathbf{1}^T h_{a,b})\mathbf{1}/N$, see (3.24), where $\mathcal{G}$ is the Green operator (or pseudoinverse of the Laplacian), to show that for every $x \in V \setminus \{a, b\}$,

$$h_{a,b}(b) - h_{a,b}(x) = h_{x,b}(b) - h_{x,b}(a).$$

(ii) For $x \in V$, let $G_{\mathcal{T}_b}(a, x)$ be the stopped Green function equal to the expected number of times a random walk, started at a, visits x before hitting b. Use (6.9) to show that,

$$\frac{G_{\mathcal{T}_b}(a, x)}{\deg_\sigma(x)} = \frac{G_{\mathcal{T}_b}(x, a)}{\deg_\sigma(a)}.$$

6.3 *Interpretations of effective resistance.* Assume $G = (V, E)$ is a graph with unit edge conductances. Fix two distinct vertices $a, b \in V$ and let $f_{a,b}$ the unit current flow from a to b.

(i) Use (5.16) and the unit flow property to show that the effective resistance $\operatorname{eff}\mathcal{R}(a, b)$ can be interpreted as the average expected number of traversals from a to one of its neighbors $x \sim a$, for a random walk starting at a, before it visits b.

(ii) Start a random walk at a and wait until the first time after $\mathcal{T}_b$, when the walk returns to a for the first time. Namely, stop the walk at time $\mathcal{T}_{a,b}$. We call this a 'round-trip'. Given a vertex $x \in V$, let $U(x)$ be the expected number of visits at x before $\mathcal{T}_{a,b}$.

(a) Show that $U(x) = G_{\mathcal{T}_b}(a, x) + G_{\mathcal{T}_a}(b, x)$;

(b) Use (6.9) to show that $\operatorname{eff}\mathcal{R}(a, b)$ can be interpreted as the average expected number of traversals out of x in a random round-trip from a to b and back.

(iii) Recall that effective resistance satisfies the triangle inequality; see Theorem 2.3. Use the symmetry from Exercise 6.2 (ii) to show the following refinement of the triangle inequality: given $a, b, x \in V$,

$$\operatorname{eff}\mathcal{R}(a, x) + \operatorname{eff}\mathcal{R}(x, b) - \operatorname{eff}\mathcal{R}(a, b) = \frac{G_{\mathcal{T}_x}(a, b)}{\deg_\sigma(b)} + \frac{G_{\mathcal{T}_x}(b, a)}{\deg_\sigma(a)}.$$

(iv) As in part (iii), use the symmetry from Exercise 6.2 (i) to also show that, given $a, b, x \in V$,

$$\operatorname{eff}\mathcal{R}(a, x) + \operatorname{eff}\mathcal{R}(x, b) - \operatorname{eff}\mathcal{R}(a, b) = h_{x,b}(a) + h_{x,a}(b).$$

(v) Use part (iii), the symmetry from Exercise 6.2 (ii) and Remark 5.3, to show that:

$$\mathbb{E}_a(\tau_b) = \frac{1}{2}\sum_{x\in V} \deg_\sigma(x)\left[\text{eff}\mathcal{R}(a,b) + \text{eff}\mathcal{R}(b,x) - \text{eff}\mathcal{R}(a,x)\right].$$

For more information see [79]

6.4 *Target time.* Assume $G = (V, E)$ is a connected graph with unit edge conductances. Consider the random walk on G with transition matrix P and stationary distribution π, as given in (5.6) and (5.7). For $x \in V$, recall the hitting times τ_x defined in (5.12). Consider the function $H(a, b) := \mathbb{E}_a(\tau_b)$, for $a, b \in V$.

(i) Fix $b \in V$, and set $h_b(\cdot) := H(\cdot, b)$. Use a first-step analysis to show that $(Ph_b)(x) = h_b(x) - 1$, for all $x \in V \setminus \{b\}$, and $(Ph_b)(b) = \frac{1}{\pi(b)} - 1$. It will be useful to recall that $\mathbb{E}_a(\tau_a^+) = 1/\pi(a)$, by (5.18).

(ii) Show that the function

$$h(a) := \mathbb{E}_\pi\left[H(a, \underline{b})\right]$$

is harmonic on V. It will be helpful to note that $H(b, b) = 0$, for all $b \in V$. (The notation $\underline{b}$ is explained in Remark 4.1).

(iii) Deduce that $h(a)$ is constant in a, and show that $h(a) = \tau_{\text{target}}$, for all $a \in V$, where

$$\tau_{\text{target}} := \sum_{x\in V}\sum_{y\in V} \pi(x)\mathbb{E}_x(\tau_y)\pi(y),$$

is the *target time* for G. The interpretation is that τ_{target} is the expected hitting time from a to b, when a and b are picked independently using the stationary distribution π. This can be done by stopping a random walk after a long time.

See [56, Chapter 10] for more information.

6.5 *Total relative resistance.* Let $G = (V, E, \sigma)$ be a simple connected graph with edge conductances $\sigma \in \mathbb{R}^E_{>0}$. For every edge $e \in E$, consider the *relative resistance* $\text{eff}\mathcal{R}_\sigma(e)/r(e) = \sigma(e)\text{eff}\mathcal{R}_\sigma(e)$. Then, for $1 \le p < \infty$, define the following total energy:

$$\mathcal{E}_{p,\sigma}(G) := \sum_{e\in E} \sigma(e)^p \text{eff}\mathcal{R}_\sigma(e)^p.$$

The goal of this exercise is to show the following claim: suppose G is bi-(edge)-connected and G admits a set of edge conductances $\sigma \in \mathbb{R}^E_{>0}$ that is energy minimizing for $p = 2$, i.e., so that

$$\mathcal{E}_{2,\sigma}(G) = \min_{\sigma' \in \mathbb{R}^E_{>0}} \mathcal{E}_{2,\sigma'}(G), \tag{P6.1}$$

Then, necessarily, we must have, for all $e \in E$,

$$\sigma(e)\mathrm{eff}\mathcal{R}_\sigma(e) \equiv Const.$$

We show this in a few steps.

(i) Use (3.26) to show that the total energy $\mathcal{E}_{p,\sigma}(G)$ is homogeneous in σ, meaning that, for every $t > 0$,

$$\mathcal{E}_{p,t\sigma}(G) = \mathcal{E}_{p,\sigma}(G).$$

In particular, given an edge $e \in E$, one can always normalize σ so that $\sigma(e) = 1$.

Now assume $\sigma \in \mathbb{R}^E_{>0}$ is an energy minimizer in (P6.1). Fix an edge $e_1 = \{s, t\} \in E$ such that

$$\sigma(e_1)\mathrm{eff}\mathcal{R}_\sigma(e_1) = \max_{e \in E} \sigma(e)\mathrm{eff}\mathcal{R}_\sigma(e). \tag{P6.2}$$

Let h_1 be the voltage potential needed to generate a unit current flow from s to t, normalized so that $\mathbf{1}^T h_1 = 0$. Also, fix an orientation $\vec{E}$ on the edges, and for every edge $\vec{e} = (x, y) \in E$, write $u_e := \delta_y - \delta_x$. Recall that the Laplacian can be written as $L_\sigma = \sum_{e \in E} \sigma(e) u_e u_e^T$, and that $h_1 = L_\sigma^+ u_{e_1}$. So, we can write the potential drop along the edge e as

$$\nabla h_1(e) := h_1(y) - h_1(x) = u_e^T h_1 = u_e^T L_\sigma^+ u_{e_1} \tag{P6.3}$$

In particular, recall that if $R_1 := \mathrm{eff}\mathcal{R}_\sigma(e_1)$, then R_1 is also equal to the potential drop along edge e_1 and to the Dirichlet energy of h_1. Namely,

$$\begin{aligned} R_1 = |\nabla h_1(e_1)| &:= |h_1(t) - h_1(s)| = u_{e_1}^T h_1 \\ &= u_{e_1}^T L_\sigma^+ u_{e_1} = \sum_{e \in E} \sigma(e) \left|\nabla h_1(e)\right|^2. \end{aligned} \tag{P6.4}$$

Next, we perturb σ by changing it on e_1 and fixing the other values. Then, get a formula for the pseudoinverse of the Laplacian.

(ii) By (I), we can assume without loss of generality that $\sigma(e_1) = 1$. For $t > -1$, define the following perturbation of the edge conductance on e_1:

$$\sigma_t := \sigma + t\sigma(e_1)\mathbb{1}_{e_1} = \sigma + t\mathbb{1}_{e_1},$$

where $\mathbb{1}_{e_1}$ is the indicator function on E for the edge e_1. Then, show that

$$L^+_{\sigma_t} = L^+_\sigma - \frac{t}{1+tR_1} L^+_\sigma u_{e_1} u_{e_1}{}^T L^+_\sigma.$$

Note that this is akin to the well-known Sherman-Morrison formula.

Next, we deduce a formula for the function $\psi_e(t)$, defined for $t > -1$, that maps $t \mapsto \text{eff}\mathcal{R}_{\sigma_t}(e)$, and then compute its derivative at $t = 0$.

(iii) Show that

$$\frac{d}{dt}\text{eff}\mathcal{R}_{\sigma_t}(e)\bigg|_{t=0} = -\left|\nabla h_1(e)\right|^2$$

(iv) Show that

$$\left.\frac{d\mathcal{E}_{2,\sigma_t}(G)}{dt}\right|_{t=0} = 2\sum_{e\in E}\Big\{R_1 - \sigma(e)\text{eff}\mathcal{R}_\sigma(e)\Big\}\sigma(e)\left|\nabla h_1(e)\right|^2. \tag{P6.5}$$

In particular, by (P6.2) and the fact that $\sigma(e_1) = 1$, deduce that $R_1 - \sigma(e)\text{eff}\mathcal{R}_\sigma(e) \geq 0$, for every edge $e \in E$ for which $|\nabla h_1(e)| \neq 0$. Therefore, show that if σ is a minimizer, the expression on the left-hand side of (P6.5) equals zero, and thus

$$\sigma(e)\text{eff}\mathcal{R}_\sigma(e) \equiv R_1,$$

for every edge $e \in E$ for which $|\nabla h_1(e)| \neq 0$.

(v) Now assume that G is *bi-(edge)-connected*, meaning that removing an edge from G does not disconnect the graph. With the notations above, use (6.5) to show that $|\nabla h_1(e)| \neq 0$ for every edge incident to s and t. Thus, conclude by induction that, in this case, $\sigma(e)\text{eff}\mathcal{R}_\sigma(e) \equiv R_1$, for every edge $e \in E$.

Graphs $G = (V, E)$ that admit a set of edge conductances σ with the property that all of the relative resistances are equal are called *strictly homogeneous* and will be studied more deeply in Section 18.5. The conclusion in part (iv) can be rephrased as saying that being strictly homogeneous for G is a necessary condition for the existence of a minimizer when minimizing $\mathcal{E}_{2,\sigma}$ over σ. Note also, that when $p = 1$, Exercise 3.13 can be used to show that $\mathcal{E}_{1,\sigma}$ is constant equal to $|V| - 1$. For more on relative resistances, see, for instance, [32].

6.6 *Laplacian random walk.* Let $G = (V, E)$ be a simple connected graph. Fix $b \in V$. Given a starting node $x_0 \in V$, we define a random path $\hat{X}$ from x_0 to b called a Laplacian walk, as follows:

- Begin by setting $\hat{X}_0 = x_0$ and $\eta = \{x_0\}$, then let h_0 solve the Dirichlet problem on $V \setminus \{x_0, b\}$ with boundary values $h_0(x_0) = 0$ and $h_0(b) = 1$.
- Pick the next step among the neighbors of x_0 using the probability

$$\mathbb{P}(\hat{X}_1 = y \mid \hat{X}_0 = x_0) := \frac{h_0(y)}{\sum_{z \sim x_0} h_0(z)} \qquad \text{for } y \sim x_0,$$

  then if $X_1 = x_1$ extend $\eta = \{x_0, x_1\}$.

- After $k \geq 1$ steps, assuming $\eta = \{x_0, \ldots, x_k\}$, let h_k solve the Dirichlet problem on $V \setminus \{x_0, \ldots, x_k, b\}$ with boundary values $h_k \equiv 0$ on η and $h_k(b) = 1$. Then,

$$\mathbb{P}(\hat{X}_{k+1} = y \mid [\hat{X}_0, \ldots, \hat{X}_k] = \eta) := \frac{h_k(y)}{\sum_{z \sim x_k} h_k(z)} \qquad \text{for } y \sim x_k. \tag{P6.6}$$

Let X_n be the simple random walk on G and given a subset $B \subset V$ recall the stopping times $\tau_B := \inf\{k \geq 0 : X_k \in B\}$ and $\tau_B^+ := \inf\{k \geq 1 : X_k \in B\}$.

First, we recall *Bayes' Theorem*.

(i) Give a proof of the following form of Bayes Theorem: given a measurable event A and finite measurable partition $\{B_j\}_{j=1}^n$ of the sample space Ω, with $\mathbb{P}(B_j) > 0$ for all j, we have

$$\mathbb{P}(B_k \mid A) = \frac{\mathbb{P}(A \mid B_k)\mathbb{P}(B_k)}{\sum_{j=1}^{n} \mathbb{P}(A \mid B_j)\mathbb{P}(B_j)}, \qquad \text{for } k = 1, \ldots, n.$$

The goal is to show that the next step in the Laplacian walk defined in (P6.6) can equivalently be obtained by starting a simple random walk X_n with $X_0 = x_k$, conditioned on the event $\{\tau_b < \tau_\eta^+\}$.

(ii) Use a first-step analysis and Bayes Theorem to show that, for every $y \in V \setminus \eta$:

$$\begin{aligned}&\mathbb{P}_{x_k}(X_1 = y \mid \tau_b < \tau_\eta^+)\\ &= \frac{\mathbb{P}_{x_k}(\tau_b < \tau_\eta^+ \mid X_1 = y)\mathbb{P}_{x_k}(X_1 = y)}{\sum_{z \in V \setminus \eta^+} \mathbb{P}_{x_k}(\tau_b < \tau_\eta^+ \mid X_1 = z)\mathbb{P}_{x_k}(X_1 = z)}\end{aligned}$$

(iii) Use Theorem 6.1 to deduce that, with the notations in (P6.6), for every $y \in V \setminus \eta$:

$$\mathbb{P}_{x_k}(X_1 = y \mid \tau_b < \tau_\eta^+) = \frac{h_k(y)}{\sum_{z \sim x_k} h_k(z)}$$

For more information on the Laplacian random walk, see [52, 53].

6.7 *Gaussian elimination for the Laplacian.* Let $G = (V, E, \sigma)$ be a simple connected graph with edge conductances $\sigma \in \mathbb{R}^E_{>0}$. Assume $s, t \in V$ are distinct nodes playing the role of source and target when passing a unit current from s to t through the graph G. Enumerate the nodes in V as $\{x_0, x_1, \ldots, x_n\}$ so that $x_0 = s$ and $x_n = t$. Let $h_{s,t}$ be the voltage potential normalized so that $h_{s,t}(s) = 0$ and $h_{s,t}(t) = \operatorname{eff}\mathcal{R}(s,t)$. Recall, as in (3.33), that the voltage potential $h = h_{s,t}$ solves the following Laplacian system, where the inhomogeneous term is given by the external current.

$$Lh = \delta_t - \delta_s = i_{\text{ext}}.$$

We treat this as a system of $n+1$ equations, where the unknowns are the values of $h(x_j)$ for $j = 1, \ldots, x_{n-1}$. Each equation is of the form

$$L(x_j, x_j)h(x_j) + \sum_{\substack{i=0 \\ i \neq j}}^{n} L(x_j, x_i)h(x_i) = i_{\text{ext}}(x_j), \qquad j = 0, \ldots, n.$$

(i) Use the $j = 1$ equation to solve for the first variable $h(x_1)$.

(ii) Eliminate $h(x_1)$ from every row $j \neq 1$. Then, identify the new matrix M for the reduced system in terms of L.

(iii) Write the off-diagonal entries of M in terms of the edge conductances σ. Show that they are non-positive and that $M\mathbf{1} = 0$. Hence, deduce a proof for the star-mesh transformation stated in (2.11).

CHAPTER 7

Mixing Times and Bottlenecks

As we have seen, irreducible Markov chains have a unique stationary distribution. We now address the question of the speed of convergence to the stationary distribution. This is a crucial question when using Monte Carlo methods. For a much more in-depth discussion, see [56].

7.1 TOTAL VARIATION DISTANCE

To get started, we want a way to measure the distance between two pmfs. Let $\mathcal{P}(\mathcal{S})$ be the set of pmfs on a finite state space $\mathcal{S}$. Given $\mu, \nu \in \mathcal{P}(\mathcal{S})$, the *total variation distance* between μ and ν measures the worst case discrepancy between the two measures:

$$\|\mu - \nu\|_{\mathrm{TV}} = \max_{S \subset \mathcal{S}} |\mu(S) - \nu(S)|. \tag{7.1}$$

The following result yields an equivalent formulation of the total variation distance in terms of the L^1-norm.

Lemma 7.1. *Given $\mu, \nu \in \mathcal{P}(\mathcal{S})$, we have*

$$\|\mu - \nu\|_{\mathrm{TV}} = \frac{1}{2}\|\mu - \nu\|_1.$$

DOI: 10.1201/9781003024866-8

Proof. Let $B = \{x \in \mathcal{S} : \mu(x) \geq \nu(x)\}$. For every event $S \subset \mathcal{S}$:

$$\begin{aligned}
&\mu(S) - \nu(S) \\
&= [\mu(S \cap B) + \mu(S \cap B^c)] - [\nu(S \cap B) + \nu(S \cap B^c)] && \text{(additivity)} \\
&= [\mu(S \cap B) - \nu(S \cap B)] - [\nu(S \cap B^c) - \mu(S \cap B^c)] && \text{(rearranging)} \\
&\leq \mu(S \cap B) - \nu(S \cap B) && ((\nu - \mu)|_{B^c} \geq 0) \\
&\leq \mu(B) - \nu(B) && ((\mu - \nu)|_{B^c} \geq 0).
\end{aligned}$$

Similarly, we can show that

$$\nu(S) - \mu(S) \leq \nu(B^c) - \mu(B^c) = \mu(B) - \nu(B),$$

where the last equality follows from the fact that

$$[\mu(B) - \nu(B)] - [\nu(B^c) - \mu(B^c)] = \mu(\mathcal{S}) - \nu(\mathcal{S}) = 0.$$

So

$$|\mu(S) - \nu(S)| \leq \mu(B) - \nu(B) \qquad \forall S \subset \mathcal{S}.$$

In particular, both $S = B$ and $S = B^c$ attain the maximum in $\max_{S \subset \mathcal{S}} |\mu(S) - \nu(S)|$, and therefore

$$\begin{aligned}
\|\mu - \nu\|_{\mathrm{TV}} &= \frac{1}{2}\left[(\mu(B) - \nu(B)) + (\nu(B^c) - \mu(B^c))\right] \\
&= \frac{1}{2}\left[\left(\sum_{x \in B} \mu(x) - \nu(x)\right) + \left(\sum_{x \in B^c} \nu(x) - \mu(x)\right)\right] \\
&= \frac{1}{2}\left[\sum_{x \in B} |\mu(x) - \nu(x)| + \sum_{x \in B^c} |\mu(x) - \nu(x)|\right] \\
&= \frac{1}{2}\|\mu - \nu\|_1
\end{aligned}$$

□

It's worth recording the following corollary from the proof.

Corollary 7.1. *We have*

$$\|\mu - \nu\|_{\mathrm{TV}} = \sum_{\substack{x \in \mathcal{S} \\ \mu(x) \geq \nu(x)}} (\mu(x) - \nu(x)).$$

The next result is a version of $L^1 - L^\infty$ duality.

Corollary 7.2. *We have:*

$$\|\mu - \nu\|_{\mathrm{TV}} = \frac{1}{2} \sup_{\|f\|_\infty \leq 1} \langle f, \mu - \nu \rangle.$$

Proof. If $\|f\|_\infty \leq 1$, then

$$\langle f, \mu - \nu \rangle = \sum_{x \in \mathcal{S}} f(x)(\mu(x) - \nu(x)) \leq \sum_{x \in \mathcal{S}} |f(x)||\mu(x) - \nu(x)| \leq \|\mu - \nu\|_1.$$

Conversely, let $f^* = \mathbb{1}_B - \mathbb{1}_{B^c}$, where $B = \{x \in \mathcal{S} : \mu(x) \geq \nu(x)\}$. Then, $\|f^*\|_\infty \leq 1$ and

$$\langle f, \mu - \nu \rangle = \sum_{x \in B} (\mu(x) - \nu(x)) + \sum_{x \in B^c} (\nu(x) - \mu(x)) = \|\mu - \nu\|_1.$$

□

7.2 A BASIC EXAMPLE

Suppose $\mathcal{S}$ has only two states $\mathcal{S} = \{1, 2\}$. Consider a Markov chain on $\mathcal{S}$ with transition matrix P. Assume that $P(1, 2) = p_1$ and $P(2, 1) = p_2$. See Figure 7.1. One can check that the distribution π given by

$$\pi(1) = \frac{p_2}{p_1 + p_2} \qquad \text{and} \qquad \pi(2) = \frac{p_1}{p_1 + p_2}, \tag{7.2}$$

satisfies the balance equation, thus the chain is reversible and π is a stationary distribution. See Exercise 7.1. Let π_0 be an arbitrary initial distribution and let $\pi_k = \pi_0 P^k$. To study convergence it is enough to look at $\delta_k := \pi_k(1) - \pi(1)$, since $\|\pi_k - \pi\|_{\mathrm{TV}} = |\delta_k|$. One can show that

$$|\delta_{k+1}| = (1 - p_1 - p_2)\|\pi_k - \pi\|_{\mathrm{TV}} = (1 - p_1 - p_2)|\delta_k|.$$

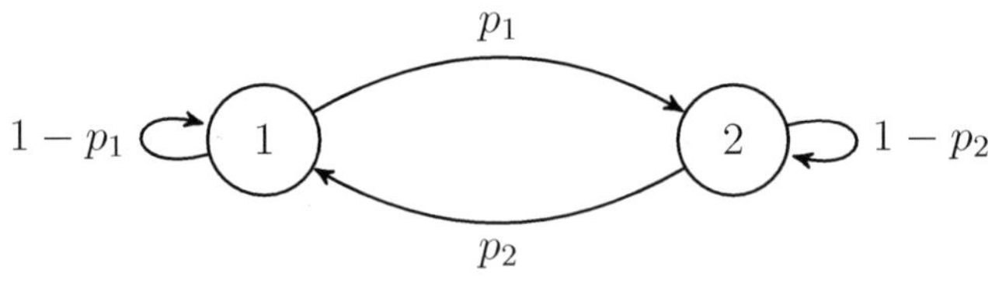

Figure 7.1 A simple Markov chain on two states.

Therefore, if $p_1, p_2 \in (0,1)$, then $1 - p_1 - p_2 \in (-1,1)$, and

$$\|\pi_k - \pi\|_{\mathrm{TV}} = |\delta_k| = |1 - p_1 - p_2|^k |\delta_0| \longrightarrow 0,$$

exponentially fast.

Also, the eigenvalues of the transition matrix P are 1 and $1-p_1-p_2$.

7.3 CONVERGENCE

Assume that a chain P is irreducible so that a stationary distribution π exists and is unique. It turns out that if P is also aperiodic, then we can find $r \in \mathbb{N}$ such that

$$P^r(x,y) > 0 \qquad \forall x, y \in \mathcal{S}. \tag{7.3}$$

A proof of this fact can be found in Proposition 1.7 of [56]. The following result shows that given any starting state $x \in \mathcal{S}$, the occupation probability $\pi_n := \mathbb{1}_x^T P^n$ after n steps converges geometrically to the stationary pmf π in the total variation distance.

Theorem 7.1. *Assume P is irreducible and aperiodic with stationary distribution π. Then there are $\alpha \in (0,1)$ and $C > 0$ such that*

$$\max_{x\in\mathcal{S}} \left\| \mathbb{1}_x^T P^n - \pi \right\|_{\mathrm{TV}} \le C\alpha^n, \qquad \forall n \in \mathbb{N}.$$

Proof. Let $\Pi = \mathbf{1}\pi$ be a matrix with all rows equal to π. Also let $\pi_{\max} = \max_{x\in\mathcal{S}} \pi(x)$. By (7.3), there is $\delta > 0$ such that:

$$\min_{x,y\in\mathcal{S}} P^r(x,y) \ge \delta > 0.$$

So for any $x, y, z \in \mathcal{S}$,

$$\frac{\pi(z)}{\pi_{\max}} \le 1 \le \frac{P^r(x,y)}{\delta}$$

i.e.,

$$P^r(x,y) \ge \frac{\delta}{\pi_{\max}} \pi(z). \tag{7.4}$$

Pick δ small enough so that

$$\theta := 1 - \frac{\delta}{\pi_{\max}} \in (0,1) \tag{7.5}$$

Define

$$Q(x,y) := \frac{P^r(x,y) - (1-\theta)\pi(y)}{\theta}. \tag{7.6}$$

The matrix Q is row-stochastic, since $Q(x,y) \geq 0$ by (7.4), and

$$\sum_{y \in \mathcal{S}} Q(x,y) = \frac{1-(1-\theta)}{\theta} = 1.$$

Note that (7.6) can be rewritten as the following convex combination:

$$P^r = (1-\theta)\Pi + \theta Q. \tag{7.7}$$

In Exercise 7.2, you will show that, for $k = 1, 2 \ldots,$ and $j = 0, \ldots, r-1$,

$$P^{rk+j} - \Pi = \theta^k (Q^k P^j - \Pi). \tag{7.8}$$

Note that, by (7.5), $0 < \theta < 1$ and Q^k is row-stochastic so it stays bounded as $k \to \infty$. So (7.8) already says that P^n tends to Π as n tends to infinity.

To conclude, we fix $x \in \mathcal{S}$ and note that multiplying a transition matrix A from the left by $\mathbb{1}_x^T$ has the effect of singling out the row in A corresponding to x, namely, the probability distribution describing the next step from x. In particular, Π is a transition matrix where every state gets sent to a random state according to the stable distribution, i.e.,

$$\mathbb{1}_x^T \Pi = (\mathbb{1}_x^T \mathbf{1})\pi = \pi. \tag{7.9}$$

Dividing $n \in \mathbb{N}$ by r yields a quotient k and a remainder $j \in \{0, 1, \ldots, r-1\}$ and we can write $n = rk + j$. Then,

$$\begin{aligned}
\left\| \mathbb{1}_x^T P^n - \pi \right\|_{\text{TV}} &= \left\| \mathbb{1}_x^T P^{rk+j} - \mathbb{1}_x^T \Pi \right\|_{\text{TV}} && \text{(by (7.9))} \\
&= \left\| \mathbb{1}_x^T \left(P^{rk+j} - \Pi \right) \right\|_{\text{TV}} && \text{(factoring)} \\
&= \theta^k \left\| \mathbb{1}_x^T \left(Q^k P^j - \Pi \right) \right\|_{\text{TV}} && \text{(by (7.8))} \\
&= \theta^k \left\| \mathbb{1}_x^T Q^k P^j - \pi \right\|_{\text{TV}} && \text{(distributing)} \\
&\leq \theta^k.
\end{aligned}$$

The last inequality follows from Exercise 7.2. In conclusion, we can let $\alpha = \theta^{1/r}$ and $C = \theta^{-(1-1/r)}$. □

7.4 MIXING TIMES

We saw in Theorem 7.1 that for irreducible aperiodic chains, the quantity

$$d(k) = \max_{x \in \mathcal{S}} \left\| \mathbb{1}_x^T P^k - \pi \right\|_{\text{TV}} \tag{7.10}$$

decays exponentially as k tends to infinity. It therefore is natural to set the chain's *mixing time* to equal the first time $d(k)$ falls below $1/4$, namely, set

$$t_{\text{mix}} := \inf \left\{ k : d(k) \leq \frac{1}{4} \right\}. \tag{7.11}$$

One way to get a lower bound on t_{mix} is in situations where n steps are insufficient to visit every other node from x. For instance, in Exercise 7.3, you will show that if $G = (V, E)$ is a d-regular graph with $d \geq 2$, then t_{mix} can be bounded below in terms of d and the size of the vertex set V.

More generally, we can use the diameter of a graph to get a lower bound on the mixing time.

Definition 7.1. Given a connected graph $G = (V, E)$. Let $d(x, y)$ be the hop length on G as defined in (1.2). Then, the *diameter* of G is

$$\text{Diam}(G) = \max_{x,y \in V} d(x, y).$$

It is also useful to consider the following related quantity:

$$\bar{d}(k) := \max_{x,y \in \mathcal{S}} \left\| \mathbb{1}_x^T P^k - \mathbb{1}_y^T P^k \right\|_{\text{TV}} \leq 2d(k), \tag{7.12}$$

where the inequality follows from the triangle inequality.

Proposition 7.1. *Suppose $G = (V, E)$ is a connected graph. Then*

$$t_{\text{mix}} \geq \left\lfloor \frac{\text{Diam}(G) - 1}{2} \right\rfloor$$

Proof. Pick $a, b \in V$ such that $d(a, b) = \text{Diam}(G)$ and set $k = \left\lfloor \frac{\text{Diam}(G)-1}{2} \right\rfloor$. Then the measures $\mathbb{1}_a^T P^k$ and $\mathbb{1}_b^T P^k$ have disjoint supports, so $\bar{d}(k) = 1$. Hence $d(k) \geq 1/2$, and this means that $t_{\text{mix}} \geq k$. □

Example 7.1 (The circuit graph). Consider the circuit graph C_N, obtained from the path graph P_N by adding the edge $\{1, N\}$. Then, since C_N is 2-regular, therefore (P7.1) gives the lower bound $t_{\text{mix}}(C_N) \geq$

$O(\log N)$, while Proposition 7.1, gives the better lower bound $t_{\text{mix}}(C_N) \geq O(N)$. Using Chernoff's bounds, one can give an even better lower bound of the order of $O(N^2)$, see Exercise 7.4. One can also establish an $t_{\text{mix}}(C_N) \leq N^2$ using the notion of coupling, see [56, Sec. 5.3.1] for the details.

We now look at more sophisticated lower bounds that depend on the geometry of the graph.

7.5 THE BOTTLENECK RATIO

Let P be an irreducible aperiodic Markov chain with stationary distribution π. Consider the *edge measure*,

$$Q(x, y) := \pi(x)P(x, y).$$

which represents the probability that the chain transitions from x to y in the distant future. Recall from (5.23) that a chain is reversible if $Q(x, y) = Q(y, x)$.

For instance, if $G = (V, E, \sigma)$ is a simple connected graph with edge weights $\sigma(e) > 0$ for every $e \in E$, then, using (6.1), we see that for every edge $\{x, y\} \in E$:

$$Q(x, y) = \frac{\sigma(x, y)}{2\sigma(E)}. \tag{7.13}$$

To interpret this as an actual probability, consider the space of all *sockets*:

$$\mathcal{S} := \{(x, e) : x \in V, e \in E, \text{ and } e \text{ is incident at } x\}.$$

There are two natural projections $q_1 : \mathcal{S} \to V$ and $q_2 : \mathcal{S} \to E$. For a fixed vertex $x \in V$, the corresponding fiber is $q_1^{-1}(x) = \{(e, x) : e \text{ is incident at } x\}$. In particular, $|q_1^{-1}(x)| = \deg(x)$. For a fixed edge $e = \{x, y\} \in E$, $q_2^{-1}(e) = \{(e, x), (e, y)\}$. In particular, $|q_2^{-1}(e)| = 2$. The fibers of q_1 and q_2 give two ways of partitioning $\mathcal{S}$ and therefore, we recover the Handshake Lemma:

$$|\mathcal{S}| = \sum_{x \in V} \deg(x) = 2|E|.$$

Moreover, there is a canonical way to "lift" a walk on the graph G to a walk on $\mathcal{S}$. Let γ be a walk on G, written as $\gamma = x_0\ e_1\ x_1\ e_2\ x_2 \cdots x_{n-1}\ e_n\ x_n$. Then the corresponding walk on $\mathcal{S}$ is

$$\tilde{\gamma} = (x_0, e_1)\ (x_1, e_2)\ \cdots (x_{n-1}, e_n)$$

where we only list the "vertices" visited by $\tilde{\gamma}$. In particular, a random walk X_k on G induces a random walk Y_k on $\mathcal{S}$.

Write $\tilde{\pi}$ for the stationary distribution of the induced random walk Y_k on $\mathcal{S}$. Suppose $e = \{x, y\} \in E$ and consider the socket $s = (x, e) \in \mathcal{S}$. Then

$$\tilde{\pi}(s) = Q(x, y) = \frac{\sigma(e)}{2\sigma(E)}.$$

To see this, let $z_1, \ldots, z_k$ be the enumerated neighbors of x and set $e_j = \{z_j, x\}$. Then the induced random walk will transition from the socket $s_j = (z_j, e_j)$ to s with probability $\tilde{P}(s_j, s) = \sigma(x, y)/\deg_\sigma(x)$. We compute

$$\sum_{j=1}^{k} \tilde{\pi}(s_j)\tilde{P}(s_j, s) = \sum_{j=1}^{k} \frac{\sigma(z_j, x)}{2\sigma(E)} \frac{\sigma(x, y)}{\deg_\sigma(x)} = \frac{\sigma(x, y)}{2\sigma(E)} = \tilde{\pi}(s)$$

One of the advantages of this induced random walk is that, in the unweighted case when $\sigma(e) \equiv 1$, $\tilde{\pi}$ becomes the uniform distribution on $\mathcal{S}$.

For two disjoint sets of states A and B, the probability of jumping from A to B in one step, in the long run, is

$$Q(A, B) := \sum_{x \in A, y \in B} Q(x, y).$$

For $S \subset V$, define the *bottleneck ratio* of S as

$$\phi(S) := \frac{Q(S, V \setminus S)}{\pi(S)}. \tag{7.14}$$

Example 7.2. Specializing to random walks on a weighted graph $G = (V, E, \sigma)$, define the *edge-boundary* of a subset $S \subset V$ as

$$\partial S := \{e = \{x, y\} \in E : x \in S,\ y \notin S\}.$$

Then,

$$\phi(S) = \frac{\sigma(\partial S)}{\sum_{x \in S} \deg_\sigma(x)}, \tag{7.15}$$

and in the unweighted case when $\sigma \equiv 1$:

$$\phi(S) = \frac{|\partial S|}{\sum_{x \in S} \deg(x)} \tag{7.16}$$

In particular, since $\sum_{x \in S} \deg(x) = |\partial S| + 2|E(S)|$, we can think of S as the ratio of the perimeter of S over the volume of S.

Finally, define the *bottleneck ratio of the chain P* as

$$\phi_*(P) := \min_{\substack{S \subset V \\ \pi(S) \leq 1/2}} \phi(S). \tag{7.17}$$

The condition $\pi(S) \leq 1/2$ is needed to avoid making $V \setminus S$ too small.

The following result yields a lower bound for the mixing time in terms of the bottleneck ratio. Namely, it shows that if the bottleneck ratio is small, the mixing time must be large. The intuition is that the chain gets stuck in a subset of states for a long time.

Theorem 7.2. *Let P be an irreducible aperiodic Markov chain. Then,*

$$t_{\text{mix}}(P) \geq \frac{1}{4\phi_*(P)}$$

We leave the proof of Theorem 7.2 to Exercise 7.6. We will see that it is helpful to think of the diffusion interpretation. Given a probability mass distribution π_0 on V, we can apply the transition matrix and obtain a new probability mass distribution $\pi_1 = \pi_0 P$. On an unweighted graph, every node x takes the mass $\pi_0(x)$ and divides it equally among its neighbors (and a portion stays at x if the walk is lazy).

Example 7.3. Take two copies of the complete graph K_N, each with a distinguished vertex, then identify the two distinguished vertices. We get a graph G with $|V| = 2N - 1$ and $|E| = N(N-1)$. Let S be equal to one of the two copies minus the distinguished vertex. Then $|\partial S| = Q(S, V \setminus S) = N - 1$ and $\sum_{x \in S} \deg(x) = (N-1)^2$. Therefore, by (7.15),

$$\phi_*(G) \leq \phi(S) = \frac{N-1}{(N-1)^2} = \frac{1}{N-1}$$

In particular, for this example, Theorem 7.2 implies that

$$t_{\text{mix}}(G) \geq \frac{N-1}{4}.$$

This is a better lower bound than the one that follows from Proposition 7.1, since $\text{Diam}(G) = 2$ in this example.

7.6 COUPLING

Here, we give a brief introduction to a fundamental technique called *coupling*. For a more in-depth treatment, see [56].

One way to code a random walk on a simple graph would be to assign to every state x a partition of the interval $[0,1]$ into $\deg(x)$ equally sized subintervals, each subinterval corresponding to a given neighbor of x. Then, one can set up a routine so that at each step, the computer samples a random variable u uniformly in $[0,1]$, then looks at the current position $X = x$ of the random walk, and uses the partition at x together with the value of u, to decide what neighbor to visit next, depending on which subinterval the variable u happens to fall in.

This routine can be modified so that, at each step, the same random variable u is used to move two different random walks. In this case, we could say that the two walkers are moving to the beat of the same drum. The partitions are set deterministically, and the two walkers share the same source of randomness. In technical terms, we called this a coupling of the two random walks.

Also, this construction shows that once the two walkers meet, they move together from that moment on.

Definition 7.2. Let X_k and Y_k be two coupled Markov chains with the same transition matrix P. Let τ_{couple} be the first time they meet:

$$\tau_{\text{couple}} = \inf\{k \geq 0 : X_k = Y_k\}.$$

The stopping time τ_{couple} can be used to estimate the total variation distance of the occupation measures for the two chains at step k. Intuitively, if the corresponding diffused probability measures are not close, then it is likely that the two chains haven't met yet.

Proposition 7.2. *Let X_k and Y_k be two coupled Markov chains with the same transition matrix P, such that $X_0 = x$ and $Y_0 = y$. Then, for $k = 1, 2, \ldots$,*

$$\|\mathbb{1}_x^T P^k - \mathbb{1}_y^T P^k\|_{\text{TV}} \leq \mathbb{P}_{x,y}\left(\tau_{\text{couple}} > k\right). \tag{7.18}$$

In particular, the convergence rate $d(k)$ defined in (7.10), can be estimated as follows, for $k = 1, 2, \ldots$:

$$d(k) \leq \max_{x,y \in \mathcal{S}} \mathbb{P}_{x,y}(\tau_{\text{couple}} > k) \tag{7.19}$$

Proof. Write $\pi_k := \mathbb{1}_x^T P^k$ and $\tilde{\pi}_k := \mathbb{1}_y^T P^k$. Also, write $\mathbb{P}_{x,y}(\cdot)$ for $\mathbb{P}(\cdot \mid X_0 = x, Y_0 = y)$.

If $S \subset \mathcal{S}$ is arbitrary, then

$$\begin{aligned}
\pi_k(S) - \tilde{\pi}_k(S) &= \mathbb{P}_{x,y}(X_k \in S) - \mathbb{P}_{x,y}(Y_k \in S) \\
&= \mathbb{P}_{x,y}(X_k \in S, Y_k \notin S) - \mathbb{P}_{x,y}(X_k \notin S, Y_k \in S) \\
&\leq \mathbb{P}_{x,y}(X_k \in S, Y_k \notin S) \\
&\leq \mathbb{P}_{x,y}(X_k \neq Y_k) \\
&\leq \mathbb{P}_{x,y}\left(\tau_{\text{couple}} > k\right).
\end{aligned}$$

The last inequality follows from the fact that once the two chains meet, they don't separate. In other words, $\{k \geq \tau_{\text{couple}}\} \subset \{X_k = Y_k\}$.

Finally, (7.18) follows by switching the role of x and y and then maximizing over all $S \subset \mathcal{S}$.

The inequality (7.19) follows from (7.18) and the fact that $d(k) \leq \bar{d}(k)$, which we leave as Exercise 7.7. □

Example 7.4 (The circuit graph revisited). Consider the circuit graph C_N, obtained from the path graph P_N by adding the edge $\{1, N\}$. Chernoff's bounds were used in Exercise 7.4 to give the lower bound $t_{\text{mix}}(C_N) \geq O(N^2)$. Using the method of coupling, one can show that this bound is sharp, i.e., that we also have $t_{\text{mix}} \leq N^2$. See Exercise 7.8.

7.7 THE SPECTRAL GAP

Mixing times and bottlenecks can be analyzed from the spectral theory point of view. Recall that, by Lemma 5.4, for an irreducible Markov chain P, the value $\lambda = 1$ is always a simple positive eigenvalue for P. Moreover, all the remaining eigenvalues satisfy $|\lambda| \leq 1$.

Example 7.5. Recall the basic example of Section 7.2. In that case, we showed that the rate of convergence to the stable distribution satisfies the geometric bound:

$$\|\pi_0 P^k - \pi\|_{\text{TV}} \leq |1 - p_1 - p_2|^k \|\pi_0 - \pi\| \leq |\lambda|^k,$$

where $\lambda := 1 - p_1 - p_2$ is the eigenvalue of the transition matrix P other than 1. In particular, in this example $d(k) \leq |\lambda|^k$, and setting $|\lambda|^k \leq 1/4$, yields $k \geq (\log 4)/(\log |\lambda|^{-1})$, i.e.,

$$t_{\text{mix}} \geq \left\lceil \frac{\log 4}{\log \frac{1}{|\lambda|}} \right\rceil$$

This shows that the closer $|\lambda|$ is to 1, the greater the mixing time will be. To make this more explicit, we could use the fact that for $|\lambda|$ close enough to 1, we have $\log(1/x) \leq (1-x)/2$, and so for this example we have $t_{\text{mix}} \geq C/(1-|\lambda|)$.

More generally, assuming that $\text{Spectrum}(P) \setminus \{1\} \neq \emptyset$, set

$$\lambda_* := \max\{|\lambda| : \lambda \in \text{Spectrum}(P),\ \lambda \neq 1\}.$$

Then the *absolute spectral gap* for P is

$$\gamma_* := 1 - \lambda_*. \tag{7.20}$$

If we assume that P is reversible, then by Exercise 5.6, the spectrum of P is contained in $[-1,1]$. First, we analyze the case when the spectral gap $\gamma_* = 0$. This can only happen if $\lambda = -1$ is an eigenvalue. For random walks on graphs, this just means that the graph is bipartite.

Recall from Proposition 5.1 that random walk on a connected undirected graph is periodic of period 2 if and only if G is bipartite. Now, we add the following observation, the proof of which is entirely analogous to Exercise 3.7 (iv).

Proposition 7.3. *Let P be a random walk on a simple connected graph G. Then $\lambda = -1$ is an eigenvalue of P if and only if G is bipartite.*

One way to simplify the analysis is to consider the lazy random walk instead

$$\tilde{P} := \frac{1}{2}I + \frac{1}{2}P \tag{7.21}$$

that stays put with 50% probability. This lazy version has the same stable distribution π. Also the symmetrized transition matrix satisfies $\tilde{S} = (I+S)/2$ and the normalized Laplacian is $\tilde{\mathcal{L}} = \mathcal{L}/2$. Moreover, as will be shown in Exercise 7.9, the eigenvalues of the matrix $\tilde{P}$ are contained in $[0,1]$.

The lazy random walk on a connected graph G is an example of an irreducible, aperiodic, and reversible Markov chain. In this case, we can order the eigenvalues of the transition matrix so that $\lambda_1 = 1 > \lambda_2 \geq \cdots \lambda_N \geq 0$, then define the *spectral gap* as

$$\gamma := 1 - \lambda_2. \tag{7.22}$$

Theorem 7.3. *Let P be an irreducible, aperiodic, reversible Markov chain on $\mathcal{S}$. Let π be the stationary distribution of P. Set*

$$\pi_{\min} = \min_{x \in \mathcal{S}} \pi(x).$$

and let γ_* be the absolute spectral gap for P. Then,

$$\left(\frac{1}{\gamma_*} - 1\right) \log 2 \leq t_{\text{mix}}(P) \leq \log\left(\frac{4}{\pi_{\text{min}}}\right) \frac{1}{\gamma_*}.$$

We leave the proof of Theorem 7.3 to Exercise 7.10.

Remark 7.1. Theorem 7.3 essentially says that the absolute spectral gap is comparable to t_{mix}^{-1}. While Theorem 7.2 gives a lower bound for the bottleneck ratio in terms of t_{mix}^{-1}. Combining these two results, we get a lower bound for the bottleneck ratio in terms of the absolute spectral gap:

$$\left(4 \log\left(4\pi_{\text{min}}^{-1}\right)\right)^{-1} \gamma_*(P) \leq \phi_*(P).$$

7.8 CHEEGER'S INEQUALITY

Cheeger's inequality provides an upper bound for the bottleneck ratio in terms of the absolute spectral gap. Instead of working in the full generality of Markov chains, we follow D. Spielman's notes and specialize to the case of a random walk on a simple connected graph.

Recall the normalized Laplacian $\mathcal{L}$ introduced in Section 6.1. Write its eigenvalues as

$$0 = \nu_1 < \nu_2 \leq \cdots \leq \nu_N \leq 2.$$

From (6.2), since $\mathcal{L} = I - S$, we deduce that $\nu_j = 1 - \lambda_j$, where λ_j are the eigenvalues for the transition matrix P. Therefore, the eigenvalue ν_2 is equal to the spectral gap γ defined in (7.22).

Also, recall the formulation of the bottleneck ratio derived in Example 7.2. Specialized to unweighted graphs, the bottleneck ratio for the graph G can be written as:

$$\phi(S) := \frac{|\partial S|}{\min\{d(S), d(V \setminus S)\}} \qquad \text{and} \qquad \phi_G := \min_{S \subset V} \phi(S),$$

where $d(A) := \sum_{x \in A} \deg(x)$, whenever $A \subset V$.

The goal is to prove the following bound known as *Cheeger's inequality*:

$$\phi_G \leq \sqrt{2\nu_2}.$$

The idea is to analyze the quadratic form of the Laplacian. First, define the vectors

$$w_1(x) := \deg(x)^{1/2} \qquad \text{and} \qquad v_1(x) := \deg(x) \qquad \forall x \in V, \qquad (7.23)$$

Then $D^{-1/2}w_1 = \mathbf{1}$. Therefore, since $L\mathbf{1} = 0$, we get that

$$\mathcal{L}w_1 = D^{-1/2}LD^{-1/2}w_1 = D^{-1/2}L\mathbf{1} = 0.$$

In other words, w_1 is the eigenvector of $\mathcal{L}$ for ν_1. In particular, letting $v = D^{-1/2}w$, we can interpret ν_2 variationally, using the Courant-Fisher Theorem 3.3, as follows:

$$\nu_2 = \min_{w \perp w_1} \frac{w^T \mathcal{L} w}{w^T w} = \min_{v \perp v_1} \frac{v^T L v}{v^T D v}.$$

Theorem 7.4. *Suppose $v \perp v_1$ and consider the level set*

$$S_t(v) := \{x \in V : v(x) < t\}. \tag{7.24}$$

Then there exists t so that

$$\phi\left(S_t(v)\right) \leq \sqrt{2\frac{v^T L v}{v^T D v}}. \tag{7.25}$$

Proof. Without loss of generality, we can relabel the nodes so that $v(1) \leq v(2) \leq \cdots \leq v(N)$. Also, we let

$$j := \min\left\{k : \sum_{i=1}^{k} \deg(i) \geq \frac{d(V)}{2}\right\}, \tag{7.26}$$

and set $u := v - v(j)\mathbf{1}$. Note that $u(j) = 0$. Also note that for any vector $m \in \mathbb{R}^V$, the function

$$s \mapsto (m + s\mathbf{1})^T D(m + s\mathbf{1}),$$

is minimized at s^* when $(m + s^*\mathbf{1})^T v_1 = 0$, as can be seen by taking the derivative in s. Applying this fact to $m = u$ and $s^* = v(j)$, we see that

$$u^T D u \geq v^T D v.$$

On the other hand, since $L\mathbf{1} = \mathbf{1}^T L = 0$, we have $u^T L u = v^T L v$. Hence

$$\frac{u^T L u}{u^T D u} \leq \frac{v^T L v}{v^T D v}.$$

So, it is enough to work with u instead of v.

Finally, since the Raleigh quotient is homogeneous, we can normalize u so that $u(1)^2 + u(N)^2 = 1$. The goal is to use a "probabilistic method" to find $t \in [u(1), u(N)]$ so that

$$\phi\left(S_t(u)\right) \leq \sqrt{2\frac{u^T L u}{u^T D u}}. \tag{7.27}$$

We leave the details of this argument for Exercise 7.11. □

EXERCISES

7.1 *A two-states example.* Suppose $\mathcal{S}$ has only two states $\mathcal{S} = \{1,2\}$. Consider a Markov chain on $\mathcal{S}$ with transition matrix P. Assume that $P(1,2) = p_1$ and $P(2,1) = p_2$. See Figure 7.1. Also, let π_0 be an arbitrary initial distribution and let $\pi_k := \pi_0 P^k$.

(i) Show that π given in (7.2) satisfies the balance equation, thus the chain is reversible and π is a stationary distribution.

(ii) Define $\delta_k := \pi_k(1) - \pi(1)$, and show that $\|\pi_k - \pi\|_{\mathrm{TV}} = |\delta_k|$.

(iii) Show that $|\delta_{k+1}| = (1 - p_1 - p_2)\|\pi_k - \pi\|_{\mathrm{TV}} = (1 - p_1 - p_2)|\delta_k|$.

(iv) Show that the eigenvalues of the transition matrix P are 1 and $1 - p_1 - p_2$.

7.2 *Convergence.* The purpose of this exercise is to complete the proof of Theorem 7.1.

(i) Use induction to obtain a formula for P^{rk}, for $k = 1, 2, \ldots$, then deduce (7.8) from (7.7).

(ii) Prove that the product of two row-stochastic matrices is also row-stochastic and deduce that, with the notations of (7.8), the expression $\mathbb{1}_x^T Q^k P^j$ is some probability distribution μ in $\mathcal{P}(\mathcal{S})$.

(iii) Conclude from (ii) that

$$\left\| \mathbb{1}_x^T Q^k P^j - \pi \right\|_{\mathrm{TV}} \leq 1.$$

7.3 *A growth-based lower bound.* Consider Random Walk on a simple connected d-regular graph $G = (V, E)$.

(i) Show that the stationary distribution π is uniform in this case.

(ii) Consider a number of steps $k \in \mathbb{N}$ such that $d^k \leq (3/4)|V|$. Let V_x^k be the set of all nodes that can be reached from x in less than k steps. Then the assumption on k implies that $|V_x^k| \leq (3/4)|V|$. Recall that if a RW starts at x, then $\pi_0 = \mathbb{1}_x^T$ and $\pi_k = \mathbb{1}_x^T P^k$. In particular, $\pi_k(S) = \mathbb{P}_x(X_k \in S)$ is the probability that the RW lands in S in k steps. Show that $d(k) \geq \frac{1}{4}$.

(iii) Choose $n \in \mathbb{N}$ so that $d^n \leq (3/4)|V| < d^{n+1}$, and deduce the following lower bound on the mixing time:

$$t_{\text{mix}} \geq \frac{\log((3/4)|V|)}{\log d} \tag{P7.1}$$

7.4 *Mixing time for the circuit graph.* The purpose of this exercise is to use Simple Random Walk on the line and Chernoff bounds to analyze the diffusion of random walk on the circuit graph C_N. More precisely, the goal is to show that there is an $\alpha > 0$ such that

$$t_{\text{mix}}(C_N) \geq \alpha N^2. \tag{P7.2}$$

For $k = 1, 2, 3, \ldots$, let X_k be independent Bernoulli variables with $X_k \in \{0, 1\}$ and $\mathbb{P}(X_k = 1) = 1/2$. We modify the X_k so that they take values 1 or -1, i.e., we let $Y_k = 2X_k - 1$. Then the random walk Z_n on the circuit graph C_N is equivalent to the usual random walk $W_n = \sum_{k=1}^{n} Y_k$ on the integer line $\mathbb{Z}$ where we identify two integers a and b, if $a - b$ is a multiple of N.

For simplicity, assume $N = 2K + 1$ is odd and let $V = \{-K, \ldots, 0, \ldots, K\}$ be the vertex set for C_N. Assume also that the random walk Z_k starts at 0 so that we can write $\pi_0 := \mathbb{1}_0$, and let $\pi_k = \pi_0 P^k$, where P is the transition matrix.

(i) Write $S_n := \sum_{k=1}^{n} X_k$. Then $W_n = 2S_n - n$, and since $\mathbb{E}(W_n) = 0$, we have $\mathbb{E}(S_n) = n/2$. Use the Chernoff bound (P4.4) with $m := n/2$ and $\delta := K/(2n)$ and show that there exists a set $U \subset V$ with $|U| = K = \lfloor N/2 \rfloor$ and a constant $\alpha > 0$, such that, for $k < \alpha N^2$,

$$\mathbb{P}_0(Z_k \in U) = \pi_k(U) < \frac{1}{4}.$$

(ii) Prove that (P7.2) holds.

7.5 *Random walk on sockets.* Show that if G is connected and not bipartite, then the random walk Y_k is irreducible and aperiodic on $\mathcal{S}$ (however, it will not be reversible in general).

7.6 *Bottlenecks and mixing times.* The goal of this exercise is to prove Theorem 7.2.

(i) We begin by relating the bottleneck ratio to the effect of taking one step in the diffusion. Given $S \subset V$, let μ_S be the stationary distribution π conditioned on S. Show that

$$\phi(S) = \|\mu_S P - \mu_S\|_{\mathrm{TV}}.$$

(ii) Next, we show that taking one step in the diffusion is a contraction of the set of all probability measures. Let P be a chain with stationary distribution π. Given $\mu, \nu \in \mathcal{P}(V)$, show that

$$\|\mu P - \nu P\|_{\mathrm{TV}} \leq \|\mu - \nu\|_{\mathrm{TV}}. \tag{P7.3}$$

In particular, deduce that, for all $k \in \mathbb{N}$,

$$\|\mu P^{k+1} - \pi\|_{\mathrm{TV}} \leq \|\mu P^k - \pi\|_{\mathrm{TV}}.$$

(iii) Combining the one-step interpretation of the bottleneck ratio in (i) and the contraction property in (ii) shows that given $S \subset V$, for all $k \in \mathbb{N}$,

$$\|\mu_S P^{k+1} - \mu_S P^k\|_{\mathrm{TV}} \leq \|\mu_S P - \mu_S\|_{\mathrm{TV}} = \phi(S).$$

Therefore, by telescoping sums

$$\|\mu_S P^k - \mu_S\|_{\mathrm{TV}} \leq k\phi(S). \tag{P7.4}$$

(iv) Finally, give an equivalent characterization of the quantity $d(k)$ from (7.10), extending its definition to all probability measures. Namely, show that

$$d(k) = \max_{x \in V} \left\| \mathbb{1}_x^T P^k - \pi \right\|_{\mathrm{TV}} = \max_{\mu \in \mathcal{P}(V)} \|\mu P^k - \pi\|_{\mathrm{TV}}.$$

(v) Finally, prove Theorem 7.2.

7.7 *The relation between $d(k)$ and $\bar{d}(k)$.* Let $d(k)$ and $\bar{d}(k)$ be defined as in (7.10) and (7.12). We will show that

$$d(k) \leq \bar{d}(k).$$

See [56, Chapter 4] for more information.

(i) Fix $x \in \mathcal{S}$. Rewrite $\left\|\mathbb{1}_x^T P^k - \pi\right\|_{\mathrm{TV}}$, using (7.1) and the fact that $P^k(x, \cdot)$ can be thought as a pmf on $\mathcal{S}$.

(ii) Fix $A \subset \mathcal{S}$. Rewrite $\left|P^k(x,A) - \pi(A)\right|$, using the fact that π is stationary. Then, use the triangle inequality to bound it by $\sum_{y\in\mathcal{S}} \pi(y)\left|P^k(x,A) - P^k(y,A)\right|$.

(iii) Now take the maximum over $A \subset \mathcal{S}$ and use the fact that the max of a sum is always less than the sum of the max to show that

$$\max_{A\subset\mathcal{S}} \left|P^k(x,A) - \pi(A)\right| \leq \sum_{y\in\mathcal{S}} \pi(y) \left\|\mathbb{1}_x^T P^k - \mathbb{1}_y^T P^k\right\|_{\mathrm{TV}}$$

(iv) Finally, use the fact that a weighted average is always less than the max to get the upper bound $\max_{y\in\mathcal{S}} \left\|\mathbb{1}_x^T P^k - \mathbb{1}_y^T P^k\right\|_{\mathrm{TV}}$, and then take the max over $x \in \mathcal{S}$.

7.8 *Coupling for the circuit graph.* Here, we use coupling to give an upper bound for the mixing time for the circuit graph. Recall from (P7.2) that $t_{\mathrm{mix}}(C_N) \geq O(N^2)$. The goal is to show that this bound is sharp. Think of C_N as $\mathbb{Z}/N\mathbb{Z}$, with $V = \{0,\ldots,N-1\}$ and the rule that moving to the right from $N-1$ leads to 0. Couple two random walks on $\mathbb{Z}/N\mathbb{Z}$ so that only one moves at every step. To make sense of this, we switch to lazy versions of the walks. With probability 1/2, the walker stays put; otherwise, with equal probabilities (1/4), the walk moves left or right. Note that the mixing time for the lazy walk is longer than that of the regular walk. So, without loss of generality, it is enough to find an upper bound for the mixing time of the lazy walk. The coupling is as follows: a fair coin decides whether X_n or Y_n moves, and another independent, fair coin decides if the move is left or right. Let $Z_n = X_n - Y_n \pmod{N}$. Then Z_n performs a regular random walk on $\mathbb{Z}/N\mathbb{Z}$ that gets absorbed when it reaches 0. Equivalently, we could look at the gambler's ruin walk on $\{1,\ldots,N-1\}$ that gets absorbed at 0 and N. Use Claim 5.1 and Markov's inequality (4.5) to prove that

$$t_{\mathrm{mix}} \leq N^2.$$

7.9 *Spectrum of the lazy random walk.* Show that the eigenvalues of the matrix $\tilde{P}$ for the lazy random walk on a simple connected graph G defined in (7.21) are contained in $[0,1]$.

7.10 *Spectral bounds for mixing time.* Here we will prove Theorem 7.3.

(i) Consider the matrix $S = D_\pi^{1/2} P D_\pi^{-1/2}$ and its eigenpairs (ϕ_j, λ_j). Recall that the eigenpairs for P are (f_j, λ_j), with $f_j(x) = \pi(x)^{-1/2}\phi_j(x)$. Show that, for $k = 1, 2, \ldots,$

$$P^k(x,y)\pi(y)^{-1} = \sum_{j=1}^{N} \lambda_j^k f_j(x) f_j(y) = 1 + \sum_{j=2}^{N} \lambda_j^k f_j(x) f_j(y),$$

and deduce that

$$\left| P^k(x,y)\pi(y)^{-1} - 1 \right| \leq \lambda_*^k \sqrt{\sum_{j=2}^{N} f_j(x)^2} \sqrt{\sum_{j=2}^{N} f_j(y)^2}.$$

(ii) Decompose the delta function δ_x in the basis $\{\phi_j\}$ to show that

$$\sum_{j=1}^{N} f_j(x)^2 = \frac{1}{\pi(x)}.$$

Then, estimate the total variation

$$\|\delta_x^T P^k - \pi\|_{\mathrm{TV}} \leq \pi_{\min}^{-1} e^{-\gamma_* k},$$

and thus deduce the upper bound

$$t_{\mathrm{mix}}(P) \leq \log\left(\frac{4}{\pi_{\min}}\right) \frac{1}{\gamma_*}.$$

(iii) Use orthogonality of $\{\phi_j\}$, to show that

$$\sum_{x \in \mathcal{S}} \pi(x) f_j(x) = 0,$$

and deduce that for $j \neq 1$,

$$\begin{aligned} |\lambda_j|^k |f_j(x)| &= \left| \sum_{y \in \mathcal{S}} \left[P^k(x,y) f_j(y) - \pi(y) f_j(y) \right] \right| \\ &\leq 2 \left(\max_{y \in \mathcal{S}} |f_j(y)| \right) d(k). \end{aligned}$$

(iv) Pick $x \in \mathcal{S}$ such that $|f_j(x)| = \max_{y \in \mathcal{S}} |f_j(y)|$, and deduce that $|\lambda_j|^k \leq 2d(k)$ for all $j \neq 1$. Finally, prove the lower bound

$$t_{\mathrm{mix}} \geq \left(\frac{1}{\gamma_*} - 1\right) \log 2.$$

7.11 *Cheeger's inequality.* Here, we will finish the proof of Theorem 7.4. Recall that the vector v_1 was defined in (7.23), while v is an arbitrary vector perpendicular to v_1. Moreover, the nodes were relabeled so that $v(1) \leq \cdots \leq v(N)$, and $u = v - v(j)\mathbf{1}$, where j is chosen as in (7.26), so that $u(j) = 0$. Finally, u was normalized so that $u(1)^2 + u(N)^2 = 1$.

(i) Consider a continuous random variable X in $[u(1), u(N)]$ with probability density function

$$f_X(t) := 2|t|.$$

Check that f_X is a probability density function, as in (4.4). Also, prove that, for $u(1) \leq a < b \leq u(N)$,

$$\mathbb{P}\left(t \in [a, b]\right) = (\operatorname{sign} b)b^2 - (\operatorname{sign} a)a^2. \tag{P7.5}$$

(ii) Orient each edge $e = \{x, y\}$ as $\vec{e} = (x, y)$ in such a way that $u(x) \leq u(y)$. Consider the random level set S_X as defined in (7.24) when t is given by X. Use the indicator function for the event $\{(x, y) \in \partial S_X\}$, and linearity of expectation, to show that

$$\mathbb{E}\left[|\partial S_X(u)|\right] \leq \sum_{(x,y)\in\vec{E}} |u(x) - u(y)| \left(|u(x)| + |u(y)|\right). \tag{P7.6}$$

(iii) Note that by our choice of j in (7.26), the quantity $\min\left\{d\left(S_X(u)\right), d\left(V \setminus S_X(u)\right)\right\}$ equals $d\left(S_X(u)\right)$, if $X < 0$, and equals $d\left(V \setminus S_X(u)\right)$, if $X \geq 0$, i.e.,

$$\begin{aligned}
&\min\left\{d\left(S_X(u)\right), d\left(V \setminus S_X(u)\right)\right\} \\
&= d\left(S_X(u)\right) \mathbb{1}_{\{X<0\}} + d\left(V \setminus S_X(u)\right) \mathbb{1}_{\{X\geq 0\}} \\
&= \sum_{x<j} \deg(x) \mathbb{1}_{\{u(x)<X<0\}} + \sum_{x\geq j} \deg(x) \mathbb{1}_{\{0\leq X<u(x)\}}.
\end{aligned}$$

Hence, deduce that

$$\mathbb{E}\left[\min\left\{d\left(S_X(u)\right), d\left(V \setminus S_X(u)\right)\right\}\right] = u^T D u. \tag{P7.7}$$

(iv) Next, use the Cauchy-Schwarz inequality and the estimate

$$2(a^2 + b^2) = (a + b)^2 + (a - b)^2 \geq (a + b)^2, \tag{P7.8}$$

to show that

$$\left(\sum_{(x,y)\in\vec{E}} |u(x)-u(y)|\,(|u(x)|+|u(y)|)\right)^2 \leq 2(u^TLu)(u^TDu).$$

Therefore, deduce that

$$\begin{aligned}&\mathbb{E}\left[|\partial S_X(u)|\right]\\ &\leq \left(\sqrt{2\frac{u^TLu}{u^TDu}}\right)\mathbb{E}\left[\min\left\{d\left(S_X(u)\right), d\left(V\setminus S_X(u)\right)\right\}\right].\end{aligned}$$

Here is where the "probabilistic" part of the method kicks in. If the mean of a random variable Y is less than a constant C, then at least one value of Y is less than C. We apply this principle to $Y = f(X) = |\partial S_X(u)|$. So there must be at least one value $X = t \in [u(1)^2, u(N)^2]$ such that

$$|\partial S_t(u)| \leq \left(\sqrt{2\frac{u^TLu}{u^TDu}}\right)\min\left\{d\left(S_t(u)\right), d\left(V\setminus S_t(u)\right)\right\}.$$

This proves (7.25) and therefore establishes Cheeger's Theorem 7.4.

II

Optimization Problems on Graphs

CHAPTER 8

Introduction to Optimization on Graphs

In Part I of this book, we encountered several problems of the form "Find the ____est ____." For example, we might seek the shortest path (Section 1.3) or the largest s-t flow (Definition 1.9). Such problems are known as optimization problems and are one of the most common problems studied on graphs. In this part of the book, we provide a brief overview of some of the essential aspects of optimization theory applied to problems on graphs.

8.1 PARTS OF AN OPTIMIZATION PROBLEM

Most optimization problems have three basic parts: a set of *decision variables*, an *objective (or goal) function*, and a set of *constraints*.

Decision variables

The decision variables are the quantities that we control in the optimization. For example, in the shortest path problem, the path is the decision variable; we seek the "best" path in the sense of the shortest length. In the maximum flow problem, the flow values on each edge are the decision variables; we wish to choose these to make the flow value as large as possible. Often, we define the decision variables by stating the set of possible choices for the variables. For example, in the shortest path problem, we might say that the decision variable is a path $\gamma \in \Gamma(a, b)$. For the maximum flow problem, on the other hand, we may consider

DOI: 10.1201/9781003024866-10

that our decision variables are the values of a function $f : \hat{E} \to \mathbb{R}$ (see Definition 1.1), or, equivalently, as the entries in a vector $f \in \mathbb{R}^{\hat{E}}$.

Objective function

The objective function in an optimization problem is the quantity that we seek to minimize or maximize by choosing the decision variables. It is a function that takes the decision variables as inputs and outputs a real number. In the shortest path problem, the path length is the objective function; in the maximum flow problem, the flow-value function is the objective.

Constraints

Constraints in an optimization problem restrict our choice of decision variables. Not all optimization problems have constraints, but many do. If we consider the decision variable for the shortest path problem as $\gamma \in \Gamma(a, b)$, no additional constraints are needed. However, if our decision variables for the maximum flow problem are $f \in \mathbb{R}^{\hat{E}}$, then we need to constrain f to be an s-t flow that satisfies the capacity constraints. In this case, the constraints on the decision variables will be those of Definition 1.6 together with the feasibility constraints of Definition 1.9.

Decision variables vs. constraints

Although the above descriptions may seem to distinguish three parts of an optimization problem clearly, we often have a significant degree of flexibility in our thinking about the problem, which can sometimes be exploited to our advantage. Consider again the problem of finding the maximum s-t flow on a network. Above, we chose to think of the decision variables as $f \in \mathbb{R}^{\hat{E}}$ with objective function $\mathrm{Val}(f)$ and with constraints that f be an s-t flow that satisfies the capacity constraints. However, we could also have considered the decision variable f to be an s-t flow, with constraints made up of the capacity constraints. This distinction may seem academic, but we'll see examples later where simply rethinking an optimization problem like this allows us to use certain mathematical tools that are unavailable with a different representation of the problem. For now, it is enough to realize that the boundary between decision variables and constraints can be fuzzier than it initially appears.

Objective function vs. constraints

A similar fuzziness exists in the boundary between constraint and objective function. For example, if we select our decision variables as the values of $f \in \mathbb{R}^{\hat{E}}$, we can avoid adding constraints by rethinking our objective function. Consider a new function Val* defined on $\mathbb{R}^{\hat{E}}$ as

$$\text{Val}^*(f) = \begin{cases} \text{Val}(f) & \text{if } f \text{ is an } s\text{-}t \text{ flow}, \\ -\infty & \text{otherwise.} \end{cases}$$

Functions that are allowed to take the values $\pm\infty$ are typically referred to as *extended-value* functions. Since the function Val* is an extension of Val, it is called an *extended-value extension* of Val. Much care must be taken with extended-value functions since not all arithmetic operations are well-defined when infinities are in play. Nevertheless, extended-value functions can greatly simplify some otherwise complicated calculations and expressions. In this case, the extension of Val to take the value $-\infty$ when its argument is not a valid s-t flow implicitly enforces the constraints of the optimization problem. Since we are trying to maximize the objective function, it will never be in our interest to choose an f that is not an s-t flow and obtain a value of $-\infty$ when we can instead choose a valid flow and obtain a finite value. Thus, we could either write the max flow problem as

$$\begin{aligned} \text{maximize} \quad & \text{Val}(f) \\ \text{subject to} \quad & f \in \mathbb{R}^{\hat{E}} \text{ is an } s\text{-}t \text{ flow}, \\ & |f(e)| \le u(e) \text{ for every } e \in E, \end{aligned}$$

or as

$$\begin{aligned} \text{maximize} \quad & \text{Val}^*(f) \\ \text{subject to} \quad & f \in \mathbb{R}^{\hat{E}}, \\ & |f(e)| \le u(e) \text{ for every } e \in E. \end{aligned}$$

8.2 CHARACTERISTICS OF OPTIMIZATION PROBLEMS

There are many different ways of characterizing optimization problems. We'll review a few important ones here. Remember, though, that this list is by no means exhaustive. To aid in the discussion, consider a generic minimization problem of the form

$$\inf\{f(x) : x \in X\}. \tag{8.1}$$

In this problem, the decision variable is x, and the objective function is f, which assigns to each choice of x a real value $f(x)$. The set X is defined by the constraints and, therefore, represents the set of all allowed choices of x for the problem (see Definition 8.2). A similar generic maximization problem can be written down by replacing the infimum with a supremum.

Discrete or continuous

Definition 8.1. If X is finite or countably infinite, the problem is called a *discrete optimization problem.* If X is a continuum (e.g., real numbers), the problem is called a *continuous optimization problem.*

The shortest path problem is a discrete problem (X is the finite set of paths connecting two vertices). In contrast, the maximum flow problem is continuous (X is a vector of real values subject to capacity and flow constraints). It is also possible to have mixed problems with some variables continuous and others discrete.

The distinction between discrete and continuous plays a vital role in the techniques we choose to solve a problem. The critical difference is that continuous variables can be subjected to infinitesimal perturbations while discrete variables cannot. For this reason, many solution methods for continuous problems rely on techniques from calculus, while solution methods for discrete problems do not.

Feasibility

Definition 8.2. Any choice of the decision variables $x \in X$, i.e. x satisfying all constraints, is referred to as a *feasible point* (or sometimes a *possible solution*) for the optimization problem. The collection of all feasible points, X, is called the *feasible set.*

Optimization aims to select the best choice among all the feasible points. A natural question to ask early about an optimization problem is "Does the problem have any feasible points?"

Definition 8.3. If $X \neq \emptyset$, the problem is called *feasible.* A problem with no feasible points is called an *infeasible* problem.

Boundedness

Sometimes, there is no bound on how well we can do. For example, consider the problem of maximizing x^2 over all real x. Since $\lim_{x\to\infty} x^2 = \infty$, there is no limit on how large we can make the objective function.

Definition 8.4. If there exists a sequence of feasible points $\{x_k\} \subset X$ such that $\lim_{k\to\infty} f(x_k) = -\infty$, then the problem is said to be *unbounded below*. Otherwise, the problem is said to be *bounded below*. In the case of a maximization problem, the problem is *unbounded above* if there is a sequence of feasible points satisfying $\lim_{k\to\infty} f(x_k) = +\infty$ and is *bounded above* otherwise.

It is common to adopt a less precise vocabulary and refer to a problem as either *bounded* or *unbounded*. The direction of boundedness (above or below) is inferred from the type of optimization problem considered. For minimization, boundedness from below is important. For maximization, the important direction is boundedness from above. For example, the function $f(x) = x^2$ over all real x is bounded from below, but not from above. So, if we maximize x^2, we have an unbounded problem, but if we are to minimize, we have a bounded problem.

Value

Definition 8.5. By the *value* of an optimization problem, we mean the value of the infimum in (8.1) (or the supremum for a maximization problem). It is conventional to assign an infeasible problem the value of $\pm\infty$; an infeasible minimization problem is assigned the value $+\infty$, while an infeasible maximization problem is assigned the value $-\infty$. It is similarly conventional to assign an unbounded problem the value of $\pm\infty$; a minimization problem that is unbounded below is assigned the value $-\infty$, while a maximization problem that is unbounded above is assigned the value $+\infty$.

In other words, using this convention, the problem in (8.1) always has a value. If there are no feasible points, we say the value is $+\infty$. If the problem is unbounded below, we say the value is $-\infty$. Otherwise, the set $\{f(x) : x \in X\}$ is non-empty and bounded below. Thus, the set has a greatest lower bound, and this bound is called the value of the problem.

Attainability

Consider the following two expressions.

$$p_1 := \inf\{\sqrt{x} : x \geq 1\}, \qquad p_2 := \inf\{\sqrt{x} : x > 1\}. \tag{8.2}$$

In the first expression, we assign the variable p_1 a value; we give it the value of the infimum of all numbers of the form $\sqrt{x}$ where $x \geq 1$. This is a minimization problem with the objective function $f(x) = \sqrt{x}$, decision variable $x \in \mathbb{R}$, and a constraint that $x \geq 1$. In this case, $p_1 = 1$.

The second expression in (8.2) is similar. Here, we assign a variable p_2 the value of an optimization problem. The problem is nearly identical to the previous one, except that $x = 1$ is not feasible (the strict inequality in the constraint disallows it). Nevertheless, the value of the optimization problem, and hence of p_2, is 1. These two optimization problems have the same value but have an important difference: attainability.

Definition 8.6. If there exists a feasible point $x^* \in X$ with the property that

$$f(x^*) = \inf\{f(x) : x \in X\},$$

then we say the problem is *attained*. In this case, the point x^* is called an *optimal* point. An optimal point is sometimes called a *solution* to the optimization problem.

Not every optimization problem is attained. An infeasible or unbounded problem is certainly not attained. Even if the problem is feasible and bounded, it may fail to be attained. The second expression in (8.2) shows an example of how this can happen. The problem is feasible and bounded with the value $p_2 = 1$. However, any feasible x must satisfy $x > 1$ and, therefore, $\sqrt{x} > 1 = p_2$. Thus, this problem is not attained.

Uniqueness

When an optimization problem is attained, there must be at least one optimal choice of the decision variables.

Definition 8.7. If an optimization problem has exactly one solution, we say that the problem has a *unique* solution.

Many optimization problems discussed earlier in this book are attained but do not necessarily have unique solutions. Generally, the shortest path, maximum flow, and minimum cut problems are all attained but are not guaranteed to have unique solutions.

Taking care with the vocabulary

In light of the preceding discussion, it is crucial to be precise in our use of language when talking about optimization problems. Before talking about the value of an optimization problem, we should consider whether this value is finite (i.e., whether the problem is feasible and bounded).

If we can verify feasibility and boundedness, we can be assured that the problem has a finite value (a real value as opposed to $\pm\infty$). Even then, we should be careful not to talk about an optimal point or solution to the problem until we have verified that the problem is attained. And even then, we should be careful not to refer to *the* solution unless we have verified that there is precisely one.

8.3 EXPRESSING OPTIMIZATION PROBLEMS

As is common in mathematics, there are many different notations to choose from when expressing an optimization problem. You have already experienced several of these in Part I of this book. We might, for example, express the maximum flow problem as

$$\sup\{\mathrm{Val}(f) : f \text{ is an } s\text{-}t \text{ flow}, \ |f(e)| \le u(e) \text{ for all } e \in E\}.$$

Here, we're using predicate notation to define a set; we are considering the set consisting of the values of all valid, capacity-respecting s-t flows, and we are asking for the supremum of these values. Compared to other ways of writing optimization problems, this is one of the most mathematically rigorous. The use of supremum here is an intentionally careful choice; it implies nothing about the feasibility, boundedness, or attainability of the problem. At times, sup will be replaced by max, but this should be done only if the problem is known to be attained (which necessarily requires that it be feasible and bounded). Sometimes, when the predicate part of the set is simple to express, you'll find a similar but more compact form of an optimization problem. For example, consider the form of (1.5). As with the case of sup vs. max, we often use inf to signal a minimization problem unless we know a minimum value exists, in which case we might use min.

The use of the operators arg min and arg max should be treated with similar care. These two operators only make sense if the underlying optimization problem is attained. Even then, their use can be problematic unless the solution is unique. In principle, they could be defined as the set of all solutions in their respective cases, but it is probably best to reserve their use for cases where the solution is unique.

It is sometimes helpful to write optimization problems in a form like that of (3.7), which may more clearly identify the objective function and constraints. When reading a problem written in this form, it is important to remember that there is generally no assumption of attainability. In this form, we write "maximize" or "minimize" even if the problem is not (or is not known to be) attained.

8.4 TRANSFORMATION AND EQUIVALENCE OF OPTIMIZATION PROBLEMS

Roughly speaking, two optimization problems are called *equivalent* if there is an easy way to obtain one solution from the other. It is possible to formalize this idea. However, a more practical way to understand equivalence is to study a few examples. When you see a claim that two problems are equivalent, ask yourself "Equivalent in what sense?" and then try to understand the connection.

Example 8.1. Let $A \in \mathbb{R}^{m\times n}$, $b \in \mathbb{R}^m$, and consider the following two problems.

$$p_1 := \inf\{\|x\|_2 : x \in \mathbb{R}^n, Ax \geq b\}$$

and

$$p_2 := \inf\{\|x\|_2^2 : x \in \mathbb{R}^n, Ax \geq b\}.$$

Note that the objective function is the only difference between the two problems. These two problems are an example of equivalent problems. Here, we'll explore the sense in which that is true.

First, note that either both problems are feasible or both are infeasible. Thus $p_1 = +\infty$ if and only if $p_2 = +\infty$. Now, suppose both problems are feasible. Since both objective functions are bounded (from below), this implies that both p_1 and p_2 have non-negative finite values. In fact, we can argue that $p_2 = p_1^2$.

To see this, let $\{x_k\} \subset \mathbb{R}^n$ be a *minimizing sequence* for the first problem. By this, we mean that each x_k is feasible and that $\lim\limits_{k\to\infty} \|x_k\|_2 \to p_1$. (Such a sequence exists since the infimum is finite.) Using the continuity of the squaring function $z \mapsto z^2$, we conclude that this implies that

$$p_2 \leq \lim_{k\to\infty} \|x_k\|_2^2 = \left(\lim_{k\to\infty} \|x_k\|_2\right)^2 = p_1^2.$$

To establish the opposite inequality, consider a minimizing sequence $\{x_k\}$ for the second problem instead. In this case, we can use the continuity of the square root function to see that

$$p_1 \leq \lim_{k\to\infty} \|x_k\|_2 = \lim_{k\to\infty} \sqrt{\|x_k\|_2^2} = \sqrt{\lim_{k\to\infty} \|x_k\|_2^2} = \sqrt{p_2},$$

so $p_2 \geq p_1^2$. Since the square root function is multi-valued, it's important to note that we are using that $\|x_k\|_2 \geq 0$ for all k.

As long as we understand $+\infty^2 = +\infty$ and $\sqrt{+\infty} = +\infty$, this gives us a sense of the equivalence between the two problems: $p_2 = p_1^2$, or $p_1 = \sqrt{p_2}$. So, if we know the value of one of the problems, we can find the value of the other. We can make the equivalence even stronger.

The first problem has a nice geometric interpretation. The set $\{x : Ax \geq b\}$ is a closed set in $\mathbb{R}^n$ and, thus, it contains a unique point x^* closest to the origin. That point is the unique minimizer for the first problem. Now, let x' be a feasible point with the property that $\|x'\|_2^2 \leq \|x^*\|_2^2$. Then, (again, since the norm is a non-negative function), it follows that $\|x'\|_2 \leq \|x^*\|_2$. By uniqueness of x^* as a minimizer of the first problem, this implies that $x' = x^*$ and, therefore, that x^* is also the unique minimizer of the second problem. Therefore, not only are the values of the two problems related by an elementary invertible formula, but also both problems have exactly the same unique minimizer.

Example 8.2. Let $A \in \mathbb{R}^{m\times n}$, $b \in \mathbb{R}^m$, and let $C \in \mathbb{R}^{n\times n}$ be a nonsingular matrix. Consider the following two problems.

$$p_1 := \inf\{\|x\|_2^2 : x \in \mathbb{R}^n, Ax \geq b\}$$

and

$$p_2 := \inf\{x^T C^T Cx : x \in \mathbb{R}^n, ACx \geq b\}.$$

These problems are also equivalent in the following sense.

Note that both problems are bounded. First, suppose the first problem is feasible and let $\{x_k\}$ be a minimizing sequence. Define a new sequence $\{y_k\}$ where $y_k = C^{-1}x_k$. Note that, for each k, $ACy_k = Ax_k \geq b$, so the y_k are feasible for the second problem. Moreover,

$$p_2 \leq \lim_{k\to\infty} y_k^T C^T C y_k = \lim_{k\to\infty} x_k^T x_k = p_1.$$

Similarly, if the second problem is feasible, let $\{x_k\}$ be a minimizing sequence and define a new sequence $y_k = Cx_k$. For each k, $Ay_k = ACx_k \geq b$, so $\{y_k\}$ is a sequence of feasible points for the first problem. Moreover,

$$p_1 \leq \lim_{k\to\infty} y_k^T y_k = \lim_{k\to\infty} x_k^T C^T C x_k = p_2.$$

The remaining cases to consider are if either of the two problems is infeasible. But, note that if x is feasible for the first problem, then $C^{-1}x$ is feasible for the second; and if x is feasible for the second problem, then Cx is feasible for the first. Thus $p_1 = +\infty$ if and only if $p_2 = +\infty$. Thus, in all cases, we see that $p_1 = p_2$.

As in the previous example, the equivalence is even more substantial than this. It is possible to show that x is optimal for the first problem if and only if $C^{-1}x$ is optimal for the second. Thus, by a relatively simple operation (multiplication by either C or C^{-1}), we can transform the solution to one of the problems into the solution of the other.

Example 8.3. The minimum-ℓ_1-norm solution problem for an underdetermined system $Ax = b$ takes the form

$$\begin{aligned} \underset{x}{\text{minimize}} \quad & \|x\|_1 \\ \text{subject to} \quad & Ax = b, \end{aligned} \tag{8.3}$$

where the matrix $A \in \mathbb{R}^{m\times n}$ and vector $b \in \mathbb{R}^n$ are given. The x under the word "minimize" explicitly indicates the decision variables. When attempting to solve this problem numerically, the non-differentiability of the objective function tends to be problematic. In this case, replacing the problem with an equivalent problem is advantageous. The equivalent formulation is created by introducing a new auxiliary vector of decision variables $s \in \mathbb{R}^n$:

$$\begin{aligned} \underset{x,s}{\text{minimize}} \quad & \sum_{i=1}^{n} s_i \\ \text{subject to} \quad & Ax = b, \\ & x_i \leq s_i \quad \text{for } i = 1, 2, \ldots, n, \\ & -x_i \leq s_i \quad \text{for } i = 1, 2, \ldots, n. \end{aligned} \tag{8.4}$$

To establish the equivalence, let p_1 be the value of (8.3) and let p_2 be the value of (8.4). Note that both problems are feasible if and only if $Ax = b$ is solvable. Indeed, for (8.4), if x is a solution to the linear system, then the choice $s_i = |x_i|$ for $i = 1, 2, \ldots, n$ provides a feasible point. Thus, $p_1 = +\infty$ if and only if $p_2 = +\infty$. Now, suppose that both problems are feasible and consider the two sets

$$S_1 := \{\|x\|_1 : Ax = b\}$$

and

$$S_2 := \left\{\sum_{i=1}^{n} s_i : Ax = b \text{ and } |x_i| \leq s_i \text{ for } i = 1, 2, \ldots, n\right\}.$$

These sets are significant because

$$p_1 = \inf S_1 \quad \text{and} \quad p_2 = \inf S_2.$$

Let x be any feasible point for the first problem and define $s_i = |x_i|$ for $i = 1, 2, \ldots, n$. Then the pair (x, s) is feasible for the second problem. Moreover,

$$\|x\|_1 = \sum_{i=1}^{n} |x_i| = \sum_{i=1}^{n} s_i.$$

This implies that $S_1 \subseteq S_2$, showing that $p_2 \leq p_1$.

On the other hand, suppose the pair (x, s) is feasible for the second problem. Then, x is necessarily feasible for the first problem. Moreover,

$$\|x\|_1 \leq \sum_{i=1}^{n} s_i.$$

This shows that for every value $v_2 \in S_2$, there exists a $v_1 \in S_1$ such that $v_1 \leq v_2$. This implies that $p_1 \leq p_2$ and, therefore, that $p_1 = p_2$. Note that in this example, we did not need to determine whether or not the problems were bounded below to establish their equivalence.

EXERCISES

8.1 *Transforming Between Maximization and Minimization.* Consider the two expressions

$$p_1 := \sup\{f(x) : x \in X\}, \qquad p_2 := \inf\{-f(x) : x \in X\}.$$

Describe the equivalence between these two problems. This exercise shows how any maximization problem can be transformed into an equivalent minimization problem and vice versa.

8.2 *Squaring and Equivalence.* Consider a generic optimization problem of the form

$$\inf\{f(x) : x \in X\} \quad \text{with} \quad f : X \to \mathbb{R}_{\geq 0}.$$

(i) Describe the equivalence between this problem and $\inf\{f(x)^2 : x \in X\}$.

(ii) Provide a simple example demonstrating why the non-negativity of f is important in part (a).

8.3 *The Knapsack Problem.* The Knapsack Problem (KP) is a famous optimization problem. We imagine we have a set of n items labeled $1, 2, \ldots, n$. Each item has a weight $w_i \geq 0$ and a value $v_i \geq 0$, which indicates how much utility this object will provide us. We need to pack a knapsack with items in such a way as to maximize the overall utility, but we cannot pack more than a maximum weight of $W \geq 0$. As an optimization problem, we can represent the Knapsack Problem as

$$\begin{aligned} \text{maximize} \quad & v^T x \\ \text{subject to} \quad & w^T x \leq W, && (1) \\ & x \in \{0,1\}^n. && (2) \end{aligned}$$

Answer the following about the mathematical formulation of the KP.

(i) Identify and interpret the decision variables.

(ii) Explain the expression for the objective function.

(iii) Explain each of the constraints.

(iv) Is the problem discrete, continuous, or mixed?

(v) Is the problem bounded?

(vi) Is the problem feasible for all choices of the parameters, $w_i \geq 0$, $v_i \geq 0$ and $W \geq 0$? If not, what additional assumptions about the parameters must be made to ensure feasibility?

8.4 *Facility Location.* Consider the following simple model of a multi-warehouse distribution system called a Facility Location Problem (FLP). In this model, there are n warehouses that can meet the demands of m customers. Warehouse i can either be open (and thus incur an operation cost $a_i \geq 0$) or closed (costing nothing in this model). If the ith warehouse is open, it can support a maximum supply of $s_i \geq 0$ units of goods. Customer j has a demand of $d_j \geq 0$ units of goods, which must be met by shipping goods from one or more warehouses to the customer. The cost of shipping one unit of goods from warehouse i to customer j is $c_{ij} \geq 0$. The company controlling the warehouses wishes to meet all customer demands while minimizing operating costs. Mathematically, we can represent this problem as

$$
\begin{aligned}
\text{minimize} \quad & \sum_{i=1}^{n}\sum_{j=1}^{m} c_{ij}x_{ij} + \sum_{i=1}^{n} a_i b_i \\
\text{subject to} \quad & \sum_{i=1}^{n} x_{ij} = d_j \quad \text{for } j = 1, 2, \ldots, m, && (1)\\
& \sum_{j=1}^{m} x_{ij} \le b_i s_i \quad \text{for } i = 1, 2, \ldots, n, && (2)\\
& x_{ij} \ge 0 \quad \text{for } i = 1, 2, \ldots, n, \ j = 1, 2, \ldots, m, && (3)\\
& b_i \in \{0, 1\} \quad \text{for } i = 1, 2, \ldots, n. && (4)
\end{aligned}
$$

Answer the following about the mathematical formulation of the FLP.

(i) Identify and interpret the decision variables.

(ii) Explain the expression for the objective function.

(iii) Explain each of the constraints.

(iv) Is the problem discrete, continuous, or mixed?

(v) Is the problem bounded?

(vi) Is the problem feasible for all choices of the parameters ($a_i \ge 0$, $s_i \ge 0$, etc.)? If not, what additional assumptions about the parameters must be made to ensure feasibility?

8.5 Consider the following minimum-ℓ_1-norm solution problem (a special case of Example (8.3)).

$$
\begin{aligned}
\underset{x}{\text{minimize}} \quad & \|x\|_1 \\
\text{subject to} \quad & \begin{bmatrix} 1 & 1 \\ 1 & a \end{bmatrix} x = \begin{bmatrix} 1 \\ b \end{bmatrix}.
\end{aligned}
$$

(The real numbers a and b are parameters of the problem.) Characterize this problem in terms of value, feasibility, boundedness, attainability, and uniqueness of solution. (Your answer must consider different cases for the parameters a and b.

CHAPTER 9

Continuous Optimization

MANY important continuous optimization problems have the form

$$\begin{aligned} \underset{x \in \mathcal{D}}{\text{minimize}} \quad & f_0(x) \\ \text{subject to} \quad & f_i(x) \leq 0 \quad \text{for } i = 1, 2, \ldots, m, \\ & h_i(x) = 0 \quad \text{for } i = 1, 2, \ldots, p. \end{aligned} \tag{9.1}$$

Here, f_i for $i = 0, 1, 2, \ldots, m$, and h_i for $i = 1, 2, \ldots, p$ are all real-valued functions on $\mathcal{D} \subseteq \mathbb{R}^n$. To keep the discussion concise, we assume that $\mathcal{D}$ is an open set, which is true for the applications in this book.

Problems in the form of (9.1) are so common that we shall follow [24] and refer to this as the *standard form* of a continuous optimization problem. A standard form problem necessarily has a finite number ($n \geq 1$) of decision variables, a finite number ($m \geq 0$) of inequality constraints, and a finite number ($p \geq 0$) of equality constraints. It is possible for m or p to be zero. If both are zero, the result is called an *unconstrained optimization problem.*

Because standard-form optimization problems are so common, many theoretical tools and numerical algorithms have been derived to treat them. This chapter briefly introduces this topic, focusing mainly on the specialized theory of Lagrangian duality and convex optimization problems. The interested reader will find a more comprehensive treatment of the topic and many applications and examples in [24]. A rigorous and detailed treatment of the finer details of convex optimization can be found in [69]. For now, we shall focus on the minimization problem. Section 9.7 contains a corresponding theory for maximization problems.

DOI: 10.1201/9781003024866-11

9.1 LAGRANGIAN DUALITY

The first step in developing the theory of Lagrangian duality is to introduce the Lagrangian function.

Definition 9.1. The *Lagrangian function* (sometimes called the *Lagrangian energy* or simply the *Lagrangian*) associated with the standard-form optimization problem (9.1) is the function

$$\mathcal{L}(x,\lambda,\nu) := f_0(x) + \sum_{i=1}^{m} \lambda_i f_i(x) + \sum_{j=1}^{p} \nu_j h_i(x). \tag{9.2}$$

The variables $\lambda \in \mathbb{R}^m$ and $\nu \in \mathbb{R}^p$ are called *Lagrangian dual variables.*

Each constraint from the optimization problem has an associated dual variable. By convention, the dual variables associated with an inequality constraint are denoted by λ_i, and the dual variables associated with an equality constraint are denoted by ν_i. It is helpful to remember this correspondence; each constraint has an associated dual variable, and vice versa.

Writing the Lagrangian in a more "vectorized" form is also sometimes useful. To do this, we combine the inequality constraints into a single, vector-valued function $f(x) = (f_1(x), f_2(x), \ldots, f_m(x))^T$ and similarly with the equality constraints, $h(x) = (h_1(x), h_2(x), \ldots, h_p(x))^T$. With this notational convention, (9.2) can be rewritten as

$$\mathcal{L}(x,\lambda,\nu) = f_0(x) + \lambda^T f(x) + \nu^T h(x).$$

The importance of the Lagrangian derives from the following lemma.

Lemma 9.1. *The standard form optimization problem in* (9.1) *is equivalent to*

$$\inf_{x\in\mathcal{D}} \sup_{\substack{\lambda\in\mathbb{R}^m_{\geq 0}\\ \nu\in\mathbb{R}^p}} \mathcal{L}(x,\lambda,\nu).$$

Proof. See Exercise 9.1. □

The next step in establishing Lagrangian duality uses a property of infima and suprema called the *max-min inequality.*

Lemma 9.2. *Suppose that X and Y are nonempty sets and that $f : X \times Y \to \mathbb{R} \cup \{-\infty, +\infty\}$. Then,*

$$\inf_{x\in X} \sup_{y\in Y} f(x,y) \geq \sup_{y\in Y} \inf_{x\in X} f(x,y).$$

Proof. See Exercise 9.2 □

In Lagrangian duality theory, the original problem (9.1) is referred to as the *primal problem*, and the decision variables are referred to as the *primal variables*. We have already introduced a set of dual variables. These will be decision variables in an associated *dual optimization problem*. The derivation of the dual problem comes from Lemmas 9.1 and 9.2. If we denote by p^* the value of the primal problem, then the lemmas imply that

$$p^* = \inf_{x\in\mathcal{D}} \sup_{\substack{\lambda\in\mathbb{R}^m_{\geq 0}\\ \nu\in\mathbb{R}^p}} \mathcal{L}(x,\lambda,\nu) \geq \sup_{\substack{\lambda\in\mathbb{R}^m_{\geq 0}\\ \nu\in\mathbb{R}^p}} \inf_{x\in\mathcal{D}} \mathcal{L}(x,\lambda,\nu) = \sup_{\substack{\lambda\in\mathbb{R}^m_{\geq 0}\\ \nu\in\mathbb{R}^p}} g(\lambda,\nu) \quad (9.3)$$

where

$$g(\lambda,\nu) := \inf_{x\in\mathcal{D}} \mathcal{L}(x,\lambda,\nu).$$

The function g is called the *dual objective function*. The right-hand side of (9.3) is the dual optimization problem and can be written as

$$\begin{aligned} &\underset{\lambda\in\mathbb{R}^n,\nu\in\mathbb{R}^p}{\text{maximize}} && g(\lambda,\nu), \\ &\text{subject to} && \lambda_i \geq 0 \text{ for } i = 1,2,\ldots,m. \end{aligned} \quad (9.4)$$

By convention, we denote the value of the dual problem by d^*. In short, (9.3) shows that $p^* \geq d^*$; that is, the value of the primal problem is bounded from below by the value of the associated dual problem. This fact is typically referred to as *weak duality*. This inequality can be extended to include the possibility that p^* and d^* may take the values $\pm\infty$.

In this setting, the variables in the vector $x \in \mathcal{D}$ in (9.1) are referred to as *primal variables*, while the variables in the pair of vectors (λ, ν) in (9.4) are referred to as *dual variables*. If, additionally, x satisfies $f_i(x) \leq 0$ for $i = 1, 2, \ldots, m$ and $h_i(x) = 0$ for $i = 1, 2, \ldots, p$, x is called *primal feasible*. Similarly, if $\lambda \geq 0$, then the pair (λ, ν) is called *dual feasible*.

Example 9.1. Consider the primal problem

$$\begin{aligned} &\underset{x\in\mathbb{R}^n}{\text{minimize}} && \|x\|_2^2 \\ &\text{subject to} && a^T x \geq b, \end{aligned}$$

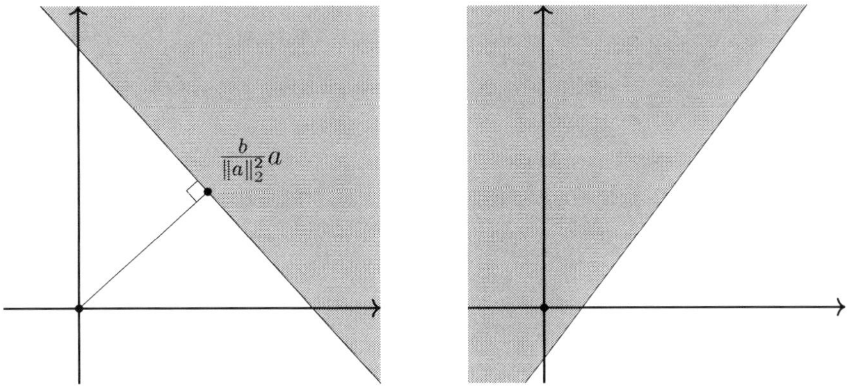

Figure 9.1 Two possible cases for Example 9.1. If the half-space $\{x : a^T x \geq b\}$ does not contain the origin (left), then the optimal x is the projection of the hyperplane onto the vector a. If the half-space does contain the origin (right), then the optimal x is the origin.

with parameters $a \in \mathbb{R}^n \setminus \{0\}$ and $b \in \mathbb{R}$. Geometrically, the problem describes the closest point to the origin in a half-space bounded by the hyperplane $\{x : a^T x = b\}$. It is straightforward to check that this plane is normal to the vector a and contains the point $(b/\|a\|_2^2)a$. There are two cases to consider, as shown in Figure 9.1. If $b \leq 0$, then the origin belongs to the half-space of feasible points, and, therefore, the value of the problem is 0. If $b > 0$, on the other hand, the closest point to the origin is the projection of the hyperplane onto the vector a, namely $x = (b/\|a\|_2^2)a$. In this case, the value of the problem is $b^2/\|a\|_2^2$.

This problem can be put into standard form, with $f_0(x) = \|x\|_x^2 = x^T x$ and $f_1(x) = b - a^T x$. (There are no equality constraints.) Thus, the Lagrangian incorporates a single dual variable λ:

$$\mathcal{L}(x, \lambda) = x^T x + \lambda(b - a^T x).$$

For a given λ, the dual objective function $g(\lambda)$ is defined as

$$g(\lambda) = \inf_{x \in \mathbb{R}^n} \{x^T x - \lambda a^T x + \lambda b\}.$$

Geometrically, $g(\lambda)$ is the "height" of a paraboloid in n-dimensional space. The location of the vertex of this paraboloid can be found by setting the gradient equal to zero. The vertex occurs where $2x - \lambda a = 0$, or $x = \frac{\lambda}{2}a$, providing the formula

$$g(\lambda) = \frac{\lambda^2}{4}\|a\|_2^2 - \frac{\lambda^2}{2}\|a\|_2^2 + \lambda b = \lambda b - \frac{\lambda^2}{4}\|a\|_2^2.$$

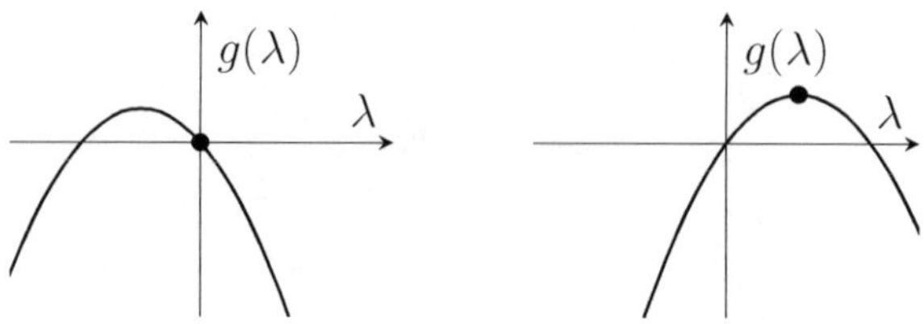

Figure 9.2 Two possible cases for the dual problem in Example 9.1. If $b \le 0$ (left), then the optimal choice of the dual variable is $\lambda = 0$. If $b > 0$ (right), on the other hand, then the maximum in (9.5) occurs at the vertex of the parabola.

Thus, the dual problem takes the form

$$\begin{aligned} \underset{\lambda}{\text{maximize}} \quad & \lambda\left(b - \frac{\|a\|_2^2}{4}\lambda\right) \\ \text{subject to} \quad & \lambda \ge 0. \end{aligned}$$

To find the value d^*, we must find the maximum of a one-dimensional parabola restricted to the right half-plane.

The vertex of this parabola occurs at $\lambda = 2b/\|a\|_2^2$, where g takes the value $b^2/\|a\|_2^2$. As in the case of the primal problem, the solution to the dual problem divides into two cases. (See Figure 9.1.) If $b \le 0$, the vertex of the parabola is in the left half-plane, and the optimal value occurs at $\lambda = 0$ yielding $d^* = 0$. If $b > 0$, on the other hand, the optimal value occurs at the vertex. So,

$$d^* = \begin{cases} 0 & \text{if } b \le 0, \\ \frac{b^2}{\|a\|_2^2} & \text{if } b > 0. \end{cases} \tag{9.5}$$

In all cases, we see that $p^* = d^*$. This is an example of strong duality, as discussed in the next section.

9.2 STRONG DUALITY

Section 9.1 established the Lagrangian duality relationship between a primal optimization problem (9.1) and its corresponding dual problem (9.4).

Definition 9.2. Given a primal optimization problem and its corresponding Lagrangian dual, we call the quantity $p^* - d^* \ge 0$ the *duality gap* for the problem. If $p^* - d^* = 0$, we say that *strong duality* holds for the problem.

There is no reason to expect strong duality to hold for general optimization problems. However, when it does, strong duality has important implications.

Definition 9.3. A point $(\tilde{x}, \tilde{\lambda}, \tilde{\nu}) \in \mathcal{D} \times \mathbb{R}^m_{\geq 0} \times \mathbb{R}^p$ is called a *saddle point of the Lagrangian* if

$$\mathcal{L}(\tilde{x}, \lambda, \nu) \leq \mathcal{L}(\tilde{x}, \tilde{\lambda}, \tilde{\nu}) \leq \mathcal{L}(x, \tilde{\lambda}, \tilde{\nu}) \text{ for all } (x, \lambda, \nu) \in \mathcal{D} \times \mathbb{R}^m_{\geq 0} \times \mathbb{R}^p.$$

Theorem 9.1. *The point, $(\tilde{x}, \tilde{\lambda}, \tilde{\nu})$ is a saddle point of the Lagrangian if and only if strong duality holds, $\tilde{x}$ is optimal for the primal problem, and $(\tilde{\lambda}, \tilde{\nu})$ is optimal for the dual problem.*

Proof. First, assume that strong duality holds and that $\tilde{x}$ and $(\tilde{\lambda}, \tilde{\nu})$ are optimal for their respective problems. This implies that

$$\begin{aligned}
\mathcal{L}(\tilde{x}, \tilde{\lambda}, \tilde{\nu}) &\leq \sup_{\substack{\lambda \in \mathbb{R}^m_{\geq 0} \\ \nu \in \mathbb{R}^p}} \mathcal{L}(\tilde{x}, \lambda, \nu) && \text{(definition of supremum)} \\
&= p^* && (\tilde{x} \text{ is primal-optimal}) \\
&= d^* && \text{(strong duality)} \\
&= \inf_{x \in D} \mathcal{L}(x, \tilde{\lambda}, \tilde{\nu}) && ((\tilde{\lambda}, \tilde{\nu}) \text{ is dual-optimal}) \\
&\leq \mathcal{L}(\tilde{x}, \tilde{\lambda}, \tilde{\nu}) && \text{(definition of infimum).}
\end{aligned}$$

Since the first and last quantities are identical, the inequalities hold as equality. This implies the saddle point property since

$$\begin{aligned}
\mathcal{L}(\tilde{x}, \lambda, \nu) &\leq \sup_{\substack{\hat{\lambda} \in \mathbb{R}^m_{\geq 0} \\ \hat{\nu} \in \mathbb{R}^p}} \mathcal{L}(\tilde{x}, \hat{\lambda}, \hat{\nu}) && \text{(definition of supremum)} \\
&= \mathcal{L}(\tilde{x}, \tilde{\lambda}, \tilde{\nu}) && \text{(equality holds above)} \\
&= \inf_{\hat{x} \in \mathcal{D}} \mathcal{L}(\hat{x}, \tilde{\lambda}, \tilde{\nu}) && \text{(equality holds above)} \\
&\leq \mathcal{L}(x, \tilde{\lambda}, \tilde{\nu}) && \text{(definition of infimum).}
\end{aligned}$$

Now, assume that $(\tilde{x}, \tilde{\lambda}, \tilde{\nu})$ is a saddle point. The first inequality in the saddle point definition implies that

$$p^* \leq \sup_{\substack{\lambda \in \mathbb{R}^m_{\geq 0} \\ \nu \in \mathbb{R}^p}} \mathcal{L}(\tilde{x}, \lambda, \nu) = \mathcal{L}(\tilde{x}, \tilde{\lambda}, \tilde{\nu}).$$

The second inequality implies that

$$d^* \geq g(\tilde{\lambda}, \tilde{\nu}) = \inf_{x \in \mathcal{D}} \mathcal{L}(x, \tilde{\lambda}, \tilde{\nu}) = \mathcal{L}(\tilde{x}, \tilde{\lambda}, \tilde{\nu}),$$

so

$$p^* \leq \mathcal{L}(\tilde{x}, \tilde{\lambda}, \tilde{\nu}) \leq d^*.$$

Combining this with weak duality ($p^* \geq d^*$) implies that strong duality holds and that the above inequalities involving p^* and d^* are satisfied as equality. In particular, this implies that $\tilde{x}$ is optimal for the primal problem and that $(\tilde{\lambda}, \tilde{\nu})$ is optimal for the dual problem. □

9.3 THE KKT CONDITIONS

Here, we further study the implications of strong duality.

Assumptions 9.1. Consider the optimization problem (9.1) satisfying the following:

1. The set $\mathcal{D}$ is open.
2. The functions $f_0, \ldots, f_m$ and $h_1, \ldots, h_p$ are differentiable on $\mathcal{D}$.
3. Strong duality holds.
4. Both the primal problem, (9.1), and the dual problem, (9.4), are attained.

These assumptions are satisfied by most of the continuous optimization problems described in this book. The first two assumptions are typically straightforward to verify. The third and fourth assumptions are often verified using tools described later in this section. In light of Theorem 9.1, these last two assumptions imply the existence of a saddle point of the Lagrangian. In what follows, we shall assume that (x^*, λ^*, ν^*) is such a saddle point.

Lemma 9.3. *Under Assumptions 9.1, if (x^*, λ^*, ν^*) is a saddle point of the Lagrangian, then*

$$\nabla f_0(x^*) + \sum_{i=1}^{m} \lambda_i^* \nabla f_i(x^*) + \sum_{i=1}^{p} \nu_i^* \nabla h_i(x^*) = 0.$$

Proof. The saddle point assumption implies that x^* is a local minimizer of the function $x \mapsto \mathcal{L}(x, \lambda^*, \nu^*)$. Since the f_i and h_i are differentiable and $\mathcal{D}$ is open, the gradient for this function vanishes at x^*. □

Lemma 9.4. *Under Assumptions 9.1, if* (x^*, λ^*, ν^*) *is a saddle point of the Lagrangian, then*

$$\lambda_i^* f_i(x^*) = 0 \quad \text{for } i = 1, 2, \ldots, m.$$

Proof. This comes from the fact that the Lagrangian (9.2) is linear in λ_i. At a saddle point, the Lagrangian is maximized with respect to λ_i, which implies that either the derivative vanishes ($f_i(x^*) = 0$) or that the inequality constraint $\lambda_i \geq 0$ is active ($\lambda_i^* = 0$). □

This leads to the Karush-Kuhn-Tucker (KKT) conditions for optimality.

Theorem 9.2 (KKT Conditions). *Under Assumptions 9.1, if* (x^*, λ^*, ν^*) *is a saddle point of the Lagrangian, then*

$$f_i(x^*) \leq 0, \ i = 1, 2, \ldots, m, \quad h_i(x^*) = 0, \ i = 1, 2, \ldots, p, \tag{9.6}$$

$$\lambda_i^* \geq 0, \ i = 1, 2, \ldots, m, \tag{9.7}$$

$$\nabla f_0(x^*) + \sum_{i=1}^{m} \lambda_i^* \nabla f_i(x^*) + \sum_{i=1}^{p} \nu_i^* \nabla h_i(x^*) = 0, \tag{9.8}$$

$$\lambda_i^* f_i(x^*) = 0, \ i = 1, 2, \ldots, m. \tag{9.9}$$

Remark 9.1. Condition (9.6) is called the *primal feasibility* condition, (9.7) is called *dual feasibility*. Condition (9.8) is called the *stationarity condition* and (9.9) is called *complementary slackness*.

Proof. The primal and dual feasibility conditions are implied by definition; a point can only be optimal if it is feasible. Stationarity and complementary slackness are Lemmas 9.3 and 9.4 respectively. □

Thus far, the KKT conditions have only been established as necessary conditions for a saddle point; if a saddle point exists, it must satisfy the KKT conditions. One may wonder about the converse: Does satisfying the KKT conditions indicate a saddle point? In other words, are the KKT conditions also sufficient? The answer is no, in general. However, there is a better answer for convex optimization problems.

9.4 CONVEXITY

We now recall two related concepts of convexity: convex sets and convex functions.

Definition 9.4. A set $\mathcal{S} \subseteq \mathbb{R}^n$ is called *convex* if, for every $x, y \in \mathcal{S}$ and every $\theta \in [0,1]$, the point $\theta x + (1-\theta)y$ is also in $\mathcal{S}$.

In other words, whenever we take two points in a convex set, the line segment connecting those two points also lies in the set.

Definition 9.5. A real-valued function f on a set $\mathcal{D} \subseteq \mathbb{R}^n$ is called *convex* if $\mathcal{D}$ is convex and if, for every $x, y \in \mathcal{D}$ and every $\theta \in [0,1]$, we have

$$f(\theta x + (1-\theta)y) \leq \theta f(x) + (1-\theta)f(y).$$

The concepts of convexity for sets and functions are closely related, as the following theorem shows.

Theorem 9.3. *Let $f : \mathcal{D} \to \mathbb{R}$ be a function on a convex set $\mathcal{D} \subseteq \mathbb{R}^n$ and let*

$$\operatorname{epi} f := \{(x,t) \in \mathcal{D} \times \mathbb{R} : f(x) \leq t\}$$

be its epigraph. Then, f is a convex function if and only if epi f *is a convex set in $\mathcal{D} \times \mathbb{R}$.*

Proof. First, assume that f is a convex function. Let

$$(x_1, t_1), (x_2, t_2) \in \operatorname{epi} f,$$

and let $\theta \in [0,1]$. By definition, $f(x_i) \leq t_i$ for $i = 1, 2$. It follows from the convexity of f that

$$f(\theta x_1 + (1-\theta)x_2) \leq \theta f(x_1) + (1-\theta)f(x_2) \leq \theta t_1 + (1-\theta)t_2$$

and, therefore, that $(\theta x_1 + (1-\theta)x_2, \theta t_1 + (1-\theta)t_2) \in \operatorname{epi} f$. This shows that epi f is a convex set.

On the other hand, suppose that epi f is convex and let $x_1, x_2 \in \mathcal{D}$ and $\theta \in [0,1]$. Since the points $(x_i, f(x_i))$ belong to epi f for $i = 1, 2$, the convexity of epi f implies that

$$(\theta x_1 + (1-\theta)x_2, \theta f(x_1) + (1-\theta)f(x_2)) \in \operatorname{epi} f.$$

By definition of the epigraph, then,

$$f(\theta x_1 + (1-\theta)x_2) \leq \theta f(x_1) + (1-\theta)f(x_2),$$

implying that f is a convex function. □

Establishing convexity by directly verifying the conditions of Definition 9.5 can be cumbersome for many functions. A better approach is to build up a toolbox of convexity properties that can be combined to establish the convexity of more complicated functions. Chapter 3 of [24] contains many such properties. Here, we review a few that will be particularly useful for this book.

Theorem 9.4. *The following is a short list of ways to establish convexity.*

(1) ***Second-order conditions.*** *A twice differentiable function f, on an open domain $\mathcal{D}$, is convex if and only if its Hessian matrix $\nabla^2 f$ is positive semi-definite everywhere in $\mathcal{D}$.*

(2) ***Affine functions.*** *Every affine function, $f(x) = a^T x + b$ is convex.*

(3) ***Quadratic functions.*** *A function of the form $f(x) = x^T P x$ (where P is a real symmetric matrix) is convex if and only if P is positive semi-definite.*

(4) ***Scaling and translating.*** *If f is convex, then so is $g = af + b$ for any constants $a \geq 0$ and b.*

(5) ***Sums.*** *If functions f_i for $i = 1, 2, \ldots, m$ are convex and share the same domain, then $f = f_1 + f_2 + \cdots + f_m$ is also convex.*

(6) ***Composition with affine functions.*** *Suppose that $f : \mathbb{R}^m \to \mathbb{R}$ is convex, $A \in \mathbb{R}^{m \times n}$ and $b \in \mathbb{R}^m$. Then, the function $g(x) = f(Ax + b)$ is convex.*

(7) ***Composition of certain convex functions.*** *Although compositions of convex functions are not necessarily convex in general, they are convex in certain special cases. For example, suppose that $f : \mathbb{R} \to [0, \infty)$ is convex and that $g : [0, \infty) \to \mathbb{R}$ is both convex and nondecreasing. Then the composition $h(x) = g(f(x))$ is convex.*

Proof. (1) is the higher-dimensional analog to the second derivative test $f''(x) \geq 0$ from calculus. We omit the proof here, but it can be reconstructed by following Exercise 3.8 of [24].

(2) is straightforward to check directly. Let $x, y \in \mathbb{R}^n$ and let $\theta \in [0,1]$. Then

$$\begin{aligned} f(\theta x + (1-\theta)y) &= a^T(\theta x + (1-\theta)y) + b \\ &= \theta a^T x + (1-\theta)a^T y + \theta b + (1-\theta)b \\ &= \theta(a^T x + b) + (1-\theta)(a^T y + b) \\ &= \theta f(x) + (1-\theta)f(y). \end{aligned}$$

Alternatively, one can observe that the Hessian matrix is identically zero and, therefore, is trivially positive semi-definite.

The proofs of (3)–(7) are left to the reader in Exercise 9.3. □

9.5 CONVEX OPTIMIZATION

Definition 9.6. A standard form optimization problem (9.1) is called a *convex optimization problem* if

(a) The functions f_i for $i = 0, 1, 2, \ldots, m$ are all convex (see Definition 9.5), and

(b) The functions h_i for $i = 1, 2, \ldots, p$ are all affine; that is, each h_i has the form $h_i(x) = a_i^T x - b_i$ for some $a_i \in \mathbb{R}^n$ and some $b_i \in \mathbb{R}$.

In linear algebra notation, (b) means that the equality constraints take the form $Ax = b$, where A is the $p \times n$ matrix with rows made up of the a_i and b is the vector in $\mathbb{R}^p$ with entries given by the b_i.

Theorem 9.5. *The set of feasible points for a convex optimization problem is convex.*

Proof. If the problem is infeasible, then the result is vacuously true. Otherwise, there is at least one feasible point. Let x_1 and x_2 be feasible points, let $\theta \in [0,1]$, and let $x = \theta x_1 + (1-\theta)x_2$. By convexity, for $i = 1, 2, \ldots, n$,

$$f_i(x) \le \theta f(x_1) + (1-\theta)f(x_2) \le \theta \cdot 0 + (1-\theta) \cdot 0 = 0.$$

Moreover,

$$Ax = \theta Ax_1 + (1-\theta)Ax_2 = \theta b + (1-\theta)b = b,$$

so x is also feasible. □

The importance of convexity in optimization stems from the following property.

Theorem 9.6. *Any local minimum of a convex optimization problem is a global minimum.*

Proof. We will prove the contrapositive. Suppose a feasible point x is *not* a global minimum. That implies that there is another feasible point y such that $f_0(y) < f_0(x)$. Now let $\theta \in (0, 1)$. By Theorem 9.5, the point $\theta x + (1 - \theta)y$ is also feasible. Moreover, by the convexity of f_0,

$$f_0(\theta x + (1 - \theta)y) \leq \theta f_0(x) + (1 - \theta)f_0(y) < f_0(x).$$

By choosing θ arbitrarily close to 1, we find feasible points arbitrarily close to x with strictly smaller objective values. Thus, x is not a local minimum. □

For convex problems, the KKT conditions can be used to guarantee optimality.

Theorem 9.7. *Under Assumptions 9.1, (x^*, λ^*, ν^*) satisfies the KKT conditions if and only if it is a saddle point of the Lagrangian.*

Proof. The necessity of the KKT conditions is established in Theorem 9.2. Conversely, suppose that (x^*, λ^*, ν^*) satisfy the KKT conditions. Stationarity implies that x^* is a critical point of the function

$$\phi_p(x) = \mathcal{L}(x, \lambda^*, \nu^*) = f_0(x) + \sum_{i=1}^{m} \lambda_i^* f_i(x) + \sum_{i=1}^{p} \nu_i^* h_i(x).$$

The λ_i^* are all nonnegative by dual feasibility, so ϕ_p is convex. Thus, the first-order stationarity conditions establish that x^* is a global minimizer, implying one-half of the saddle point inequalities:

$$\mathcal{L}(x^*, \lambda^*, \nu^*) \leq \mathcal{L}(x, \lambda^*, \nu^*) \text{ for all } x \in \mathcal{D}.$$

For the other inequality, consider the function

$$\phi_d(\lambda, \nu) = \mathcal{L}(x^*, \lambda, \nu) = f_0(x^*) + \sum_{i=1}^{m} \lambda_i f_i(x^*) + \sum_{i=1}^{p} \nu_i h_i(x^*).$$

Assuming $\lambda_i \geq 0$, we can use primal feasibility to obtain

$$\phi_d(\lambda, \nu) = f_0(x^*) + \sum_{i=1}^{m} \lambda_i f_i(x^*) \leq f_0(x^*).$$

On the other hand, complementary slackness implies that (λ^*, ν^*) attains this upper bound, establishing the other inequality:

$$\mathcal{L}(\lambda, \nu, x^*) \leq \mathcal{L}(\lambda^*, \nu^*, x^*) \text{ for all } (\lambda, \nu) \in \mathbb{R}^m_{\geq 0} \times \mathbb{R}^p.$$

□

9.6 OPTIMALITY AND SLATER'S CONDITION

The KKT conditions provide an algebraic way to establish optimality (and strong duality), at least for differentiable convex problems. An alternative method we shall use frequently is establishing coincident upper and lower bounds. The motivation behind this method arises from weak duality. Suppose x is feasible for the primal problem and that (λ, ν) is feasible for the dual problem. By weak duality, we have that

$$f_0(x) \geq p^* \geq d^* \geq g(\lambda, \nu).$$

Any feasible choices of x and (λ, ν) provide such a bound. Now, suppose we are able to find x, λ and ν so that $f_0(x) = g(\lambda, \nu)$. This would force the inequalities above to hold as equality:

$$f_0(x) = p^* = d^* = g(\lambda, \nu).$$

This immediately implies that x is primal optimal, (λ, ν) is dual optimal, and that strong duality holds. An example of how such an argument can be used is found in Exercise 9.6.

Of course, such a technique can only work if strong duality holds and if both the primal and dual optima are attained. Before searching for coincident upper and lower bounds, it can be helpful to know if this is even possible. That is where "constraint qualifications" like Slater's condition come in.

As previously mentioned, strong duality does not generally hold. However, if the optimization problem is convex and satisfies an additional, relatively mild condition, then strong duality is guaranteed. This additional condition is known as a *constraint qualification.* Slater's condition is one of the most widely used.

Definition 9.7. Problem (9.1) is said to satisfy *Slater's condition* if there exists a feasible point, x, that satisfies all inequality constraints strictly:

$$f_i(x) < 0 \text{ for } i = 1, 2, \ldots, m.$$

If some inequality constraints are affine (i.e., $f_i(x) = c_i^T x + d_i$) and there is a feasible point that strictly satisfies all non-affine inequality constraints, then the problem is said to satisfy *the weak form of Slater's condition*.

Theorem 9.8. *If a standard form optimization problem is convex and satisfies the weak form of Slater's condition, then strong duality holds.*

While not particularly difficult, the proof of this theorem uses concepts of convex analysis (separating and supporting hyperplanes, for example) that are outside the scope of this book. Moreover, Slater's condition implies additional beneficial properties; for instance, if the primal problem is bounded, then the dual problem is attained. For details, we refer the reader to [24, 69].

For many problems in this book, the objective function is convex, and all constraints are affine. For such problems, the weak form of Slater's condition is satisfied trivially, and we immediately know that strong duality holds.

9.7 MAXIMIZATION PROBLEMS

The standard form for a maximization problem is

$$\begin{aligned} \underset{x\in\mathcal{D}}{\text{maximize}} \quad & f_0(x) \\ \text{subject to} \quad & f_i(x) \geq 0 \quad \text{for } i = 1, 2, \ldots, m, \\ & h_i(x) = 0 \quad \text{for } i = 1, 2, \ldots, p. \end{aligned} \tag{9.10}$$

(Note the difference in the direction of inequality when compared with (9.1).)

One way to handle such a problem is to convert it into a minimization problem using the fact that maximizing $f_0(x)$ is equivalent to minimizing $-f_0(x)$. We can treat the minimization problem as described above and only need to remember to take care of the minus sign once we've found the value.

An equivalent approach is to mimic the analysis of the minimization problem, using the max-min inequality (Lemma 9.2) in the opposite direction. Conceptually, that looks like this. We define the Lagrangian function as in (9.2) and observe that

$$p^* = \sup_{x\in\mathcal{D}} \inf_{\substack{\lambda\in\mathbb{R}^m_{\geq 0} \\ \nu\in\mathbb{R}^p}} \mathcal{L}(x,\lambda,\nu) \leq \inf_{\substack{\lambda\in\mathbb{R}^m_{\geq 0} \\ \nu\in\mathbb{R}^p}} \sup_{x\in\mathcal{D}} \mathcal{L}(x,\lambda,\nu) = d^*.$$

Notice that we have essentially reversed the roles of infimum and supremum here. The dual objective function is

$$g(\lambda, \nu) = \sup_{x \in \mathcal{D}} \mathcal{L}(x, \lambda, \nu)$$

and the Lagrangian dual problem is

$$\begin{aligned} \underset{\lambda \in \mathbb{R}^n, \nu \in \mathbb{R}^p}{\text{minimize}} \quad & g(\lambda, \nu), \\ \text{subject to} \quad & \lambda_i \geq 0 \text{ for } i = 1, 2, \ldots, m. \end{aligned} \tag{9.11}$$

EXERCISES

9.1 *Lagrangian representation.* Prove Lemma 9.1 using the following steps.

(i) Beginning with the supremum, define the function

$$\phi(x) := \sup_{\substack{\lambda \in \mathbb{R}^m_{\geq 0} \\ \nu \in \mathbb{R}^p}} \mathcal{L}(x, \lambda, \nu).$$

Show that, for $x \in \mathcal{D}$,

$$\phi(x) = \begin{cases} f_0(x) & \text{if } x \text{ is feasible}, \\ +\infty & \text{otherwise}. \end{cases}$$

(ii) Show that (9.1) is equivalent to $\inf\{\phi(x) : x \in \mathcal{D}\}$. Be sure to address the possibility that the problem may be infeasible or unbounded.

9.2 *The max-min inequality.* Prove Lemma 9.2. To get started, consider that if $\tilde{x} \in X$ and $y \in Y$ then

$$f(\tilde{x}, y) \geq \inf_{x \in X} f(x, y)$$

and consider performing a supremum on both sides.

9.3 *Establishing convexity.* Prove statements (3) through (7) from Theorem 9.4.
Hint: For (3), write the function using sigma notation and indices and show that $\nabla^2 f = 2P$.

9.4 *Quadratic program (QP).* Consider an optimization problem of the form

$$\begin{aligned} \underset{x \in \mathbb{R}^n}{\text{minimize}} \quad & \frac{1}{2} x^T P x + q^T x \\ \text{subject to} \quad & Ax \leq b, \end{aligned}$$

where P is a symmetric, positive definite $n \times n$ matrix, $q \in \mathbb{R}^n$, $A \in \mathbb{R}^{m \times n}$ and $b \in \mathbb{R}^m$.

(i) Prove that the problem is convex.

(ii) Prove that strong duality holds.

(iii) Derive the Lagrangian dual problem.

(iv) Derive the KKT conditions for optimality.

9.5 *Quadratically constrained quadratic program (QCQP).* Consider the primal problem

$$\begin{aligned} \underset{x \in \mathbb{R}^n}{\text{minimize}} \quad & \|x\|_2^2 \\ \text{subject to} \quad & \|x - a\|_2^2 \leq b, \end{aligned}$$

with parameters $a \in \mathbb{R}^n$ and $b \in \mathbb{R}$.

(i) Provide a geometric interpretation for this problem.

(ii) Use geometric reasoning to find the value, p^* of the problem. There will be three cases. (Hints: See Example 9.1. Don't forget to consider the possibility that $b < 0$.)

(iii) Derive the Lagrangian and dual optimization problem.

(iv) Evaluate the value, d^*, of the dual problem and show that weak duality holds. Does strong duality also hold?

9.6 *Establishing upper and lower bounds.* As an example of using the method described in Section 9.6, consider the quadratic optimization problem

$$\begin{aligned} \underset{x \in \mathbb{R}^3}{\text{minimize}} \quad & \|x\|_2^2 \\ \text{subject to} \quad & x_1 + x_2 \geq 1, \\ & x_1 + x_3 \geq 1, \\ & x_2 + x_3 \geq 1. \end{aligned}$$

(i) Notice that the objective function and constraints are invariant with respect to permuting the variables. This suggests that the minimum should be attained for an x of the form $x = (c, c, c)$. Use this observation to produce an upper bound on p^*. (I.e., choose the "best" c that makes x feasible.)

(ii) Use the result of Exercise 9.4 to write the Lagrangian dual problem.

(iii) Now notice that the dual objective function and constraints are invariant with respect to permuting the dual variables. This again suggests a constant choice $\lambda = (t, t, t)$. Use this guess to obtain a lower bound on d^*.

(iv) Explain why your work on the preceding parts of this problem proves that strong duality holds and that the guesses you obtained for x and λ are indeed optimal.

III

The Basic Theory of Modulus

CHAPTER 10

Modulus on Graphs

MODULUS IS A WELL-ESTABLISHED TOOL in complex analysis and geometric function theory [2, 68]. The goal in this chapter is to systematically develop the theory of modulus on graphs and networks, the pioneering work along these lines being Duffin's 1962 paper [36], see also [25, 48, 65, 72]. In doing so, we will rely heavily on the previous chapters on convex optimization.

10.1 FAMILIES OF OBJECTS

For simplicity, we will work with finite undirected and possibly weighted graphs. However, much of the theory extends to more general settings, such as directed graphs, multigraphs, infinite graphs, and beyond. To fix notation, let $G = (V, E, \sigma)$ be a *graph*, where the vertex set V and the edge set E are finite. Also, to every edge $e \in E$ we may correspond a weight $0 < \sigma(e) < \infty$, which we will think of as a "edge conductance".

We want to study *families* Γ of objects γ, such as walks, trees, etc, but the only property that truly concerns us is the usage that each object makes of each given edge in E. This is a choice on our part and thus technically we will only be studying so-called (edge)-modulus. There are similar definitions where the usage is focused instead on the nodes, (vertex)-modulus, and possibly also situations where one is interested in the usage on the union of edges and nodes. But we won't pursue these questions here. Therefore, consider families of objects Γ such that for each $\gamma \in \Gamma$ we can associate a function

$$\mathcal{N}(\gamma, \cdot) : E \to \mathbb{R}_{\geq 0}, \tag{10.1}$$

i.e. a vector $\mathcal{N}(\gamma, \cdot)^T \in \mathbb{R}^E_{\geq 0}$, that measures the *usage of edge e by* γ.

DOI: 10.1201/9781003024866-13

Remark 10.1. Sometimes we will identify the family Γ with the associated vectors $\mathcal{N}(\gamma,\cdot)^T \in \mathbb{R}^E_{\geq 0}$. In that case we will write simply $\Gamma \subset \mathbb{R}^E_{\geq 0}$. Conversely, a set of vectors in $\mathbb{R}^E_{\geq 0}$ can always be thought of as a family of objects on G by identifying each object with its corresponding usage vector.

Here are some examples of possible objects.

- A walk $\gamma = x_0\ e_1\ x_1\ \cdots\ e_n\ x_n$ can be given the usage

$$\mathcal{N}(\gamma, e) = \text{ number of times } \gamma \text{ traverses } e.$$

  In this case $\mathcal{N}(\gamma,\cdot)^T \in \mathbb{Z}^E_{\geq 0}$. Also, note that two distinct walks could have the same usage function $\mathcal{N}(\gamma,\cdot)$.

- When $T \subset E$ is a set of edges, such as a spanning tree or a simple cycle, we can associate to it the indicator function $\mathcal{N}(T,e) = \mathbb{1}_T(e)$, which is equal to 1, if $e \in T$ and 0 otherwise. In this case, $\mathcal{N}(\gamma,\cdot)^T \in \{0,1\}^E$.

- A flow f (see Definition 1.6) can be given the usage function $\mathcal{N}(f,e) := |f(e)|$. Then, in this case, $\mathcal{N}(\gamma,\cdot)^T \in \mathbb{R}^E_{\geq 0}$.

- A potential $u \in \mathbb{R}^V$ (i.e., a function defined on V) may be associated with the usage $\mathcal{N}(u,e) = |u(x) - u(y)|$, whenever $e = \{x,y\} \in E$.

In other words, to each family of objects Γ we associate a $|\Gamma| \times |E|$ *usage matrix* $\mathcal{N}$ such that each row of $\mathcal{N}$ corresponds to an object $\gamma \in \Gamma$ and records the usage of edge e by γ. Note, however, that the families Γ under consideration may very well be infinite. For instance, given a graph $G = (V,E)$ and two nodes $a \neq b \in V$, the *connecting family* $\Gamma(a,b)$ is the family of all walks in G that start at a and end at b:

$$\Gamma(a,b) := \{\gamma : \gamma \text{ is a walk } a \rightsquigarrow b\}.$$

Such connecting families are always infinite families of walks.

10.2 ADMISSIBLE DENSITIES

We will represent the *cost of using* an edge e with a density $\rho \in \mathbb{R}^E_{\geq 0}$. Given an object γ,

$$\ell_\rho(\gamma) := \sum_{e\in E} \mathcal{N}(\gamma,e)\rho(e) = (\mathcal{N}\rho)(\gamma),$$

represents the *ρ-length* (or total usage cost) of γ with respect to the density ρ. Given a family of objects Γ, a density $\rho \in \mathbb{R}^E_{\geq 0}$ is *admissible for the family* Γ, if

$$\ell_\rho(\gamma) \geq 1 \qquad \forall \gamma \in \Gamma.$$

Equivalently, ρ is admissible for Γ if

$$\ell_\rho(\Gamma) := \inf_{\gamma \in \Gamma} \ell_\rho(\gamma) \geq 1. \tag{10.2}$$

In matrix notations, we want

$$\mathcal{N}\rho \geq \mathbf{1}.$$

A good way to remember the property of admissibility is via the motto *"Everybody pays a dollar"*. Namely, when ρ is an admissible density for Γ, then every object $\gamma \in \Gamma$ must pay a cost greater or equal than one.

Define the set of *all admissible densities* $\operatorname{Adm}\Gamma$ as

$$\left\{\rho \in \mathbb{R}^E_{\geq 0} : \mathcal{N}\rho \geq \mathbf{1}\right\} = \bigcap_{\gamma \in \Gamma} \left\{\rho \in \mathbb{R}^E_{\geq 0} : \sum_{e \in E} \mathcal{N}(\gamma, e)\rho(e) \geq 1\right\} \tag{10.3}$$

Note that $\operatorname{Adm}\Gamma$ is an intersection of closed half-spaces, hence it is a closed convex set in $\mathbb{R}^E$.

We say that γ corresponds to a *face* of $\operatorname{Adm}\Gamma$ if there is $\rho \in \mathbb{R}^E_{\geq 0}$ such that $\ell_\rho(\gamma) < 1$, while $\ell_\rho(\gamma') \geq 1$ for every $\gamma' \in \Gamma \setminus \{\gamma\}$. See Section 13.2 for a more general definition of 'face'. In other words, the constraint corresponding to γ is not implied by the other constraints. Notice that $\operatorname{Adm}\Gamma$ might also have faces corresponding to the additional inequalities $\rho(e) \geq 0$ for each $e \in E$. The admissible set is a *polyhedron*, if it has finitely many faces. In particular, if Γ is a finite family, then $\operatorname{Adm}\Gamma$ is a polyhedron. However, we will see later, in Theorem 14.2, that families of walks, while infinite, also have a polyhedral admissible set.

Borrowing from the theory of convex sets we say that ρ is an *extreme point* (or *vertex*) of $\operatorname{Adm}\Gamma$ if $\rho = t\rho_1 + (1-t)\rho_2$ for some $\rho_1, \rho_2 \in \operatorname{Adm}\Gamma$ and some $t \in (0,1)$, implies that $\rho_1 = \rho_2$.

Remark 10.2. Computationally, one way to find extreme points is to choose $|E|$ rows of $\mathcal{N}$ and check if the square submatrix A thus obtained is non-singular. Then, if A is invertible, set $\rho = A^{-1}\mathbf{1}$ and check if ρ is admissible.

Theorem 10.1 (Minkowski). *Suppose K is a compact convex set in $\mathbb{R}^N$. Then K is the convex hull of its extreme points.*

Proof. We will only sketch a proof here. For more details, see [17, Theorem II.3.3 and III.4.1]. Proceed by induction on the dimension of K. In $\mathbb{R}$, every compact convex set with non-empty interior is a bounded interval and in this case the extreme points are the endpoints of the interval. Assume that the result holds for compact convex sets with non-empty interior in $\mathbb{R}^n$ for $n = 1, \dots, N-1$. Let K be a compact convex set in $\mathbb{R}^N$ and assume without loss of generality that $0 \in \operatorname{int}(K)$. If $x_0 \in K$, then either

(a) $x_0 \in \partial K$ or

(b) $x_0 \in \operatorname{int}(K)$.

In case (a), there exists a hyperplane H containing x_0, such that K is all on one side of H (this is called a *supporting hyperplane*) . More specifically, there is a vector $h \in \mathbb{R}^N$ and a scalar α so that $h^T x_0 = \alpha$ and $h^T x \le \alpha$ for all $x \in K$. With these notations $H = \{y \in \mathbb{R}^N : h^T y = \alpha\}$. Taking the span of $K_1 := K \cap H$ about x_0, note that K_1 is a compact convex set with a non-empty interior in some $\mathbb{R}^n$ with $n \le N-1$. Therefore, by induction hypothesis, x_0 is a convex combination of the extreme points of K_1. The key observation is that the extreme points of K_1 are also extreme points of K. To see this, suppose z is an extreme point for K_1 and suppose that $z = tx_1 + (1-t)x_2$ for points $x_1, x_2 \in K$ and some $t \in (0,1)$. Note that $h^T x_j \le \alpha$ for $j = 1, 2$, but by linearity $th^T x_1 + (t-1)h^T x_2 = h^T z = \alpha$, which is only possible if $h^T x_j = \alpha$ for $j = 1, 2$, i.e., if $x_j \in K_1$ for $j = 1, 2$. This proves the claim in case (a).

In case (b), draw a line through x_0, then x_0 is the convex combination of two points on ∂K. Now use part (a). □

Remark 10.3. The set of admissible densities $\operatorname{Adm}\Gamma$ is not only convex but also closed under addition of vectors taken from $\mathbb{R}^E_{\ge 0}$. A variant of Minkowski's theorem holds in this setting, see [69, Theorem 18.5]. Namely,

$$\operatorname{Adm}\Gamma = \operatorname{co}(\operatorname{ext}(\operatorname{Adm}\Gamma)) + \mathbb{R}^E_{\ge 0}.$$

See Proposition 13.1 for an exposition of this fact.

Example 10.1. The path graph P_3 with two edges in series, $e_1 = \{a, b\}$ and $e_2 = \{b, c\}$, is ideal for visualizing the admissible set, because in this case, $\mathbb{R}^E_{\ge 0} = \mathbb{R}^2_{\ge 0}$ is the first quadrant in the plane.

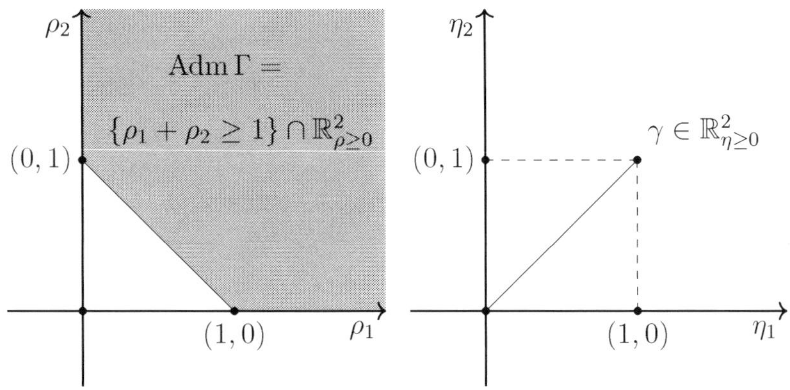

Figure 10.1

Figure 10.1 shows a picture of $\mathrm{Adm}\,\Gamma$ when $\Gamma = \Gamma_{\text{path}}(a,c)$ is the family of all paths from a to c in P_3. In this case, $\Gamma = \{\gamma\}$ consists of a single object $\gamma = a\ e_1\ b\ e_2\ c$. To make sure γ pays at least a dollar, it is enough to require that

$$\rho(e_1) + \rho(e_2) \geq 1. \tag{10.4}$$

Writing $\rho_i = \rho(e_i)$, for $i = 1,2$, it is useful to think of the quadrant $\mathbb{R}^E_{\rho\geq 0} = \{\rho = (\rho_1,\rho_2) : \rho_1,\rho_2 \geq 0\}$ as *ρ-space*. While the usage vectors $\mathcal{N}(\gamma,\cdot)^T \in \mathbb{R}^E_{\eta\geq 0}$ should be thought as belonging to *η-space*. This terminology will become clear later when we describe the theory of Fulkerson duality.

Exercise 10.1 explores the connecting family $\Gamma(a,c)$ of all walks from a to c in P_3, and Exercise 10.2 leads to the construction of a subfamily $\Gamma \subset \Gamma(a,c)$ with the property that $\mathrm{Adm}\,\Gamma$ has an arbitrarily large number of faces.

10.3 THE DEFINITION OF MODULUS

For $1 \leq p < \infty$, the *p-energy* of ρ is defined as

$$\mathcal{E}_{p,\sigma}(\rho) := \sum_{e\in E} \sigma(e)|\rho(e)|^p. \tag{10.5}$$

For $p = \infty$, we also define the *∞-energy* as

$$\mathcal{E}_{\infty,\sigma}(\rho) := \max_{e\in E} \sigma(e)|\rho(e)|$$

We omit the subscript σ when $\sigma \equiv 1$.

The next result shows the behavior of p-energy in the limit, as $p \to \infty$.

Lemma 10.1. *The following holds:*

$$\lim_{p\to\infty} (\mathcal{E}_{p,\sigma}(\rho))^{\frac{1}{p}} = \mathcal{E}_\infty(\rho) \qquad \text{and} \qquad \lim_{p\to\infty} (\mathcal{E}_{p,\sigma^p}(\rho))^{\frac{1}{p}} = \mathcal{E}_{\infty,\sigma}(\rho)$$

Proof. Assume ρ is not identically 0. Find $e_0 \in E$ with $\rho(e_0) = \mathcal{E}_\infty(\rho)$. Then

$$\begin{aligned}
&\mathcal{E}_\infty(\rho)\sigma(e_0)^{1/p} \\
&\le \left(\sum_{e\in E} \sigma(e)|\rho(e)|^p\right)^{1/p} && \text{(add terms for } e \neq e_0) \\
&= (\mathcal{E}_{p,\sigma}(\rho))^{\frac{1}{p}} && \text{(by definition (10.5))} \\
&= \mathcal{E}_\infty(\rho)\left(\sum_{e\in E} \sigma(e)\left(\frac{|\rho(e)|}{\mathcal{E}_\infty(\rho)}\right)^p\right)^{1/p} && \text{(factoring } \mathcal{E}_\infty(\rho) \text{ out)} \\
&\le \mathcal{E}_\infty(\rho)\sigma(E)^{1/p} && \text{(since } |\rho(e)| \le \mathcal{E}_\infty(\rho))
\end{aligned}$$

where $\sigma(E) = \sum_{e\in E}\sigma(e)$.

Now let $p \to \infty$. Then, both $\sigma(e_0)^{1/p}$ and $\sigma(E)^{1/p}$ tend to 1, and the first claim follows.

The second claim follows similarly and is left for the reader in Exercise 10.3. □

Given a graph $G = (V, E, \sigma)$ and a family Γ with usage matrix $\mathcal{N}$, let $1 \le p \le \infty$. Then, the *p-modulus* of Γ is:

$$\mathrm{Mod}_{p,\sigma}(\Gamma) := \inf_{\rho\in \mathrm{Adm}\,\Gamma} \mathcal{E}_{p,\sigma}(\rho) \tag{10.6}$$

Equivalently, p-modulus is the following optimization problem

$$\begin{aligned}
\underset{\rho\in\mathbb{R}^E}{\text{minimize}} \quad & \mathcal{E}_{p,\sigma}(\rho) \\
\text{subject to} \quad & \mathcal{N}\rho \ge \mathbf{1},\ \rho \ge 0.
\end{aligned}$$

where each object γ adds one constraint. Since each constraint is affine and the objective function is convex, we see that p-modulus is a convex optimization problem.

Example 10.2. Recall Example 10.1. Consider the ℓ_p-norm balls centered at the origin

$$B_p(r) = \{\rho \in \mathbb{R}^2 : \|\rho\|_p \leq r\}.$$

Inflate the radius r until the first time $B_p(r)$ touches $\operatorname{Adm}\Gamma$ as defined in (10.4), see Figure 10.1. Depending on the shape of $B_p(r)$ the contact point should be different. In this case, however, by symmetry, the closest point to the origin is always $(1/2, 1/2)$, independent of p.

10.4 BASIC PROPERTIES

There are two special cases. If some object has zero usage, then the modulus is infinite. On the other hand, the modulus is zero, if and only if, the family of objects is empty.

Proposition 10.1. *Let Γ be a family of objects on a graph $G = (V, E, \sigma)$. Then, for $1 \leq p \leq \infty$:*

1. *If for some $\gamma \in \Gamma$, the usage is $\mathcal{N}(\gamma, \cdot) \equiv 0$, then*

$$\operatorname{Mod}_{p,\sigma}(\Gamma) = \infty.$$

2. *$\Gamma = \emptyset$ if and only if $\operatorname{Mod}_{p,\sigma}(\Gamma) = 0$.*

Proof. For 1., if γ has zero usage, then no density can make γ pay a dollar. Hence, no density can be admissible for Γ, and the minimization is infeasible. So, the infimum is infinite.

For 2., if $\Gamma = \emptyset$, then every density $\rho \in \mathbb{R}^E_{\geq 0}$ is admissible. So $\operatorname{Mod}_{p,\sigma}(\Gamma) = 0$. Conversely, suppose $\Gamma \neq \emptyset$ and $\operatorname{Mod}_{p,\sigma}(\Gamma) = 0$. Then, there is a sequence $\rho_n \in \operatorname{Adm}\Gamma$ such that $\mathcal{E}_{p,\sigma}(\rho_n) \to 0$, as $n \to \infty$. Fixing an edge $e \in E$, since $\sigma(e) > 0$, we have

$$\rho_n(e) \leq \frac{1}{\sigma(e)} \mathcal{E}_{p,\sigma}(\rho_n)$$

and hence,

$$\sup_{e \in E} \rho_n(e) \leq \left(\sup_{e \in E} \frac{1}{\sigma(e)} \right) \mathcal{E}_{p,\sigma}(\rho_n) \to 0,$$

as $n \to \infty$.

Since $\Gamma \neq \emptyset$, we can pick an object $\gamma_0 \in \Gamma$. Then,

$$\begin{aligned} \ell_{\rho_n}(\gamma_0) &= \sum_{e \in E} \mathcal{N}(\gamma_0, e) \rho_n(e) \\ &\leq \left(\sup_{e \in E} \rho_n(e) \right) \sum_{e \in E} \mathcal{N}(\gamma_0, e) \to 0, \end{aligned}$$

as $n \to \infty$. This contradicts the fact that ρ_n is admissible. So the modulus can't be zero. □

Remark 10.4. Note that even if no usage vector is identically zero, the modulus could still be infinite. Indeed, assume that there is a sequence $\gamma_n \in \Gamma$ with usage $\mathcal{N}(\gamma_n, e) = \epsilon_n \mathbb{1}_{e_n}(e)$, where $e_n \in E$ and $\epsilon_n \to 0$. Since E is finite, we can assume that $e_n \equiv e_0$, by passing to a subsequence. So, if $\rho \in \operatorname{Adm}\Gamma$, then

$$\ell_\rho(\gamma_n) = \mathcal{N}(\gamma_n, e_0)\rho(e_0) = \epsilon_n \rho(e_0) \geq 1.$$

Hence, $\rho(e_0) \geq \epsilon_n^{-1} \to \infty$. So again, the modulus problem is infeasible in this case.

In order to avoid these trivial cases, we make the following standing assumptions, which will always be assumed to hold, unless otherwise stated, for the remainder of this book.

Standing Assumptions. If Γ is a family of objects on $G = (V, E, \sigma)$, we will assume

(1) **Non-emptiness:**

$$\Gamma \neq \emptyset. \tag{10.7}$$

(2) **Non-triviality:** Every object has non-trivial usage:

$$\mathcal{N}(\gamma, \cdot) \not\equiv 0 \qquad \forall \gamma \in \Gamma. \tag{10.8}$$

(3) **Usage gap:** There is a gap $\delta > 0$, so that given an object $\gamma \in \Gamma$ and an edge $e \in E$, either $\mathcal{N}(\gamma, e) = 0$ or $\mathcal{N}(\gamma, e) \geq \delta$.

These three assumptions are equivalent to requiring that

$$\mathcal{N}_{\min} := \inf_{\gamma \in \Gamma} \min_{e \in E : \mathcal{N}(\gamma, e) \neq 0} \mathcal{N}(\gamma, e) > 0. \tag{10.9}$$

Remark 10.5. The usage gap property (3) above follows from (2) if Γ is finite, and follows from (1) if Γ is a family of walks. Under this assumption the modulus is finite, as the following corollary shows. Also, note that most common families of objects, such as path families and families with 0/1 usage will have $\mathcal{N}_{\min} = 1$.

Corollary 10.1. *Let $G = (V, E, \sigma)$ be a weighted graph, with $\sigma \in \mathbb{R}^E_{>0}$, and let Γ be a non-empty non-trivial family of objects on G, satisfying $\mathcal{N}_{\min} > 0$ as in (10.9). Then,* $\operatorname{Adm}\Gamma \neq \emptyset$ *and*

$$0 < \operatorname{Mod}_{p,\sigma}(\Gamma) < \infty,$$

for all $p \in [1, \infty]$.

Proof. We prove the case $p < \infty$. The case $p = \infty$ is similar.

Set $\rho \equiv C$. Then, for every $\gamma \in \Gamma$,

$$\ell_\rho(\gamma) = \sum_{e \in E} \mathcal{N}(\gamma, e)\rho(e) \geq C\mathcal{N}_{\min}.$$

So, as soon as $C \geq \mathcal{N}_{\min}^{-1}$, we have have $\rho \in \operatorname{Adm}\Gamma$. This shows that

$$\operatorname{Mod}_{p,\sigma}(\Gamma) \leq \sigma(E)\mathcal{N}_{\min}^{-p} < \infty.$$

The fact that $\operatorname{Mod}_{\sigma,p}(\Gamma) > 0$ follows from (2) in Proposition 10.1 and the fact that Γ is non-empty. □

10.5 EXISTENCE AND UNIQUENESS OF AN EXTREMAL DENSITY

Proposition 10.2. *Let Γ be as in Corollary 10.1. Then*

1. *For $1 \leq p \leq \infty$ a minimizer ρ^* for (10.6) always exists. We call such an optimal ρ^* an* extremal density.

2. *For $1 \leq p < \infty$, all extremal densities satisfy*

$$\begin{aligned} &0 \leq \rho^* \leq 1, && \text{if } \mathcal{N}_{\min} = 1, \text{ and} \\ &0 \leq \rho^* \leq \mathcal{N}_{\min}^{-1}, && \text{in general,} \end{aligned} \tag{10.10}$$

   coordinate-wise. While, for $p = \infty$, there always is an extremal density satisfying (10.10).

3. *For $1 < p < \infty$, the extremal density ρ^* is unique.*

Proof. For 1., note that, as in the proof of Corollary 10.1, setting $\rho \equiv C$ for $C \geq \mathcal{N}_{\min}^{-1}$, yields $\rho \in \operatorname{Adm}\Gamma$. Therefore, we can choose $R > 0$ large enough so that $K := \operatorname{Adm}\Gamma \cap \{\rho \in \mathbb{R}^E_{\geq 0} : \mathcal{E}_{p,\sigma}(\rho) \leq R\} \neq \emptyset$. Note that K is a closed and bounded set and $\mathbb{R}^E$ is finite dimensional. So K is compact. Therefore, since the p-norms are continuous, there always is a minimizer.

For 2., suppose ρ^* is extremal for $\mathrm{Mod}_{p,\sigma}(\Gamma)$ and assume that $\rho^*(e_0) > \mathcal{N}_{\min}^{-1}$ on some edge $e_0 \in E$. Define, a new density $\rho(e) := \rho^*(e)$ for $e \neq e_0$ and $\rho(e_0) := \mathcal{N}_{\min}^{-1}$. We claim that ρ is admissible for Γ. Indeed, if an object γ does not use e_0, then $\ell_\rho(\gamma) = \ell_{\rho^*}(\gamma) \geq 1$. On the other hand, if $\mathcal{N}(\gamma, e_0) > 0$, then

$$\ell_\rho(\gamma) = \sum_{e \neq e_0} \mathcal{N}(\gamma, e)\rho(e) + \mathcal{N}(\gamma, e_0)\mathcal{N}_{\min}^{-1} \geq 1.$$

Hence, for $1 \leq p < \infty$,

$$\mathcal{E}_{p,\sigma}(\rho) = \sum_{e \neq e_0} \sigma(e)|\rho(e)|^p + \sigma(e_0)\mathcal{N}_{\min}^{-p} < \mathcal{E}_{p,\sigma}(\rho^*)$$

which contradicts the extremality of ρ^*. When $p = \infty$, we only get

$$\mathcal{E}_{\infty,\sigma}(\rho) = \max_{e \in E} \sigma(e)\rho(e) \leq \mathcal{E}_{\infty,\sigma}(\rho^*),$$

so ρ is extremal as well and satisfies (10.10).

For 3., assume that $\rho_1 \neq \rho_2$ are two extremal densities. By convexity of $\mathrm{Adm}\,\Gamma$,

$$\rho_0 := \frac{\rho_1 + \rho_2}{2}$$

is also admissible. However, for $p \in (1, \infty)$ the functional $\mathcal{E}_{p,\sigma}(\rho)$ is strictly convex. So

$$\mathcal{E}_{p,\sigma}(\rho_0) < \frac{\mathcal{E}_{p,\sigma}(\rho_1) + \mathcal{E}_{p,\sigma}(\rho_2)}{2} = \mathrm{Mod}_{p,\sigma}(\Gamma)$$

which contradicts the extremality of ρ_1 and ρ_2. □

Next, we discuss a very useful classical criterion for deciding whether a density is extremal [2, Theorem 4-4, p. 61].

10.6 BEURLING'S CRITERION

If $\Gamma' \subset \Gamma$, we write $\mathcal{N}(\Gamma')$ for the submatrix of $\mathcal{N}$ consisting of those rows that correspond to objects $\gamma' \in \Gamma'$.

Theorem 10.2 (Beurling's Criterion for Extremality). *Let $G = (V, E, \sigma)$ be a weighted graph, and let Γ be as in Corollary 10.1. Also, assume $1 < p < \infty$. Then, $\rho \in \mathbb{R}^E_{\geq 0}$ is the extremal density for $\mathrm{Mod}_{p,\sigma}(\Gamma)$ if*

1. *$\rho \in \mathrm{Adm}(\Gamma)$;*
2. *there is a subfamily $\tilde{\Gamma} \subset \Gamma$ such that*
   (a) *$\ell_\rho(\gamma) = \sum_{e\in E} \mathcal{N}(\gamma, e)\rho(e) = 1$ for all $\gamma \in \tilde{\Gamma}$, in other words, $\mathcal{N}(\tilde{\Gamma})\rho = \mathbf{1}$;*
   (b) *whenever $h \in \mathbb{R}^E$ and $\mathcal{N}(\tilde{\Gamma})h \geq 0$, then*
$$\sum_{e\in E} \sigma(e)h(e)\rho^{p-1}(e) \geq 0.$$

Furthermore, for such a subfamily $\tilde{\Gamma}$ we have

$$\mathrm{Mod}_{p,\sigma}(\tilde{\Gamma}) = \mathrm{Mod}_{p,\sigma}(\Gamma). \tag{10.11}$$

Definition 10.1. If $\tilde{\Gamma} \subset \Gamma$ is a subfamily satisfying 2(a) and 2(b) in Theorem 10.2, for some admissible density ρ, we say that $\tilde{\Gamma}$ is a *Beurling subfamily* for Γ.

Remark 10.6. Although, the extremal density ρ is unique when $1 < p < \infty$, Beurling subfamilies for Γ may not be unique, see Exercise 10.8.

Note that the function h in 2(b) in Theorem 10.2, is in $\mathbb{R}^E$ and not $\mathbb{R}^E_{\geq 0}$. The proof is well-known, but we reproduce it here for completeness.

Proof of Theorem 10.2. By 1., we have that ρ is admissible. So it's enough to show that it minimizes the energy in $\mathrm{Adm}\,\Gamma$. To that end, let ρ' be an arbitrary admissible density for Γ. Set $h := \rho' - \rho$. If $\gamma \in \tilde{\Gamma}$, then, by 2(a),

$$\ell_h(\gamma) = \ell_{\rho'}(\gamma) - \ell_\rho(\gamma) = \ell_{\rho'}(\gamma) - 1 \geq 0.$$

Thus, assumption 2(b) implies that $\sum_{e\in E} h(e)\rho^{p-1}(e)\sigma(e) \geq 0$. In other words,

$$\sum_{e\in E} \rho'(e)\rho^{p-1}(e)\sigma(e) - \sum_{e\in E} \rho^p(e)\sigma(e) \geq 0. \tag{10.12}$$

Let q be the Hölder conjugate exponent of p, so that $pq = p + q$. Then,

$$\begin{aligned}
&\mathcal{E}_{p,\sigma}(\rho) \\
&= \sum_{e\in E} \sigma(e)\rho(e)^p \\
&\leq \sum_{e\in E} \rho'(e)\rho^{p-1}(e)\sigma(e) && \text{(by (10.12))} \\
&= \sum_{e\in E} \left(\rho'(e)\sigma(e)^{1/p}\right)\left(\rho^{p-1}(e)\sigma(e)^{1/q}\right) \\
&\leq \left(\sum_{e\in E}(\rho'(e))^p\sigma(e)\right)^{1/p}\left(\sum_{e\in E}\rho(e)^p\sigma(e)\right)^{1-1/p} && \text{(Hölder's inequality.)}
\end{aligned}$$

Dividing both sides $\mathcal{E}_{p,\sigma}(\rho)^{1-1/p}$ and raising to the p-th power, we get that $\mathcal{E}_{p,\sigma}(\rho) \leq \mathcal{E}_{p,\sigma}(\rho')$. This shows that ρ is extremal.

To prove (10.11), note that assumption 2(b) only requires $\mathcal{N}(\tilde{\Gamma})h \geq 0$, so ρ also satisfies Beurling's criterion for $\mathrm{Mod}_{p,\sigma}(\tilde{\Gamma})$, with $\tilde{\Gamma}$ as its own Beurling subfamily. Therefore, $\mathrm{Mod}_{p,\sigma}(\Gamma) = \mathcal{E}_{p,\sigma}(\rho) = \mathrm{Mod}_{p,\sigma}(\tilde{\Gamma})$. □

Example 10.3. Consider the graph in Figure 10.2 and let $\Gamma = \Gamma(a, c)$ be the family of all walks from a to c. We want to compute $\mathrm{Mod}_2(\Gamma)$. Here we make an educated guess that $\rho(ab) = \rho(bc) = \rho(ad) = \rho(dc) = 1/2$ and $\rho(bd) = 0$. To check that this ρ is extremal we use Beurling's criterion.

First note that $\rho \in \mathrm{Adm}(\Gamma)$ since every walk from a to c must use one of ab or ad as well as one of bc or dc. Now define the subfamily $\tilde{\Gamma} := \{\gamma_1 = a\ b\ c,\ \gamma_2 = a\ d\ c\}$. Note that $\ell_\rho(\gamma_1) = \ell_\rho(\gamma_2) = 1$. So we only need to check assumption (2)(b) in Theorem 10.2. Let $h \in \mathbb{R}^E$ satisfy $\ell_h(\gamma_j) \geq 0$ for $j = 1, 2$. This means that $h(ab) + h(bc) \geq 0$ and

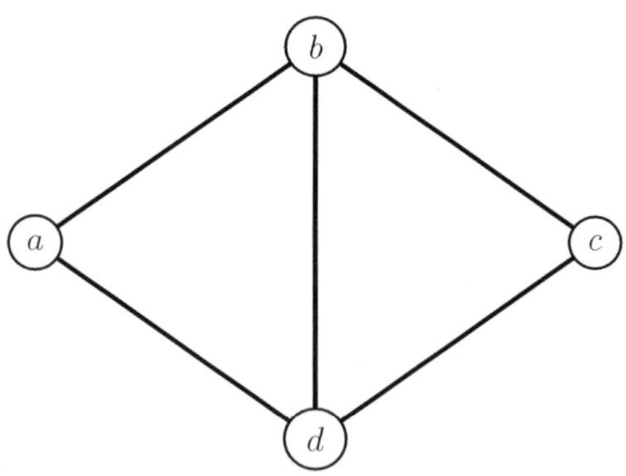

Figure 10.2 The diamond graph.

$h(ad)+h(dc) \geq 0$. We want to show that this implies $\sum_{e\in E} h(e)\rho(e) \geq 0$. Indeed,

$$\begin{aligned}\sum_{e\in E} h(e)\rho(e) &= h(ab)\rho(ab) + h(bc)\rho(bc) + h(ad)\rho(ad) + h(dc)\rho(dc) \\ &= \frac{1}{2}\left[h(ab) + h(bc)\right] + \frac{1}{2}\left[h(ad) + h(dc)\right] \geq 0.\end{aligned}$$

10.7 AN ALTERNATIVE DEFINITION

In the complex analysis literature, modulus is usually defined via its reciprocal, which is known as the *extremal length* of Γ. This approach focuses more on the shortest ρ-length $\ell_\rho(\Gamma)$ defined in (10.2). In particular, the admissible set $\operatorname{Adm}\Gamma$ is replaced by the set

$$S(\Gamma) := \left\{\rho \in \mathbb{R}^E_{\geq 0} : \ell_\rho(\Gamma) > 0\right\}, \tag{10.13}$$

of densities ρ that give Γ non-zero shortest length.

Proposition 10.3. *Let $G = (V, E, \sigma)$ be a weighted graph, and let Γ be as in Corollary 10.1. Let $S(\Gamma)$ be the set of densities defined in (10.13). Then, for $1 \leq p < \infty$,*

$$\operatorname{Mod}_{p,\sigma}(\Gamma) = \inf_{\rho\in S(\Gamma)} \frac{\mathcal{E}_{p,\sigma}(\rho)}{\ell_\rho(\Gamma)^p}; \tag{10.14}$$

and for $p = \infty$,

$$\operatorname{Mod}_\infty(\Gamma) = \inf_{\rho\in S(\Gamma)} \frac{\mathcal{E}_\infty(\rho)}{\ell(\Gamma)},$$

where $\ell(\Gamma)$ is $\ell_\rho(\Gamma)$ for $\rho \equiv 1$.

Proof. Suppose $\rho \in \operatorname{Adm}\Gamma$. Then, $\ell_\rho(\Gamma) \geq 1$, hence $\rho \in S(\Gamma)$. Namely, $\operatorname{Adm}\Gamma \subset S(\Gamma)$. So

$$\begin{aligned}\inf_{\rho\in S(\Gamma)} \frac{\mathcal{E}_{p,\sigma}(\rho)}{\ell_\rho(\Gamma)^p} &\leq \inf_{\rho\in \operatorname{Adm}\Gamma} \frac{\mathcal{E}_{p,\sigma}(\rho)}{\ell_\rho(\Gamma)^p} && \text{(since } \operatorname{Adm}\Gamma \subset S(\Gamma)\text{)} \\ &\leq \inf_{\rho\in \operatorname{Adm}\Gamma} \mathcal{E}_{p,\sigma}(\rho) && \text{(since } \ell_\rho(\Gamma) \geq 1\text{)} \\ &= \operatorname{Mod}_{p,\sigma}(\Gamma). && \text{(by (10.6))}\end{aligned}$$

Conversely, suppose $\rho \in S(\Gamma)$. Then, $\ell_\rho(\Gamma) > 0$. Set

$$\tilde{\rho} := \frac{\rho}{\ell_\rho(\Gamma)}.$$

Then $\tilde{\rho} \in \operatorname{Adm}\Gamma$. So

$$\operatorname{Mod}_{p,\sigma}(\Gamma) \leq \mathcal{E}_{p,\sigma}(\tilde{\rho}) = \frac{\mathcal{E}_{p,\sigma}(\rho)}{\ell_\rho(\Gamma)^p}.$$

Since $\rho \in S(\Gamma)$ is arbitrary, we get (10.14). □

10.8 MODULUS IS A CAPACITY

We show that p-modulus is Γ-monotone and subadditive. These two properties often define a notion of capacity.

Proposition 10.4. *Let Γ, Γ' and Γ_j, $j = 1, 2, \ldots$, be families of objects. Then, for $1 \leq p \leq \infty$, we have the following:*

1. **Γ-monotonicity:** *If $\Gamma \subset \Gamma'$, then*

$$\operatorname{Mod}_{p,\sigma}(\Gamma) \leq \operatorname{Mod}_{p,\sigma}(\Gamma'). \tag{10.15}$$

2. **Countable subadditivity:**

$$\operatorname{Mod}_{p,\sigma}\left(\bigcup_{j=1}^{\infty} \Gamma_j\right) \leq \sum_{j=1}^{\infty} \operatorname{Mod}_{p,\sigma}(\Gamma_j).$$

Proof. For 1., note that if $\Gamma \subset \Gamma'$, then $\operatorname{Adm}\Gamma \supset \operatorname{Adm}\Gamma'$.

For 2., assume that $1 \leq p < \infty$ and that $\operatorname{Mod}_{p,\sigma}(\Gamma_j) < \infty$ for all j's. Fix $\epsilon > 0$ and pick $\rho_j \in \operatorname{Adm}(\Gamma_j)$ such that

$$\mathcal{E}_{p,\sigma}(\rho_j) \leq \operatorname{Mod}_{p,\sigma}(\Gamma_j) + \epsilon 2^{-j},$$

for $j = 1, \ldots$. Set $\rho := \left(\sum_{j=1}^{\infty} \rho_j^p\right)^{1/p}$, and let $\Gamma = \cup_j \Gamma_j$. Pick $\gamma \in \Gamma$. Then $\gamma \in \Gamma_j$ for some j. Note that $\rho \geq \rho_j$, so $\ell_\rho(\gamma) \geq \ell_{\rho_j}(\gamma) \geq 1$. Therefore $\rho \in \operatorname{Adm}\Gamma$ and

$$\begin{aligned}
\operatorname{Mod}_{p,\sigma}(\Gamma) &\leq \mathcal{E}_{p,\sigma}(\rho) \\
&= \sum_{e \in E} \sum_{j=1}^{\infty} \sigma(e) \rho_j(e)^p \\
&= \sum_{j=1}^{\infty} \mathcal{E}_{p,\sigma}(\rho_j) \\
&\leq \sum_{j=1}^{\infty} \operatorname{Mod}_{p,\sigma}(\Gamma_j) + \epsilon.
\end{aligned}$$

□

A similar argument works when $p = \infty$, and in fact we have the stronger estimate

$$\mathrm{Mod}_{\infty}(\Gamma) \leq \max_{j} \mathrm{Mod}_{\infty}(\Gamma_j). \tag{10.16}$$

We call this an ultrametric property, see Exercise 10.5.

10.9 LESS USAGE PROPERTY

Definition 10.2. We say that an object γ_1 is *subordinate* to an object γ_2, and write $\gamma_1 \preceq \gamma_2$ if

$$\mathcal{N}(\gamma_1, e) \leq \mathcal{N}(\gamma_2, e) \qquad \forall e \in E.$$

Also a family of objects Γ_1 is *subordinate* to Γ_2, and write $\Gamma_1 \preceq \Gamma_2$ if for every $\gamma_2 \in \Gamma_2$, there is an object $\gamma_1 \in \Gamma_1$ such that $\gamma_1 \preceq \gamma_2$.

Proposition 10.5. *Let Γ_1, Γ_2 be two families of objects.*

$$\textit{If } \Gamma_1 \preceq \Gamma_2, \textit{ then } \mathrm{Mod}_{p,\sigma}(\Gamma_1) \geq \mathrm{Mod}_{p,\sigma}(\Gamma_2).$$

Proof. First note that if γ_1 and γ_2 are objects with $\gamma_1 \preceq \gamma_2$, and $\rho \in \mathbb{R}^E_{\geq 0}$ is an arbitrary density, then

$$\ell_\rho(\gamma_1) = \sum_{e \in E} \mathcal{N}(\gamma_1, e)\rho(e) \leq \sum_{e \in E} \mathcal{N}(\gamma_2, e)\rho(e) = \ell_\rho(\gamma_2).$$

Therefore,

$$\ell_\rho(\gamma_1) \leq \ell_\rho(\gamma_2). \tag{10.17}$$

Now, suppose $\rho \in \mathrm{Adm}(\Gamma_1)$. For every object $\gamma_2 \in \Gamma_2$, there is an object $\gamma_1 \in \Gamma_1$ such that $\gamma_1 \preceq \gamma_2$. Hence, by (10.17), $1 \leq \ell_\rho(\gamma_1) \leq \ell_\rho(\gamma_2)$. Therefore, $\rho \in \mathrm{Adm}(\Gamma_2)$. this shows that $\mathrm{Adm}(\Gamma_1) \subset \mathrm{Adm}(\Gamma_2)$ and $\mathrm{Mod}_{p,\sigma}(\Gamma_1) \geq \mathrm{Mod}_{p,\sigma}(\Gamma_2)$. □

Example 10.4. Consider the "via" family $\Gamma(a, b; c)$ consisting of all walks that start at a, end at b, and visit c. Then if $\Gamma(a, b)$ is the family of all walks that start at a and end at b, we have

$$\Gamma(a, b) \preceq \Gamma(a, b; c).$$

And therefore $\mathrm{Mod}_{p,\sigma}(\Gamma(a, b)) \geq \mathrm{Mod}_{p,\sigma}(\Gamma(a, b; c)$.

10.10 THE BASIC ESTIMATE

The general philosophy is that good upper bounds for modulus are easier to obtain because it is enough to find an admissible density and then compute its energy. In some cases, the geometry of the problem can help choose appropriate admissible densities that yield tight upper bounds for modulus. Here we give a very standard upper bound.

Given a family of objects Γ as in Corollary 10.1, define the smallest row sum of the usage matrix as

$$\ell(\Gamma) := \inf_{\gamma\in\Gamma} \sum_{e\in E} \mathcal{N}(\gamma, e). \tag{10.18}$$

Then consider the constant density

$$\rho_0 \equiv \frac{1}{\ell(\Gamma)},$$

which is admissible, since

$$\sum_{e\in E} \mathcal{N}(\gamma, e)\rho_0(e) = \frac{1}{\ell(\Gamma)} \sum_{e\in E} \mathcal{N}(\gamma, e) \geq 1. \tag{10.19}$$

This choice of density yields the following upper bound:

$$\text{Mod}_{p,\sigma}(\Gamma) \leq \mathcal{E}_{p.\sigma}(\rho_0) = \frac{\sigma(E)}{\ell(\Gamma)^p}. \tag{10.20}$$

Example 10.5. Consider Γ_{spt}, the family of all spanning trees of a simple graph G. In this case, $\ell(\Gamma) = N - 1$. Hence, assuming $p = 2$ and $\sigma \equiv 1$, we get

$$\text{Mod}_2(\Gamma_{\text{spt}}) \leq \frac{|E|}{(N-1)^2} \leq \frac{1}{2}\frac{N}{N-1}.$$

10.11 SYMMETRY

Given a graph $G = (V, E, \sigma)$ a *symmetry* is an involution $T : E \to E$ such that

$$\sigma(T(e)) = \sigma(e) \qquad \forall e \in E. \tag{10.21}$$

We say that a family of objects Γ is *T-invariant* if for every object $\gamma_1 \in \Gamma$ there is an object $\gamma_2 \in \Gamma$ such that

$$\mathcal{N}(\gamma_2, T(e)) = \mathcal{N}(\gamma_1, e) \qquad \forall e \in E. \tag{10.22}$$

Finally, we set

$$\mathrm{Adm}_T\,\Gamma := \{\rho \in \mathrm{Adm}\,\Gamma : \rho(e) = \rho(T(e)),\ \forall e \in E\}, \tag{10.23}$$

as the set of T-invariant admissible densities.

Proposition 10.6. *Let T be a symmetry for the graph $G = (V, E, \sigma)$. Assume that Γ is a family of objects on G, as in Corollary 10.1, that is also T-invariant. Then, for $1 \le p < \infty$,*

$$\mathrm{Mod}_{p,\sigma}(\Gamma) = \inf_{\rho \in \mathrm{Adm}_T(\Gamma)} \mathcal{E}_{p.\sigma}(\rho).$$

Proof. Suppose ρ is extremal for $\mathrm{Mod}_{p,\sigma}(\Gamma)$, so that in particular $\rho \in \mathrm{Adm}\,\Gamma$. Define $\rho_1 := \rho \circ T$. Then by (10.22), for every $\gamma_1 \in \Gamma$ there is $\gamma_2 \in \Gamma$ such that

$$\ell_{\rho_1}(\gamma_1) = \sum_{e \in E} \mathcal{N}(\gamma_1, e)\rho_1(e) = \sum_{e \in E} \mathcal{N}(\gamma_2, T(e))\rho(T(e)) = \ell_\rho(\gamma_2) \ge 1.$$

Therefore, $\rho_1 \in \mathrm{Adm}\,\Gamma$. Moreover, by (10.21),

$$\begin{aligned}\mathcal{E}_{p,\sigma}(\rho_1) &= \sum_{e \in E} \sigma(e)|\rho_1(e)|^p \\ &= \sum_{e \in E} \sigma(T(e))|\rho(T(e))|^p \\ &= \mathcal{E}_{p,\sigma}(\rho).\end{aligned}$$

Define $\rho_2 := (\rho + \rho_1)/2$. Then $\rho_2 \in \mathrm{Adm}\,\Gamma$. Also, since T is an involution, we have $\rho_2 \circ T = \rho_2$. So $\rho_2 \in \mathrm{Adm}_T(\Gamma)$. By convexity,

$$\mathcal{E}_{p,\sigma}(\rho_2) \le \frac{\mathcal{E}_{p.\sigma}(\rho) + \mathcal{E}_{p,\sigma}(\rho_1)}{2} = \mathrm{Mod}_{p,\sigma}(\Gamma).$$

Therefore

$$\inf_{\rho \in \mathrm{Adm}_T(\Gamma)} \mathcal{E}_{p.\sigma}(\rho) \le \mathrm{Mod}_{p,\sigma}(\Gamma).$$

The other direction follows from the fact that $\mathrm{Adm}_T(\Gamma) \subset \mathrm{Adm}\,\Gamma$. $\square$

For an example using symmetry, see Exercise 10.7.

EXERCISES

10.1 *The connecting family on the Path graph.* Here we consider the setting of Example 10.1, where G is the path graph P_3, with two edges $e_1 = \{a, b\}$ and $e_2 = \{b, c\}$ in series. Let $\Gamma = \Gamma(a, c)$ be the connecting family of all walks that start at a and end at c.

(i) Show that every walk $\gamma \in \Gamma$ uses edge e_1 an odd number n of times and the edge e_2 another odd number m of times, where $n, m \geq 1$.
Hint: Consider all walks that take an even number of steps starting at a and use induction on the number of steps.

(ii) Show that for every pair of odd integers $m, n \in \mathbb{Z}_{\geq 1}$ the vector $[n\ m]^T$ is the usage vector for some walk $\gamma \in \Gamma$. Also, show that the corresponding line

$$n\rho_1 + m\rho_2 = 1$$

intersects the coordinate axis at $(1/n, 0)$ and $(0, 1/m)$.

(iii) Deduce that the constraint corresponding to the usage vector $[1\ 1]^T$ implies every other constraint in the family Γ, hence

$$\operatorname{Adm}\Gamma = \left\{\rho \in \mathbb{R}^2_{\geq 0} : \rho_1 + \rho_2 \geq 1\right\}.$$

(iv) Finally, show that the path with usage vector $[1\ 1]^T$ is a face of $\operatorname{Adm}\Gamma$.

10.2 *Admissible sets with an arbitrarily large number of faces.* The goal of this exercise is to construct an example of an admissible set with an arbitrarily large number of extreme points and faces.

Consider again the unweighted path graph P_3 with two edges e_1 and e_2 in series. Let $n = 1, 3, 5, ...$ be an odd integer. Define the family of walks $\Gamma_n = \{\gamma_{n,k}\}_{k=0}^{(n-1)/2}$ with usages

$$\mathcal{N}(\gamma_{n,k}, e_1) = n - 2k \quad \text{and} \quad \mathcal{N}(\gamma_{n,k}, e_2) = f_n(n - 2k) \qquad \text{(P10.1)}$$

where $f_n(n) = n$ and $f_n(n - 2k) := n + 2\sum_{j=1}^{k}(j + 1)$, for $k = 1, \ldots, (n - 1)/2$.

(i) Show that $f_n(n - 2k)$ is odd for all k's. In particular, deduce from Exercise 10.1 (ii) that there exist walks $\gamma_{n,k}$ with usage vector $\mathcal{N}(\gamma_{n,k}, \cdot)$ as defined in (P10.1).

(ii) Consider the lines $L_{n,k} = \left\{\rho : \sum_{i=1}^{2} \mathcal{N}(\gamma_{n,k}, e_i)\rho(e_i) = 1\right\}$ and find the point (x_k, y_k) where $L_{n,k}$ and $L_{n,k+1}$ meet by solving the system

$$\begin{array}{l} (1) \\ (2) \end{array} \left\{ \begin{array}{lcll} x(n-2k) & + & y f_n(n-2k) & = 1 \\ x(n-2k-2) & + & y\left[f_n(n-2k) + 2(k+2)\right] & = 1 \end{array} \right.$$

where $x = \rho(e_1)$ and $y = \rho(e_2)$.

(iii) Show that each line $L_{n,k}$ becomes more and more horizontal as k varies, and that y_k decreases. Deduce that the admissible set $\mathrm{Adm}(\Gamma_n)$ must have at least $(n+1)/2$ faces.

10.3 *The behavior of p-energy as p tends to infinity.* Complete the proof of Lemma 10.1, by showing that for $\rho \in \mathbb{R}^E$:

$$\lim_{p\to\infty} \left(\mathcal{E}_{p,\sigma^p}(\rho)\right)^{\frac{1}{p}} = \mathcal{E}_{\infty,\sigma}(\rho).$$

10.4 *Beurling criterion.* In Exercise 10.2, we defined a family of walks $\Gamma_n = \{\gamma_{n,k}\}_{k=0}^{(n-1)/2}$ with n odd, and showed that the polyhedron of admissible densities $\mathrm{Adm}(\Gamma_n)$ has at least $(n+1)/2$ faces. Here, we will compute $\mathrm{Mod}_p(\Gamma_n)$ using Beurling's criterion and the candidate extremal density $\rho_n \equiv \frac{1}{2n}$.

(i) Show that by (P10.1) the walks in Γ_n have edge usage

$$\mathcal{N}(\gamma_{n,k}, e_1) = n-2k \qquad \text{and} \qquad \mathcal{N}(\gamma_{n,k}, e_2) = n+2k+k(k+1),$$

for $k = 0, \ldots, (n-1)/2$.

(ii) Show that ρ_n is admissible for Γ_n and $\ell_{\rho_n}(\gamma_{n,0}) = 1$.

(iii) Let the subfamily $\tilde{\Gamma}_n$ consist of a single walk $\{\gamma_{n,0}\}$. Assume that $h \in \mathbb{R}^E$ and $\ell_h(\rho_n) \geq 0$. Show that

$$\sum_{e\in E} h(e)\rho_n(e)^{p-1} \geq 0.$$

and deduce that ρ_n is extremal for Γ_n.

(iv) Finally, compute the value of $\mathrm{Mod}_p(\Gamma_n)$.

10.5 *Ultrametric property of infinity modulus.* Prove the inequality in (10.16).

10.6 *Modulus bound for the shortest path distance.* Let $G = (V, E)$ be an unweighted graph, and let $\Gamma(a, b)$ be the family all walks connecting a and b in G. Let $d_{\mathrm{SP}}(a, b)$ be the shortest path distance from a to b defined in (1.3). Use the basic modulus estimate (10.20) to show that

$$d_{\mathrm{SP}}(a,b) \leq |E|^{1/p}\,\mathrm{Mod}_p(\Gamma(a,b))^{-1/p}, \qquad \text{for all } p \in [1,\infty).$$

10.7 *Spanning tree modulus for the diamond graph.* Let G be the diamond graph, as in Figure 1.1 (b), and consider the family Γ_{spt} of all spanning trees.

(i) Draw the 8 spanning trees in Γ_{spt}. There should be 4 that include the diagonal and 4 that don't.

(ii) Let T_1 be the symmetry that flips the two vertices on the given diagonal and T_2 the symmetry that flips the other two vertices. Check that Γ_{spt} is invariant under both T_1 and T_2, and deduce that the densities that are invariant with respect to both T_1 and T_2 (in the sense of (10.23)) must be constant on the sides of the square.

(iii) Show that computing $\mathrm{Mod}_2(\Gamma_{\mathrm{spt}})$ is equivalent to solving the following optimization:

$$\begin{aligned} \text{minimize} \quad & 4\alpha^2 + \beta^2 \\ \text{subject to} \quad & 3\alpha \geq 1, \\ & 2\alpha + \beta \geq 1. \end{aligned}$$

Then solve the problem using Lagrangian duality and the KKT conditions.

10.8 *Non-uniqueness of Beurling subfamilies* Consider the plane graph G in Figure 10.3. Assume that the center node is at $(0, 0)$ and any two nodes are one unit away on the integer lattice. Let $A = \{(0,-2),(-1,-1),(-2,0)\}$ and $B = \{(2,0),(1,1),(0,2)\}$ and let $\Gamma = \Gamma(A, B)$, the family of all walks in G starting on A and ending on B.

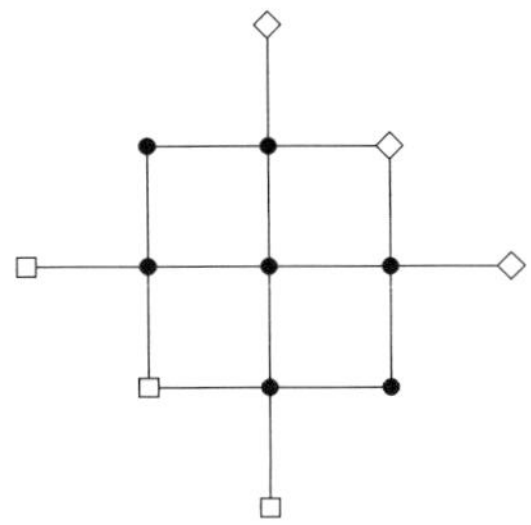

Figure 10.3 Graph G with the vertices of A indicated by squares, and the vertices of B indicated by diamonds.

Guess the extremal density ρ for $\mathrm{Mod}_2\,\Gamma$ and show that there are two different Beurling families for Γ and ρ in this case.

CHAPTER 11

Dependence on the Parameters

THE PARAMETER p in p-modulus can be thought of as a continuous "tuning parameter", which balances the sensitivity of modulus between various aspects of the family Γ. A second parameter consists of the edge weights σ that may be defined on the graph G. Using varying values of the parameter p is important in analysis on metric spaces, for instance to study the conformal dimension of fractals, [23,51]. As we will see later, when $\Gamma = \Gamma(a, b)$ is a connecting family of walks on a graph, then varying p can recover classical graph quantities such as shortest path distance, effective resistance and max-flow min-cut, see also [3]. In the case of the edge weights σ, we will often think of them as edge conductances. The dependence on σ is then useful in order to maximize modulus, see for instance [42].

11.1 INFINITY MODULUS

We begin with the special case of infinity modulus. First, we refine the basic modulus estimate we obtained in (10.20).

Lemma 11.1. *Let $G = (V, E, \sigma)$ be a graph with $\sigma \in \mathbb{R}^E_{>0}$ and let Γ be a family of objects on G as in Corollary 10.1. Let $\ell(\Gamma)$ be the smallest row sum in the usage matrix of Γ defined in (10.18). Then, for $1 \leq p < \infty$,*

$$\frac{\sigma_{\min}}{\ell(\Gamma)^p} \leq \operatorname{Mod}_{p,\sigma}(\Gamma) \leq \frac{\sigma(E)}{\ell(\Gamma)^p},$$

where $\sigma_{\min} := \min_{e \in E} \sigma(e)$ and $\sigma(E) := \sum_{e \in E} \sigma(e)$.

DOI: 10.1201/9781003024866-14

Proof. Since Γ is non-empty and satisfies (10.9), we may assume that $0 < \ell(\Gamma) < \infty$. The upper bound is just our basic estimate (10.20). For the lower bound, fix $\epsilon > 0$ and find $\gamma_0 \in \Gamma$ such that

$$\sum_{e\in E} \mathcal{N}(\gamma_0, e) \leq \ell(\Gamma) + \epsilon.$$

If $\rho \in \mathrm{Adm}(\Gamma)$, then

$$1 \leq \sum_{e\in E} \mathcal{N}(\gamma_0, e)\rho(e) \leq (\ell(\Gamma) + \epsilon) \max_{e\in E} \rho(e).$$

In particular, there exists $e_0 \in E$ such that

$$\rho(e_0) \geq \frac{1}{\ell(\Gamma) + \epsilon}.$$

And therefore,

$$\mathcal{E}_{p,\sigma}(\rho) = \sum_{e\in E} \sigma(e)|\rho(e)|^p \geq \sigma_{\min}|\rho(e_0)|^p \geq \frac{\sigma_{\min}}{(\ell(\Gamma) + \epsilon)^p}.$$

Taking the infimum in ρ and letting ϵ tend to zero gives the claim. □

Proposition 11.1. *Let $G = (V, E, \sigma)$ be a graph with $\sigma \in \mathbb{R}^E_{>0}$ and let Γ be a family of objects on G as in Corollary 10.1. Then*

$$\lim_{p\to\infty} \mathrm{Mod}_{p,\sigma}(\Gamma)^{1/p} = \frac{1}{\ell(\Gamma)} = \mathrm{Mod}_\infty(\Gamma).$$

Proof. By Lemma 11.1, we have

$$\frac{\sigma_{\min}^{1/p}}{\ell(\Gamma)} \leq \mathrm{Mod}_{p,\sigma}(\Gamma)^{1/p} \leq \frac{\sigma(E)^{1/p}}{\ell(\Gamma)},$$

and letting p tend to infinity we find that $\lim_{p\to\infty} \mathrm{Mod}_{p,\sigma}(\Gamma)^{1/p} = \ell(\Gamma)^{-1}$.

To show the other identity, fix $\epsilon > 0$. Let $\gamma_0 \in \Gamma$ satisfy

$$\sum_{e\in E} \mathcal{N}(\gamma_0, e) \leq \ell(\Gamma) + \epsilon.$$

As seen in (10.19), the constant density $\rho_0 \equiv \ell(\Gamma)^{-1}$ is admissible for Γ. Therefore, $\mathrm{Mod}_\infty(\Gamma) \leq \ell(\Gamma)^{-1}$. On the other hand, if $\rho \in \mathrm{Adm}(\Gamma)$, then

$$\begin{aligned} 1 &\leq \ell_\rho(\gamma_0) \\ &= \sum_{e\in E} \mathcal{N}(\gamma_0, e)\rho(e) \leq \max_{e\in E} \rho(e) \sum_{e\in E} \mathcal{N}(\gamma_0, e) \leq (\ell(\Gamma) + \epsilon) \max_{e\in E} \rho(e). \end{aligned}$$

Taking the infimum over $\rho \in \mathrm{Adm}(\Gamma)$, we obtain that $\mathrm{Mod}_\infty(\Gamma) \geq (\ell(\Gamma) + \epsilon)^{-1}$. Now let ϵ tend to zero. □

11.2 MONOTONICITY AND CONTINUITY IN p

Here, we state the monotonicity properties of modulus with respect to p, in full generality but also in the fairly common special case when $\mathcal{N}_{\min} = 1$ and $\sigma(E) = 1$. This makes it easier to see that the p-modulus is decreasing in p, while the p-th roots of p-modulus are increasing in p.

Theorem 11.1. *Let $G = (V, E, \sigma)$ be a graph with $\sigma \in \mathbb{R}^E_{>0}$ and let Γ a family of objects on G as in Corollary 10.1, with $\mathcal{N}_{\min}$ defined as in (10.9). Then, assuming that $\mathcal{N}_{\min} = 1$ and $\mathbf{1}^T\sigma = 1$, we have*

$$1 \le p < q < \infty \quad \Longrightarrow \quad \mathrm{Mod}_{p,\sigma}(\Gamma) \ge \mathrm{Mod}_{q,\sigma}(\Gamma), \tag{11.1}$$

and

$$1 \le p < q < \infty \quad \Longrightarrow \quad \mathrm{Mod}_{p,\sigma}(\Gamma)^{1/p} \le \mathrm{Mod}_{q,\sigma}(\Gamma)^{1/q} \tag{11.2}$$

More generally, $1 \le p < q < \infty$ implies that

$$\mathcal{N}_{\min}^p \,\mathrm{Mod}_{p,\sigma}(\Gamma) \ge \mathcal{N}_{\min}^q \,\mathrm{Mod}_{q,\sigma}(\Gamma), \tag{11.3}$$

and

$$\sigma(E)^{-1/p}\,\mathrm{Mod}_{p,\sigma}(\Gamma)^{1/p} \le \sigma(E)^{-1/q}\,\mathrm{Mod}_{q,\sigma}(\Gamma)^{1/q}. \tag{11.4}$$

Moreover, the map $F : [1, \infty) \to \mathbb{R}_{\ge 0}$, given by $F(p) := \mathrm{Mod}_{p,\sigma}(\Gamma)$, is continuous.

Proof. Equations (11.1) and (11.2) follow from (11.3) and (11.4) respectively.

For (11.3), assume that $\rho \in \mathrm{Adm}(\Gamma)$ is extremal for $\mathrm{Mod}_{p,\sigma}(\Gamma)$, then by part 2. of Proposition 10.2, we have $\rho \le \mathcal{N}_{\min}^{-1}$. Therefore, $\rho(e)\mathcal{N}_{\min} \le 1$ for every $e \in E$, and

$$\begin{aligned}
\mathcal{N}_{\min}^p \,\mathrm{Mod}_{p,\sigma}(\Gamma) = \mathcal{N}_{\min}^p \mathcal{E}_{p,\sigma}(\rho) &= \sum_{e\in E} \sigma(e)\left(\rho(e)\mathcal{N}_{\min}\right)^p \\
&\ge \sum_{e\in E} \sigma(e)\left(\rho(e)\mathcal{N}_{\min}\right)^q \\
&= \mathcal{N}_{\min}^q \mathcal{E}_{q,\sigma}(\rho) \ge \mathcal{N}_{\min}^q \,\mathrm{Mod}_{q,\sigma}(\Gamma).
\end{aligned}$$

For (11.4), we use Hölder's inequality with exponents $r = q/p$ and $s = q/(q-p)$. Note that $r > 1$ and $r + s = q^2/(p(q-p)) = rs$ so that r

and s are conjugate exponents. Then, for any $\rho \in \mathbb{R}^E_{\geq 0}$ we have:

$$\begin{aligned}\mathcal{E}_{p,\sigma}(\rho) = \sum_{e\in E}\sigma(e)\rho(e)^p &= \sum_{e\in E}\left(\sigma(e)^{1/q}\rho(e)\right)^p \sigma(e)^{1-p/q}\\ &\leq \left(\sum_{e\in E}\sigma(e)\rho(e)^q\right)^{p/q}\left(\sum_{e\in E}\sigma(e)\right)^{1-p/q}.\end{aligned}$$

Hence,

$$\left(\sigma(E)^{-1}\mathcal{E}_{p,\sigma}(\rho)\right)^{1/p} \leq \left(\sigma(E)^{-1}\mathcal{E}_{q,\sigma}(\rho)\right)^{1/q}. \tag{11.5}$$

Now take the infimum over $\rho \in \mathrm{Adm}(\Gamma)$.

For the continuity, fix $p \in [1, \infty)$. Taking the limit supremum in (11.3), as $q \downarrow p$, we get

$$\limsup_{q\downarrow p} \mathrm{Mod}_{q,\sigma}(\Gamma) \leq \mathrm{Mod}_{p,\sigma}(\Gamma)\limsup_{q\downarrow p}\left(\mathcal{N}_{\min}\right)^{p-q} = \mathrm{Mod}_{p,\sigma}(\Gamma).$$

Also, taking the limit inferior in (11.4), as $q \downarrow p$, we get

$$\liminf_{q\downarrow p} \mathrm{Mod}_{q,\sigma}(\Gamma) \geq \liminf_{q\downarrow p} \sigma(E)^{1-q/p}\,\mathrm{Mod}_{p,\sigma}(\Gamma)^{q/p} = \mathrm{Mod}_{p,\sigma}(\Gamma).$$

Therefore, the right limit $\lim_{q\downarrow p}\mathrm{Mod}_{q,\sigma}(\Gamma)$ exists and equals $\mathrm{Mod}_{p,\sigma}(\Gamma)$.

Now, fix $p \in (1, \infty)$. Taking the limit inferior in (11.3), as $q \uparrow p$, we get

$$\liminf_{q\downarrow p} \mathrm{Mod}_{q,\sigma}(\Gamma) \leq \mathrm{Mod}_{p,\sigma}(\Gamma)\liminf_{q\downarrow p}\left(\mathcal{N}_{\min}\right)^{p-q} = \mathrm{Mod}_{p,\sigma}(\Gamma),$$

Also, taking the limit superior in (11.4), as $q \uparrow p$, we get

$$\limsup_{q\downarrow p} \mathrm{Mod}_{q,\sigma}(\Gamma) \geq \limsup_{q\downarrow p} \sigma(E)^{1-q/p}\,\mathrm{Mod}_{p,\sigma}(\Gamma)^{q/p} = \mathrm{Mod}_{p,\sigma}(\Gamma).$$

Therefore, the left limit $\lim_{q\uparrow p}\mathrm{Mod}_{q,\sigma}(\Gamma)$ exists and equals $\mathrm{Mod}_{p,\sigma}(\Gamma)$.

Since the right limit and the left limit are the same, the limit exists and we have shown continuity. □

11.3 CLARKSON INEQUALITIES

In order to study the dependence of the extremal density on p, it is useful to recall the property of uniform convexity of L^p-spaces. The presentation of this material is inspired by some written class notes of Sadahiro Saeki [71]. We begin with a result about real numbers (in fact the same holds for complex numbers).

Lemma 11.2. *If $u, v \in \mathbb{R}$ and $p \geq 2$, then*

$$|u+v|^p + |u-v|^p \leq 2\left(|u|^q + |v|^q\right)^{p/q},$$

where $p+q = pq$.

We will often use the fact that $p+q = pq$ is equivalent to $(p-1)(q-1) = 1$.

Proof. Without loss of generality we can assume that $0 < v < u$. To see this, note that

$$|u+v|^p + |u-v|^p = ||u|+v|^p + ||u|-v|^p = ||u|+|v||^p + ||u|-|v||^p.$$

Also the inequality holds if $v = 0$ or $v = u$.

So consider the auxiliary function

$$\phi(x) := (x+v)^{p-1}(u^p x^{1-p} + v) + (x-v)^{p-1}(u^p x^{1-p} - v),$$

for $x \geq v$. A computation gives

$$\phi'(x) = (p-1)v(1 - u^p x^{-p})\left\{(x+v)^{p-2} - (x-v)^{p-2}\right\}.$$

Therefore, $\phi'(u) = 0$ and, since $p \geq 2$, we have $\phi'(x) < 0$ if and only if $x < u$. Namely, ϕ is minimized at $x = u$ on $[v, +\infty)$.

Consider $x_0 := u^q v^{1-q} = (u/v)^q v$. Since $x_0 > v$, we have $\phi(u) \leq \phi(x_0)$. Moreover, $\phi(u) = (u+v)^p + (u-v)^p$. And,

$$u^p x_0^{1-p} = u^p (u^q v^{1-q})^{1-p} = v.$$

So,

$$\phi(x_0) = 2v(x_0+v)^{p-1} = 2\left(v^{q-1}u^q v^{1-q} + v^q\right)^{p-1} = 2\left(u^q + v^q\right)^{p/q}.$$

□

Next we generalize the inequality about numbers in Lemma 11.2 to an inequality about L^p-norms. In this section L^p stands for $L^p(\mu)$ for some arbitrary measure μ.

Theorem 11.2 (Clarkson's Inequalities)**.** *Let $1 < p < \infty$. Let q be the conjugate exponent of p, so that $p+q = pq$. Set $M := \max\{p, q\}$ and $m = \min\{p, q\}$. Then, for any $f, g \in L^p$,*

$$\|f+g\|_p^M + \|f-g\|_p^M \leq 2\left(\|f\|_p^m + \|g\|_p^m\right)^{M/m}. \tag{11.6}$$

Proof. We need to prove two cases:

- If $p \geq 2$:

$$\|f+g\|_p^p + \|f-g\|_p^p \leq 2\left(\|f\|_p^q + \|g\|_p^q\right)^{p/q}. \tag{11.7}$$

- If $1 < p \leq 2$;

$$\|f+g\|_p^q + \|f-g\|_p^q \leq 2\left(\|f\|_p^p + \|g\|_p^p\right)^{q/p}. \tag{11.8}$$

Interestingly, (11.7) is easier to establish and follows from the regular Minkowski's Inequality, see equation (6.5) in [40]. Indeed, if $p \geq 2$,

$$\begin{aligned}
&\|f+g\|_p^p + \|f-g\|_p^p \\
&= \int \left(|f+g|^p + |f-g|^p\right) d\mu \\
&\leq \int 2\left(|f|^q + |g|^q\right)^{p/q} d\mu && \text{(by Lemma 11.2)} \\
&= 2\,\||f|^q + |g|^q\|_{p/q}^{p/q} && \text{(note } p/q \geq 1\text{)} \\
&\leq 2\left(\||f|^q\|_{p/q} + \||g|^q\|_{p/q}\right)^{p/q} && \text{(by Minkowski's inequality)} \\
&= 2\left(\|f\|_p^q + \|g\|_p^q\right)^{p/q}
\end{aligned}$$

To prove (11.8) we need instead Minkowski's Inequality for Integrals, see equation (6.19) in [40].

Theorem 11.3 (Minkowski's inequality for integrals). *Assume that $1 \leq p < \infty$. Let α and β be σ-finite measures on X and Y respectively. Let $f : X \times Y \to \mathbb{R}$ be jointly measurable. Then*

$$\left(\int \left|\int f(x,y)d\beta(y)\right|^p d\alpha(x)\right)^{1/p} \leq \int \left(\int |f(x,y)|^p d\alpha(x)\right)^{1/p} d\beta(y).$$

In other words, the p-norm of an integral is less than the integral of the p-norms. This gives rise to the following reverse triangle inequality.

Corollary 11.1. *Let $0 < r < 1$. If $f_1, f_2 \geq 0$ are measurable, then*

$$\|f_1\|_r + \|f_2\|_r \leq \|f_1 + f_2\|_r.$$

Proof of Corollary 11.1. This is an application of Minkowski's inequality for integrals with $p = 1/r$, $\beta = \mu$ and α the counting measure on $\{1, 2\}$.

$$
\begin{aligned}
(\|f_1\|_r + \|f_2\|_r)^r &= \left\{\left(\int f_1^r d\mu\right)^p + \left(\int f_2^r d\mu\right)^p\right\}^{1/p} \\
&\leq \int (f_1^{rp} + f_2^{rp})^{1/p}\, d\mu = \int (f_1 + f_2)^r d\mu.
\end{aligned}
$$

□

We are now ready to prove (11.8). Suppose $1 < p \leq 2$. Then, $p/q < 1$ and

$$
\begin{aligned}
\|f+g\|_p^q + \|f-g\|_p^q &= \||f+g|^q\|_{p/q} + \||f-g|^q\|_{p/q} \\
&\leq \||f+g|^q + |f-g|^q\|_{p/q} && \text{(by Corollary 11.1)} \\
&\leq 2\left\|(|f|^p + |g|^p)^{q/p}\right\|_{p/q} && \text{(by Lemma 11.2)} \\
&= 2\,\||f|^p + |g|^p\|_1^{q/p} \\
&= 2\left(\|f\|_p^p + \|g\|_p^p\right)^{q/p}
\end{aligned}
$$

□

For practical purposes it is often enough to appeal to the following weaker version of Clarkson's Inequalities.

Corollary 11.2. *Let $1 < p < \infty$. Set $M := \max\{p, q\}$, where $p+q = pq$. Then, for any $f, g \in L^p$,*

$$
\left\|\frac{f+g}{2}\right\|_p^M + \left\|\frac{f-g}{2}\right\|_p^M \leq \frac{\|f\|_p^M + \|g\|_p^M}{2}. \tag{11.9}
$$

Proof. Note that (11.6) can be written as

$$
\left\|\frac{f+g}{2}\right\|_p^M + \left\|\frac{f-g}{2}\right\|_p^M \leq \left(\frac{\|f\|_p^m + \|g\|_p^m}{2}\right)^{M/m}.
$$

Finally, we can use the convexity of the map $x \mapsto x^{M/m}$ to get (11.9). □

In particular, assuming that $\|f\|_p = \|g\|_p = 1$, we obtain

$$\left\|\frac{f+g}{2}\right\|_p^M \leq 1 - \left\|\frac{f-g}{2}\right\|_p^M .$$

In other words, the midpoint of f and g falls into the unit ball by an amount that only depends on how far apart f and g are. This is what is meant by uniform convexity. A moment of reflection shows that the L^1 and L^∞ norms do not have this property.

11.4 CONTINUITY IN p FOR EXTREMAL DENSITIES

Here we translate the language of p-norms into facts about p-modulus. First, we show that if $\rho \in \operatorname{Adm}\Gamma$ is almost a minimizer, then ρ must be close to the extremal density ρ^*.

Lemma 11.3. *Let $G = (V, E, \sigma)$ be a graph with $\sigma \in \mathbb{R}^E_{>0}$ and let Γ be a family of objects on G as in Corollary 10.1. Let $1 < p < \infty$ and set $M := \max\{p, q\}$, where $pq = p + q$. Let ρ^* denote the (unique) extremal density for $\operatorname{Mod}_{p,\sigma}(\Gamma)$. Then, for every $\rho \in \operatorname{Adm}\Gamma$:*

$$\|\rho - \rho^*\|_p^M \leq 2^{M-1}\sigma_{\min}^{-M/p}\left(\mathcal{E}_{p,\sigma}(\rho)^{M/p} - \operatorname{Mod}_{p,\sigma}(\Gamma)^{M/p}\right) \tag{11.10}$$

Proof. Let $f(e) := \sigma(e)^{1/p}\rho(e)$ and $f^*(e) = \sigma(e)^{1/p}\rho^*(e)$. Then

$$\|f\|_p^p = \mathcal{E}_{p,\sigma}(\rho) \qquad \text{and} \qquad \|f^*\|_p^p = \operatorname{Mod}_{p,\sigma}(\Gamma).$$

Also

$$\|f - f^*\|_p^p = \sum_{e\in E} \sigma(e)(\rho(e) - \rho^*(e))^p \geq \sigma_{\min}\|\rho - \rho^*\|_p^p,$$

and, since $\operatorname{Adm}(\Gamma)$ is convex,

$$\left\|\frac{f+f^*}{2}\right\|_p^p = \sum_{e\in E} \sigma(e)\left|\frac{\rho(e)+\rho^*(e)}{2}\right|^p = \mathcal{E}_{p,\sigma}\left(\frac{\rho+\rho^*}{2}\right) \geq \operatorname{Mod}_{p,\sigma}(\Gamma).$$

Applying (11.9) to f and f^* and substituting, we obtain

$$\operatorname{Mod}_{p,\sigma}(\Gamma)^{M/p} + \sigma_{\min}^{M/p}2^{-M}\|\rho - \rho^*\|_p^M \leq \frac{1}{2}\left(\mathcal{E}_{p,\sigma}(\rho)^{M/p} + \operatorname{Mod}_{p,\sigma}(\Gamma)^{M/p}\right).$$

Hence, (11.10) follows. □

We are ready to show the continuity of extremal densities in p.

Theorem 11.4. *Let $G = (V, E, \sigma)$ be a graph with $\sigma \in \mathbb{R}^E_{>0}$ and let Γ be a family of objects on G as in Corollary 10.1. For $1 < p < \infty$, let ρ_p denote the (unique) extremal density for $\text{Mod}_{p,\sigma}(\Gamma)$. Then, given $1 < t < \infty$,*

$$\lim_{p \to t} \|\rho_p - \rho_t\|_* = 0.$$

(The specific norm used is not important).

Proof. By Lemma 11.3, since $\rho_p \in \text{Adm}(\Gamma)$ for all $1 < p < \infty$, it is enough to show that, given $1 < t < \infty$,

$$\mathcal{E}_{t,\sigma}(\rho_p) \longrightarrow \text{Mod}_{t,\sigma}(\Gamma) \qquad \text{as } p \to t. \tag{11.11}$$

First, let $p \downarrow t$. Then, by Hölder's Inequality,

$$\begin{aligned}
\text{Mod}_{t,\sigma}(\Gamma) \le \mathcal{E}_{t,\sigma}(\rho_p) &= \sum_{e \in E} \sigma(e)\rho_p(e)^t \\
&\le \left(\sum_{e \in E} \sigma(e)\rho_p(e)^p\right)^{t/p} \left(\sum_{e \in E} \sigma(e)\right)^{1-t/p} \\
&= \mathcal{E}_{p,\sigma}(\rho_p)^{t/p}\sigma(E)^{1-t/p} = \text{Mod}_{p,\sigma}(\Gamma)^{t/p}\sigma(E)^{1-t/p}.
\end{aligned}$$

Moreover, by (11.3)

$$\mathcal{N}_{\min}^{p-t}\,\text{Mod}_{p,\sigma}(\Gamma) \le \text{Mod}_{t,\sigma}(\Gamma),$$

with $\mathcal{N}_{\min}$ defined as in (10.9). So,

$$\mathcal{N}_{\min}^{p-t}\,\text{Mod}_{p,\sigma}(\Gamma) \le \mathcal{E}_{t,\sigma}(\rho_p) \le \text{Mod}_{p,\sigma}(\Gamma)^{t/p}\sigma(E)^{1-t/p}, \tag{11.12}$$

and (11.11) follows from the continuity of modulus in p (Theorem 11.1).

Now, let $p \uparrow t$. Then, by (11.4),

$$\sigma(E)^{1-t/p}\,\text{Mod}_{p,\sigma}(\Gamma)^{t/p} \le \text{Mod}_{t,\sigma}(\Gamma)$$

Also, by part 2. of Proposition 10.2, we have $\mathcal{N}_{\min}\rho_p(e) \le 1$ for every $e \in E$. So,

$$\begin{aligned}
\text{Mod}_{t,\sigma}(\Gamma) \le \mathcal{E}_{t,\sigma}(\rho_p) &= (\mathcal{N}_{\min})^{-t} \sum_{e \in E} \sigma(e)\left(\mathcal{N}_{\min}\rho_p(e)\right)^t \\
&\le (\mathcal{N}_{\min})^{-t} \sum_{e \in E} \sigma(e)\left(\mathcal{N}_{\min}\rho_p(e)\right)^p && \text{(since } 1 < p < t\text{)} \\
&= (\mathcal{N}_{\min})^{p-t}\,\text{Mod}_{p,\sigma}(\Gamma)
\end{aligned}$$

Therefore,

$$\sigma(E)^{1-t/p}\,\mathrm{Mod}_{p,\sigma}(\Gamma)^{t/p} \le \mathcal{E}_{t,\sigma}(\rho_p) \le (\mathcal{N}_{\min})^{p-t}\,\mathrm{Mod}_{p,\sigma}(\Gamma).$$

And again the conclusion follows by the continuity of the p-modulus in p. □

Remark 11.1. Theorem 11.4 shows that the unique extremal densities ρ_p are continuous for $p \in (1,\infty)$. One may wonder if the limit as $p \downarrow 1$ always exists, even though the extremal density for 1-modulus is not necessarily unique. The answer is yes and it is Exercise 11.4.

11.5 DEPENDENCE ON THE EDGE WEIGHTS

When $G = (V, E, \sigma)$ is weighted, we think of $\sigma(e)$ as a "conductance". First note that p-modulus is homogeneous in σ, meaning that, for a given family Γ and a given parameter p,

$$\mathrm{Mod}_{p,t\sigma}(\Gamma) = t\,\mathrm{Mod}_{p,\sigma}(\Gamma), \quad \forall t > 0.$$

Next, we show some basic properties of the dependence on σ.

Theorem 11.5. *Let $G = (V, E, \sigma)$ be a weighted graph with $\sigma \in \mathbb{R}^E_{>0}$ and let Γ be a family of objects on G, as in Corollary 10.1. Fix $1 < p < \infty$ and let ρ^*_σ be the extremal density for $\mathrm{Mod}_{p,\sigma}(\Gamma)$. Then:*

1. *the map $\phi : \mathbb{R}^E_{>0} \to \mathbb{R}$ given by $\phi(\sigma) := \mathrm{Mod}_{p,\sigma}(\Gamma)$ is Lipschitz continuous;*
2. *the extremal density ρ^*_σ is continuous in σ;*
3. *the map ϕ is concave;*
4. *the map ϕ is differentiable and the partial derivatives of ϕ satisfy*

$$\frac{\partial \phi}{\partial \sigma(e)} = \rho^*_\sigma(e)^p \qquad \forall e \in E. \tag{11.13}$$

Proof. To show part 1., fix $\sigma_1, \sigma_2 \in \mathbb{R}^E_{>0}$. Assume, without loss of generality, that $\mathrm{Mod}_{p,\sigma_2}(\Gamma) \le \mathrm{Mod}_{p,\sigma_1}(\Gamma)$. Then, we have

$$\begin{aligned}
&|\mathrm{Mod}_{p,\sigma_1}(\Gamma) - \mathrm{Mod}_{p,\sigma_2}(\Gamma)| \\
&= \mathrm{Mod}_{p,\sigma_1}(\Gamma) - \mathrm{Mod}_{p,\sigma_2}(\Gamma) \\
&\le \mathcal{E}_{p,\sigma_1}(\rho^*_{\sigma_2}) - \mathcal{E}_{p,\sigma_2}(\rho^*_{\sigma_2}) \\
&= \sum_{e\in E}(\sigma_1(e) - \sigma_2(e))\rho^*_{\sigma_2}(e)^p \\
&\le \mathcal{N}_{\min}^{-p}\|\sigma_1 - \sigma_2\|_1. \qquad \text{(by part 2. of Proposition 10.2)}
\end{aligned}$$

Recall that since all norms are equivalent on $\mathbb{R}^E$, being Lipschitz does not depend on the norm chosen. This establishes the Lipschitz continuity of the map ϕ.

To show part 2., by Lemma 11.3 with $M := \max\{p, q\} \geq 2$,

$$\|\rho^*_{\sigma_2} - \rho^*_{\sigma_1}\|_p^M \leq 2^{M-1}\sigma_{1,\min}^{-M/p}\left(\mathcal{E}_{p,\sigma_1}(\rho^*_{\sigma_2})^{M/p} - \mathrm{Mod}_{p,\sigma_1}(\Gamma)^{M/p}\right).$$

So it's enough to show that

$$\mathcal{E}_{p,\sigma_1}(\rho^*_{\sigma_2}) \longrightarrow \mathrm{Mod}_{p,\sigma_1}(\Gamma) \qquad \text{as } \sigma_2 \to \sigma_1$$

But, using part 2. of Proposition 10.2 again,

$$\begin{aligned}\mathrm{Mod}_{p,\sigma_1}(\Gamma) \leq \mathcal{E}_{p,\sigma_1}(\rho^*_{\sigma_2}) &= \sum_{e\in E}\sigma_1(e)\rho^*_{\sigma_2}(e)^p \\ &= \sum_{e\in E}\sigma_2(e)\rho^*_{\sigma_2}(e)^p + \sum_{e\in E}(\sigma_1(e) - \sigma_2(e))\rho^*_{\sigma_2}(e)^p \\ &\leq \mathrm{Mod}_{p,\sigma_2}(\Gamma) + \mathcal{N}_{\min}^{-p}\|\sigma_1 - \sigma_2\|_1.\end{aligned}$$

And, as $\sigma_2 \to \sigma_1$, the last line converges to $\mathrm{Mod}_{p,\sigma_1}(\Gamma)$ by continuity of the map ϕ in part 1.

To show part 3., fix $\sigma_0, \sigma_1 \in \mathbb{R}^E_{>0}$ and $t \in [0, 1]$. Let ρ^*_t be extremal for $\mathrm{Mod}_{p,\sigma_t}(\Gamma)$, with $\sigma_t := t\sigma_1 + (1-t)\sigma_0$. Then, since $\rho^*_t \in \mathrm{Adm}(\Gamma)$,

$$\begin{aligned}t\,\mathrm{Mod}_{p,\sigma_1}(\Gamma) + (1-t)\,\mathrm{Mod}_{p,\sigma_0}(\Gamma) &\leq t\mathcal{E}_{p,\sigma_1}(\rho^*_t) + (1-t)\mathcal{E}_{p,\sigma_0}(\rho^*_t) \\ &= \sum_{e\in E}\sigma_t(e)\rho^*_t(e)^p = \mathrm{Mod}_{p,\sigma_t}(\Gamma).\end{aligned}$$

This proves concavity.

To show part 4., fix $\tau \in \mathbb{R}^E$ and let $\epsilon > 0$. Set $\sigma_\epsilon := \sigma + \epsilon\tau$. Note that for ϵ small enough $\sigma_\epsilon \in \mathbb{R}^E_{>0}$. Let ρ^*_ϵ be the extremal density corresponding to σ_ϵ. For any $\rho \in \mathbb{R}^E_{\geq 0}$:

$$\mathcal{E}_{p,\sigma_\epsilon}(\rho) = \sum_{e\in E}(\sigma(e) + \epsilon\tau(e))\rho(e)^p = \mathcal{E}_{p,\sigma}(\rho) + \epsilon\mathcal{E}_{p,\tau}(\rho).$$

So

$$\mathrm{Mod}_{p,\sigma_\epsilon}(\Gamma) = \mathcal{E}_{p,\sigma_\epsilon}(\rho^*_\epsilon) = \mathcal{E}_{p,\sigma}(\rho^*_\epsilon) + \epsilon\mathcal{E}_{p,\tau}(\rho^*_\epsilon) \geq \mathrm{Mod}_{p,\sigma}(\Gamma) + \epsilon\mathcal{E}_{p,\tau}(\rho^*_\epsilon),$$

and

$$\mathrm{Mod}_{p,\sigma_\epsilon}(\Gamma) \leq \mathcal{E}_{p,\sigma_\epsilon}(\rho^*_0) = \mathrm{Mod}_{p,\sigma}(\Gamma) + \epsilon\mathcal{E}_{p,\tau}(\rho^*_0),$$

Thus

$$\mathcal{E}_{p,\tau}(\rho_\epsilon^*) \le \frac{\phi(\sigma+\epsilon\tau)-\phi(\sigma)}{\epsilon} \le \mathcal{E}_{p,\tau}(\rho_0^*).$$

But by part 2., $\rho_\epsilon^* \to \rho_0^*$, as $\epsilon \to 0$. So the directional derivative of ϕ in the direction of τ is :

$$D_\tau(\phi) = \sum_{e\in E} \tau(e)\rho_\sigma^*(e)^p.$$

Also by Part 2., this implies that all directional derivatives are continuous, and it follows from [70, Theorem 9.21] that ϕ is differentiable. Finally, (11.13) is obtained by choosing $\tau := \delta_e$ to be the indicator function for an edge e. □

EXERCISES

11.1 *Weighted infinity modulus.* In Proposition 11.1 the dependence on the weights σ disappear for ∞-modulus. However, it is often interesting to consider weighted shortest path distance. In order to recover this with p-modulus one needs to modify Lemma 11.1.

Let $G = (V, E, \sigma)$ be a graph and Γ a family of objects on G as in Corollary 10.1. Let $r(e) := 1/\sigma(e)$ for $e \in E$, and define $\ell_r(\Gamma) := \inf_{\gamma\in\Gamma}\sum_{e\in E}\mathcal{N}(\gamma,e)r(e)$. Also, define the *weighted ∞-modulus* $\mathrm{Mod}_{\infty,\sigma}(\Gamma)$ to be the minimum energy

$$\mathcal{E}_{\infty,\sigma}(\rho) := \max_{e\in E} \sigma(e)\rho(e), \tag{P11.1}$$

over all $\rho \in \mathrm{Adm}(\Gamma)$.

(i) First, show that, for $1 \le p < \infty$,

$$\frac{1}{\ell_r(\Gamma)^p} \le \mathrm{Mod}_{p,\sigma^p}(\Gamma) \le \frac{|E|}{\ell_r(\Gamma)^p}.$$

(ii) Then, show that

$$\lim_{p\to\infty} \mathrm{Mod}_{p,\sigma^p}(\Gamma)^{1/p} = \frac{1}{\ell_r(\Gamma)} = \mathrm{Mod}_{\infty,\sigma}(\Gamma). \tag{P11.2}$$

11.2 *Continuity of modulus.* Consider the graph shown in Figure 11.1 consisting of k parallel simple paths of ℓ hops connecting node a to node b. Let $\Gamma = \Gamma(a, b)$ be the connecting family of walks for a to b.

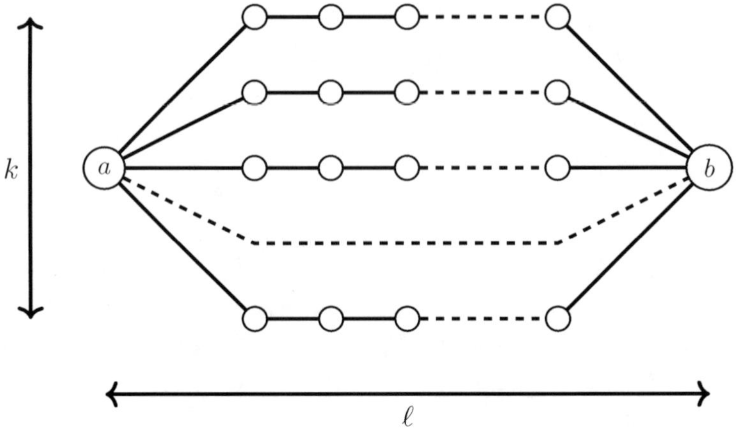

Figure 11.1 A graph consisting of k simple paths with ℓ hops each, connecting a to b.

(i) Show that $\rho^* \equiv \frac{1}{\ell}$ is extremal for all $1 \leq p \leq \infty$.

(ii) Next, show that, for $1 \leq p < \infty$, the modulus of Γ is

$$\operatorname{Mod}_p(\Gamma) = \frac{k}{\ell^{p-1}},$$

and for $p = \infty$,

$$\operatorname{Mod}_\infty(\Gamma) = \frac{1}{\ell}.$$

This shows that, as a function of p, the modulus is continuous, monotone decreasing, and decays to zero as $p \to \infty$, if $\ell > 1$, while the p-th root of modulus tends to infinity modulus.

11.3 *More monotonicity properties of modulus.* Let $G = (V, E, \sigma)$ be a graph with $\sigma \in \mathbb{R}^E_{>0}$ and let Γ a family of objects on G as in Corollary 10.1, with $\mathcal{N}_{\min}$ defined as in (10.9). Prove the following variation of inequalities (11.3) and (11.4): for $1 \leq p < q < \infty$,

$$\left(\frac{\mathcal{N}_{\min}}{\sigma(E)}\right)^p \operatorname{Mod}_{p,\sigma^p}(\Gamma) \geq \left(\frac{\mathcal{N}_{\min}}{\sigma(E)}\right)^q \operatorname{Mod}_{q,\sigma^q}(\Gamma), \tag{P11.3}$$

where $\sigma^p := (\sigma(e)^p)_{e\in E}$; and, for $1 \leq p < q < \infty$,

$$|E|^{-1/p} \operatorname{Mod}_{p,\sigma^p}(\Gamma)^{1/p} \leq |E|^{-1/q} \operatorname{Mod}_{q,\sigma^q}(\Gamma)^{1/q} \tag{P11.4}$$

11.4 *Convergence of extremal densities.* Let $G = (V, E, \sigma)$ be a graph with $\sigma \in \mathbb{R}^E_{>0}$ and let Γ be a family of objects on G as in Corollary 10.1. For $1 < p < \infty$, let ρ_p denote the (unique) extremal density for $\mathrm{Mod}_{p,\sigma}(\Gamma)$. The goal of this exercise is to show that the limit $\rho_1 = \lim_{p\downarrow 1} \rho_p$ exists and is an extremal density for $\mathrm{Mod}_{1,\sigma}(\Gamma)$.

(i) Let $t, p \to 1$ with $1 < t < p$ in (11.12) and show that

$$\lim_{p\downarrow 1} \mathcal{E}_{1,\sigma}(\rho_p) = \mathrm{Mod}_{1,\sigma}(\Gamma).$$

(ii) Deduce that we can extract a convergent subsequent from $\{\rho_p\}_{p>1}$. In particular, show that there is $\rho_1 \in \mathbb{R}^E_{\geq 0}$ and a sequence $p_n \downarrow 1$ so that $\lim_{n\to\infty} \rho_{p_n} = \rho_1$, and argue that $\rho_1 \in \mathrm{Adm}\,\Gamma$.

(iii) Show that the energies of ρ_{p_n} converge to the energy of ρ_1 and deduce that ρ_1 is extremal for $\mathrm{Mod}_{1,\sigma}(\Gamma)$.

11.5 *Dependence on the edge weights.* Let $G = (V, E, \sigma)$ be a graph with $\sigma \in \mathbb{R}^E_{>0}$ and let Γ a family of objects on G as in Corollary 10.1, with $\mathcal{N}_{\mathrm{min}}$ defined as in (10.9). The goal of this exercise is to extend parts of Theorem 11.5 to the case $p = 1$ and $p = \infty$.

(i) Show that the functions $\phi_1(\sigma) = \mathrm{Mod}_{1,\sigma}(\Gamma)$ and $\phi_\infty(\sigma) = \mathrm{Mod}_{\infty,\sigma}(\Gamma)$ are Lipschitz continuous.

(ii) Show that the function ϕ_1 in part (i) is concave.

(iii) Show that the function ϕ_∞ in part (i) is not concave in general by considering the case when G is a triangle abc and $\Gamma = \Gamma_{\mathrm{path}}(a, b)$. Let $\sigma_1 \equiv 1$ and get σ_0 by modifying σ_1 and setting $\sigma_0(a, b) = 1/3$. Now, use (P11.2) to compute ϕ_∞ at σ_1, σ_0, and $(\sigma_1 + \sigma_0)/2$.

CHAPTER 12

Duality for Modulus

SINCE MODULUS is a convex optimization problem, we now reexamine it in view of the theory of convex duality developed in Chapters 10 and 11.

12.1 THE PRIMAL PROBLEM

Fix a finite graph $G = (V, E, \sigma)$ and a finite family of objects Γ as in Corollary 10.1. For simplicity we will assume at first that $1 < p < \infty$. Recall that, as we saw in (10.6), the modulus $\mathrm{Mod}_{p,\sigma}(\Gamma)$ is the value of the following convex optimization problem

$$\begin{aligned} &\underset{\rho\in\mathbb{R}^E}{\text{minimize}} \qquad \mathcal{E}_{p,\sigma}(\rho) = \sum_{e\in E}\sigma(e)|\rho(e)|^p \\ &\text{subject to} \qquad \mathcal{N}\rho \geq \mathbf{1} \quad \text{and} \quad \rho \geq 0. \end{aligned} \tag{12.1}$$

Here, the density $\rho \in \mathbb{R}^E$ is the decision variable, the energy $\mathcal{E}_{p,\sigma}(\rho)$ is the objective, and admissibility of ρ forms the affine constraints.

Note that when $1 < p < \infty$ the objective is strictly convex. This implies that, in this case, we can remove the constraint $\rho \geq 0$. Indeed, if the extremal density ρ^* has a negative value $\rho^*(e_0)$ for some $e_0 \in E$, then defining $\tilde{\rho}^*(e_0) := -\rho^*(e_0)$ and else $\tilde{\rho}^*(e) := \rho^*(e)$ for all $e \in E \setminus \{e_0\}$, yields another admissible density with the same energy, which contradicts the uniqueness of the extremal density. Therefore, we may consider a dual variable $\lambda \in \mathbb{R}^\Gamma$ and define the *Lagrangian for p-modulus* as

$$\mathcal{L}(\rho, \lambda) := \sum_{e\in E}\sigma(e)|\rho(e)|^p + \sum_{\gamma\in\Gamma}\lambda(\gamma)\left(1 - \sum_{e\in E}\mathcal{N}(\gamma, e)\rho(e)\right). \tag{12.2}$$

DOI: 10.1201/9781003024866-15

Introducing the Lagrangian is a way to embed the constraints into the objective function. Indeed, for $\rho \in \mathbb{R}^E$, define $f(\rho) := \sup_{\lambda \geq 0} \mathcal{L}(\rho, \lambda)$. Then

$$f(\rho) = \begin{cases} \sum_{e \in E} \sigma(e)|\rho(e)|^p & \text{if } \mathcal{N}\rho \geq \mathbf{1} \\ +\infty & \text{else} \end{cases}$$

To see this note that, if $\rho \in \mathbb{R}^E$ is not admissible, then $\ell_\rho(\gamma_0) < 1$ for some $\gamma_0 \in \Gamma$. Therefore

$$1 - \sum_{e \in E} \mathcal{N}(\gamma, e)\rho(e) > 0,$$

and letting $\lambda(\gamma_0)$ tend to $+\infty$, implies that $f(\rho) = +\infty$. On the other hand, if ρ is admissible, then $1 - \sum_{e \in E} \mathcal{N}(\gamma, e)\rho(e) \leq 0$ for all $\gamma \in \Gamma$, so it's advantageous to let $\lambda = 0$ for all $\gamma \in \Gamma$.

Consequently, modulus can be thought as an unconstrained optimization by writing

$$\text{Mod}_{p,\sigma}(\Gamma) = \inf_{\rho \in \mathbb{R}^E} f(\rho) = \inf_{\rho \in \mathbb{R}^E} \sup_{\lambda \geq 0} \mathcal{L}(\rho, \lambda). \tag{12.3}$$

12.2 WEAK DUALITY

Next, we recall a very general fact. Suppose $F : X \times Y \to \mathbb{R}$. Then

$$\inf_{x \in X} \sup_{y \in Y} F(x, y) \geq \sup_{y \in Y} \inf_{x \in X} F(x, y). \tag{12.4}$$

Applying (12.4) to (12.3) we obtain

$$\text{Mod}_{p,\sigma}(\Gamma) = \inf_{\rho \in \mathbb{R}^E} \sup_{\lambda \in \mathbb{R}^\Gamma_{\geq 0}} \mathcal{L}(\rho, \lambda) \geq \sup_{\lambda \in \mathbb{R}^\Gamma_{\geq 0}} \inf_{\rho \in \mathbb{R}^E} \mathcal{L}(\rho, \lambda).$$

This inequality goes under the name of *weak duality*. The dual objective functions is

$$g(\lambda) := \inf_{\rho \in \mathbb{R}^E} \mathcal{L}(\rho, \lambda), \tag{12.5}$$

and the dual problem is

$$\begin{aligned} \underset{\lambda \in \mathbb{R}^\Gamma}{\text{maximize}} \quad & g(\lambda) \\ \text{subject to} \quad & \lambda \geq 0. \end{aligned} \tag{12.6}$$

12.3 STRONG DUALITY

To borrow notations from convex optimization let P^* be the value of the primal problem in (12.3) and let D^* be the value of the dual problem (12.6). Then *strong duality* is said to hold if $P^* = D^*$ and both the primal and dual problem are attained.

There are well-known sufficient conditions for strong duality to hold. One condition that applies in our case is Slater's condition (see also [69, Theorem 28.2]).

Fact 12.1 (Slater). *Strong duality holds if the primal feasible region has a point for which all inequality constraints are satisfied strictly.*

For p-modulus it is enough to find ρ such that $\ell_\rho(\gamma) > 1$ for all $\gamma \in \Gamma$. Thus, for families of objects such as walks or subsets of E, it is enough to let $\rho \equiv 2$. More generally, if Γ is a finite family of objects that are allowed to make fractional use of an edge, it's enough to set $\rho(e) > \mathcal{N}_{\min}^{-1}$, for all $e \in E$, with $\mathcal{N}_{\min}$ defined as in (10.9).

Assume that (ρ^*, λ^*) is an optimal pair. Then by (12.3) and (12.5):

$$\begin{aligned} P^* = \operatorname{Mod}_{p,\sigma}(\Gamma) = \mathcal{L}(\rho^*, \lambda^*) &= \min_{\rho \in \mathbb{R}^E} \mathcal{L}(\rho, \lambda^*) \\ = g(\lambda^*) &= \max_{\lambda \in \mathbb{R}^\Gamma_{\geq 0}} \mathcal{L}(\rho^*, \lambda) = D^*. \end{aligned}$$

In other words, for every $\rho \in \mathbb{R}^E$ and every $\lambda \in \mathbb{R}^\Gamma_{\geq 0}$, we have the following *saddle property*:

$$\mathcal{L}(\rho^*, \lambda) \leq \mathcal{L}(\rho^*, \lambda^*) \leq \mathcal{L}(\rho, \lambda^*). \tag{12.7}$$

12.4 THE DUAL MODULUS PROBLEM

Theorem 12.1. *On a finite graph $G = (V, E, \sigma)$, consider a finite family of objects Γ as in Corollary 10.1. Assume that $1 < p < \infty$ and let q be the conjugate of p, so that $(p-1)(q-1) = 1$. Then, the dual problem for $\operatorname{Mod}_{p,\sigma}(\Gamma)$ is*

$$\max_{\lambda \in \mathbb{R}^\Gamma_{\geq 0}} \left\{ \lambda^T \mathbf{1} - \frac{(q-1)^{q-1}}{q^q} \sum_e \sigma(e)^{q-1} \left[\left(\mathcal{N}^T \lambda \right)(e) \right]^q \right\} \tag{12.8}$$

Moreover, optimal solutions λ^* always exist (although they may not be unique) and are related to the unique extremal density ρ^* via the formula

$$\rho^*(e) = \left(\frac{1}{p\sigma(e)} \left(\mathcal{N}^T \lambda^* \right)(e) \right)^{q-1} \qquad \forall e \in E. \tag{12.9}$$

Proof. As a consequence of strong duality, we obtain a new formula for the modulus $\text{Mod}_{p,\sigma}(\Gamma)$, which must equal

$$\sup_{\lambda\geq 0}\inf_{\rho\in\mathbb{R}^E}\left\{\sum_{e\in E}\sigma(e)\,|\rho(e)|^p+\sum_{\gamma\in\Gamma}\lambda(\gamma)\left[1-\sum_{e\in E}\mathcal{N}(\gamma,e)\rho(e)\right]\right\} \tag{12.10}$$

We now rewrite the dual problem using standard convex optimization techniques. In particular, we use the fact that for $1<p<\infty$ the function $x\mapsto|x|^p$ is C^1-smooth and its derivative is

$$\frac{d}{dx}|x|^p=px|x|^{p-2}. \tag{12.11}$$

Therefore, for a fixed λ, the Lagrangian $\mathcal{L}$ in (12.2) is minimized by a density ρ_λ which satisfies the following stationarity condition:

$$p\sigma(e)\rho_\lambda(e)\,|\rho_\lambda(e)|^{p-2}-\sum_{\gamma}\lambda(\gamma)\mathcal{N}(\gamma,e)=0, \tag{12.12}$$

for every $e\in E$.

Dual feasibility implies that $\lambda\geq 0$, so (12.12) implies that $\rho_\lambda\geq 0$ as well, and therefore

$$\rho_\lambda(e)=\left(\frac{1}{p\sigma(e)}\sum_{\gamma}\lambda(\gamma)\mathcal{N}(\gamma,e)\right)^{\frac{1}{p-1}}=\left(\frac{1}{p\sigma(e)}(\mathcal{N}^T\lambda)(e)\right)^{\frac{1}{p-1}} \tag{12.13}$$

Substituting (12.13) into (12.10) we get the following dual problem: maximize the function

$$g(\lambda)=\sum_{\gamma}\lambda(\gamma)-(p-1)\sum_{e}\sigma(e)\left(\frac{1}{p\sigma(e)}\sum_{\gamma}\lambda(\gamma)\mathcal{N}(\gamma,e)\right)^{\frac{p}{p-1}} \tag{12.14}$$

subject to $\lambda(\gamma)\geq 0$, for all $\gamma\in\Gamma$.

Equation (12.9) follows from (12.13). □

Remark 12.1. Note that λ appears in (12.14) composed with two linear operations, which may have a kernel. So in general the solutions to the dual problem are not unique. However, if λ^* is a solution to (12.14), then (12.13) gives a solution ρ^* to the primal problem (12.1) written in terms of λ^* (and we already know that ρ^* is unique).

12.5 LAGRANGIAN DUALITY FOR $p = 1$

Given a finite graph $G = (V, E, \sigma)$ and a finite family of objects Γ with usage matrix $\mathcal{N}$. It follows from (10.6) that $\mathrm{Mod}_{1,\sigma}(\Gamma)$ is the value of the linear problem:

$$\begin{aligned} \underset{\rho\in\mathbb{R}^E}{\text{minimize}} \quad & \sum_{e\in E}\sigma(e)\rho(e) = \sigma^T\rho \\ \text{subject to} \quad & \mathcal{N}\rho \geq \mathbf{1},\ \rho \geq 0; \end{aligned} \tag{12.15}$$

Let $\mathrm{Adm}\,\Gamma = \{\rho \in \mathbb{R}^E_{\geq 0} : \mathcal{N}\rho \geq \mathbf{1}\}$ be the polyhedron of admissible densities as defined in (10.3). Recall from Proposition 10.2 that minimizers always exist. Note, however, that uniqueness of the minimizers is not necessary anymore because the norm is not strictly convex in this case.

Introducing the dual variables $\lambda \in \mathbb{R}^\Gamma_{\geq 0}$ and $\mu \in \mathbb{R}^E_{\geq 0}$, the Lagrangian for problem (12.15) is

$$\begin{aligned} \mathcal{L}_1(\rho,\lambda,\mu) &:= \sum_{e\in E}\sigma(e)\rho(e) \\ &\quad + \sum_{\gamma\in\Gamma}\lambda(\gamma)\left(1 - \sum_{e\in E}\mathcal{N}(\gamma,e)\rho(e)\right) - \sum_{e\in E}\mu(e)\rho(e) \\ &= \sum_{\gamma\in\Gamma}\lambda(\gamma) + \sum_{e\in E}\rho(e)\left(\sigma(e) - \sum_{\gamma\in\Gamma}\mathcal{N}(\gamma,e)\lambda(\gamma) - \mu(e)\right). \end{aligned}$$

Since the Lagrangian is affine in ρ, taking the infimum in ρ, we see that it equals $-\infty$, unless $\sigma(e) - \sum\limits_{\gamma\in\Gamma}\mathcal{N}(\gamma,e)\lambda(\gamma) - \mu(e) = 0$ for all $e \in E$, in which case we have,

$$g_1(\lambda,\mu) := \inf_{\rho\in\mathbb{R}^E} L_1(\rho,\lambda,\mu) = \sum_{\gamma\in\Gamma}\lambda(\gamma).$$

The dual problem involves maximizing g_1 over feasible λ and μ, and therefore it will never be advantageous to choose the dual variables such that $g_1(\lambda,\mu) = -\infty$ (assuming this can be avoided). In fact, provided λ satisfies the inequalities $\sum_{\gamma\in\Gamma}\mathcal{N}(\gamma,e)\lambda(\gamma) \leq \sigma(e)$ (e.g., $\lambda \equiv 0$), a suitable nonnegative μ can always be defined to make $g_1(\lambda,\mu)$ finite, whereas, if any of the aforementioned inequalities is violated in the choice of λ, there will be no nonnegative μ for which $g_1(\lambda,\mu)$ does not evaluate to $-\infty$.

Thus, the dual modulus problem when $p = 1$ can be written as follows:

$$\begin{aligned} \underset{\lambda\in\mathbb{R}^\Gamma}{\text{maximize}} \quad & \sum_{\gamma\in\Gamma}\lambda(\gamma) \\ \text{subject to} \quad & \sum_{\gamma\in\Gamma}\mathcal{N}(\gamma,e)\lambda(\gamma) \le \sigma(e) \quad \forall e\in E \\ & \lambda(\gamma)\ge 0 \quad \forall\gamma\in\Gamma. \end{aligned} \tag{12.16}$$

This can be interpreted as a *generalized max flow* problem. Indeed, the weights $\sigma(e)$ can be thought as edge capacities, and then the task is to distribute non-negative weights $\lambda(\gamma)$ on the objects of Γ so as to maximize the total weight while respecting the edge constraints. When the objects in Γ are connecting paths, then this is a path flow formulation of a max flow problem.

12.6 LAGRANGIAN DUALITY FOR $p = \infty$

Recall that $\text{Mod}_{\infty,\sigma}(\Gamma)$ is the value of the problem:

$$\begin{aligned} \underset{\rho\in\mathbb{R}^E}{\text{minimize}} \quad & \max_{e\in E}\sigma(e)\rho(e) \\ \text{subject to} \quad & \mathcal{N}\rho \ge \mathbf{1},\ \rho\ge 0. \end{aligned} \tag{12.17}$$

In this case, an appropriate equivalent problem is the following:

$$\begin{aligned} \underset{\rho\in\mathbb{R}^E,\ t\in\mathbb{R}}{\text{minimize}} \quad & t \\ \text{subject to} \quad & \sum_{e\in E}\mathcal{N}(\gamma,e)\rho(e) \ge 1 \quad \forall\gamma\in\Gamma \\ & 0\le\sigma(e)\rho(e)\le t \quad \forall e\in E. \end{aligned} \tag{12.18}$$

The associated Lagrangian takes the form

$$\begin{aligned} &\mathcal{L}_\infty(t,\rho,\lambda,\mu,\beta) \\ &:= t + \sum_{\gamma\in\Gamma}\lambda(\gamma)\left(1-\sum_{e\in E}\mathcal{N}(\gamma,e)\rho(e)\right) - \sum_{e\in E}\mu(e)\sigma(e)\rho(e) \\ &+ \sum_{e\in E}\beta(e)\left(\sigma(e)\rho(e)-t\right) \\ &= t\left(1-\sum_{e\in E}\beta(e)\right) + \sum_{e\in E}\sigma(e)\rho(e)\left(-\sum_{\gamma\in\Gamma}\mathcal{N}(\gamma,e)\lambda(\gamma)-\mu(e)+\beta(e)\right) \\ &+ \sum_{\gamma\in\Gamma}\lambda(\gamma). \end{aligned}$$

As in the $p = 1$ case, the Lagrangian is affine in the primal variables t and ρ, so the dual objective function $g_\infty(\lambda, \mu, \beta)$ is $-\infty$, unless $\sum_{e\in E} \beta(e) = 1$ and $\sum_{\gamma\in\Gamma} \mathcal{N}(\gamma, e)\lambda(\gamma) + \mu(e) - \beta(e) = 0$, for all $e \in E$, in which case,

$$g_\infty(\lambda, \mu, \beta) := \sum_{\gamma\in\Gamma} \lambda(\gamma).$$

Similarly, this produces the following form of the dual problem.

$$\begin{aligned}
\underset{\lambda\in\mathbb{R}^\Gamma,\beta\in\mathbb{R}^E}{\text{maximize}} \quad & \sum_{\gamma\in\Gamma} \lambda(\gamma) \\
\text{subject to} \quad & \sum_{\gamma\in\Gamma} \mathcal{N}(\gamma, e)\lambda(\gamma) \le \beta(e) \quad \forall e \in E \\
& \sum_{e\in E} \beta(e) = 1 \\
& \lambda(\gamma) \ge 0 \quad \forall \gamma \in \Gamma \\
& \beta(e) \ge 0 \quad \forall e \in E.
\end{aligned} \tag{12.19}$$

This can also be interpreted as a generalized max-flow problem, where the edge capacities can be freely assigned under a budget constraint.

12.7 THE KKT OPTIMALITY CONDITIONS

When $1 < p < \infty$ the modulus problem (12.1) is C^1-smooth, hence the following Karush-Kuhn-Tucker necessary and sufficient conditions for optimality hold.

Proposition 12.1. *Let $G = (V, E, \sigma)$ be a weighted graph, Γ a finite family of objects on G as in Corollary 10.1, and let $1 < p < \infty$. A pair $(\rho^*, \lambda^*) \in \mathbb{R}^E \times \mathbb{R}^\Gamma$ is optimal for the primal problem (12.1) and the dual problem (12.14) respectively if and only if*

1. **Primal and dual feasibility:** $\rho^* \in \mathrm{Adm}(\Gamma)$ *and* $\lambda^* \ge 0$.

2. **Complementary slackness:** *For every* $\gamma \in \Gamma$

$$\lambda^*(\gamma)\left(1 - \sum_{e\in E} \mathcal{N}(\gamma, e)\rho^*(e)\right) = 0.$$

3. **Stationarity:** *If $\mathcal{L}$ is the Lagrangian as in* (12.2), *then*

$$(\nabla_\rho \mathcal{L})(\rho^*, \lambda^*) = 0.$$

The proof of this proposition is omitted.

12.8 LAGRANGIAN SUBFAMILIES

Definition 12.1. If λ^* is optimal for the dual problem (12.14), we say that

$$\Gamma_{\lambda^*} := \{\gamma \in \Gamma : \lambda^*(\gamma) > 0\}$$

is the *Lagrangian subfamily* defined by λ^*.

In the next proposition, we use the KKT conditions to show that every Lagrangian subfamily is a Beurling subfamily; see Definition 10.1. The converse does not always hold, see, for instance, Exercise 12.3.

Proposition 12.2. *Let $G = (V, E, \sigma)$ be a weighted graph, Γ a finite family of objects on G as in Corollary 10.1, and let $1 < p < \infty$. Let $(\rho^*, \lambda^*) \in \mathbb{R}^E \times \mathbb{R}^\Gamma$ be an optimal pair for (12.1) and (12.14) respectively. Let Γ_{λ^*} be the Lagrangian subfamily defined by λ^*. Then Γ_{λ^*} is a Beurling subfamily for Γ.*

Proof. First note that $\rho^* \in \mathrm{Adm}(\Gamma)$ by primal feasibility. Also by complementary slackness, for every $\gamma \in \Gamma_{\lambda^*}$, we have $\lambda^*(\gamma) > 0$, therefore $\ell_{\rho^*}(\gamma) = 1$.

We are left to verify 2(b) in Theorem 10.2. By the saddle property (12.7), for every $\epsilon > 0$ and every $h \in \mathbb{R}^E$, we have

$$\begin{aligned}
0 &\leq \mathcal{L}(\rho^* + \epsilon h, \lambda^*) - \mathcal{L}(\rho^*, \lambda^*) \\
&= \mathcal{E}_{p,\sigma}(\rho^* + \epsilon h) + \sum_{\gamma\in\Gamma} \lambda^*(\gamma)\left(1 - \ell_{\rho^*+\epsilon h}(\gamma)\right) - \mathcal{E}_{p,\sigma}(\rho^*) \\
&= \sum_{e\in E} \sigma(e)\left[|\rho^*(e) + \epsilon h(e)|^p - |\rho^*(e)|^p\right] - \epsilon \sum_{\gamma\in\Gamma_{\lambda^*}} \lambda^*(\gamma)\ell_h(\gamma)
\end{aligned}$$

Dividing by ϵ and letting $\epsilon \to 0$, using (12.11), we obtain

$$p\sum_{e\in E} \sigma(e)\rho^*(e)^{p-1}h(e) \geq \sum_{\gamma\in\Gamma_{\lambda^*}} \lambda^*(\gamma)\ell_h(\gamma) \geq 0.$$

□

12.9 MINIMAL SUBFAMILIES

Definition 12.2. Let Γ be a family of objects on a weighted graph G and let $p \in (1, \infty)$. We say that a subfamily $\tilde{\Gamma} \subset \Gamma$ is a *minimal subfamily for* $\mathrm{Mod}_{p,\sigma}(\Gamma)$ if

1. $\mathrm{Mod}_{p,\sigma}(\tilde{\Gamma}) = \mathrm{Mod}_{p,\sigma}(\Gamma)$,

2. $\text{Mod}_{p,\sigma}(\tilde{\Gamma} \setminus \{\gamma\}) < \text{Mod}_{p,\sigma}(\Gamma)$ for all $\gamma \in \tilde{\Gamma}$.

Theorem 12.2 below will show that minimal subfamilies are Lagrangian and hence, by Proposition 12.2, also Beurling. Moreover, minimal subfamilies are "small" in the sense that their cardinality is bounded above by $|E|$. In fact, in specific situations, minimal subfamilies might even be much smaller than that.

Assume that Γ is a finite family of objects. By definition, the matrix $\mathcal{N} = [\mathcal{N}(\gamma, e)]_{\gamma\in\Gamma, e\in E}$ has rank at most $\min\{|\Gamma|, |E|\}$. It will be important in designing algorithms for computing modulus to understand when $\mathcal{N}$ has full rank. The following is the main result in [13, Theorem 3.5].

Theorem 12.2. *Let $G = (V, E, \sigma)$ be a weighted graph and let $p \in (1, \infty)$. Assume Γ is a finite family of objects on G as in Corollary 10.1. Then:*

(i) A minimal subfamily $\tilde{\Gamma} \subset \Gamma$ for $\text{Mod}_{p,\sigma}(\Gamma)$ always exists.

(ii) If $\tilde{\Gamma}$ is a minimal subfamily for $\text{Mod}_{p,\sigma}(\Gamma)$, then there exists a unique optimal λ^* for $\text{Mod}_{p,\sigma}(\Gamma)$ so that $\tilde{\Gamma} = \Gamma_{\lambda^*}$, as in Definition 12.1. In particular, $\tilde{\Gamma}$ is a Beurling subfamily.

(iii) The objects in a minimal subfamily $\tilde{\Gamma}$ are "linearly independent", in the sense that

$$\text{rank}\,\mathcal{N}(\tilde{\Gamma}) = |\tilde{\Gamma}|. \tag{12.20}$$

(iv) In particular, this implies that the cardinality of minimal subfamilies is bounded above by the number of edges $|E|$.

(v) Moreover, if Γ is itself minimal, then the optimizer for the dual problem is unique and positive.

The proof of Theorem 12.2 we give here is an extended and more detailed version of the one in [13].

Proof. To see that minimal families always exist, start from a finite family Γ and enumerate it as $\Gamma = \{\gamma_1, \dots, \gamma_m\}$. By Corollary 10.1, $\text{Mod}_{p,\sigma}(\Gamma)$ is finite and non-zero. Next, remove the objects of Γ one at a time. By monotonicity, see part 1. of Proposition 10.4, the modulus can only decrease, and the empty family has zero modulus, y part 2. of Proposition 10.1. Therefore, there is a smallest index $1 \le k \le m$ such that

$$\text{Mod}_{p,\sigma}(\Gamma) = \text{Mod}_{p,\sigma}(\Gamma \setminus \{\gamma_1, \dots, \gamma_{k-1}\}) > \text{Mod}_{p,\sigma}(\Gamma \setminus \{\gamma_1, \dots, \gamma_k\}),$$

with the understanding that $\{\gamma_1, \ldots, \gamma_{k-1}\} = \emptyset$ if $k = 1$. Then, $\Gamma \setminus \{\gamma_1, \ldots, \gamma_{k-1}\}$ is a minimal subfamily of Γ, and (i) is established.

Now assume that $\tilde{\Gamma}$ is a minimal subfamily of Γ for $\text{Mod}_{p,\sigma}(\Gamma)$. Consider the extremal density $\tilde{\rho} \in \mathbb{R}^E_{\geq 0}$ and a set of optimal dual variables $\tilde{\lambda} \in \mathbb{R}^{\tilde{\Gamma}}_{\geq 0}$, for the modulus problem $\text{Mod}_{p,\sigma}(\tilde{\Gamma})$. Extend $\tilde{\lambda}$ by setting it equal to zero outside of $\tilde{\Gamma}$, i.e., define $\lambda^*(\gamma) = \tilde{\lambda}(\gamma)$, for $\gamma \in \tilde{\Gamma}$ and set $\lambda^* = 0$ otherwise. Then, λ^* gives a solution to the dual problem (12.14) for Γ, since $\text{Mod}_{p,\sigma}(\tilde{\Gamma}) = \text{Mod}_{p,\sigma}(\Gamma)$. In particular, by (12.9), we see that $\tilde{\rho}$ is equal to the extremal density ρ^* for $\text{Mod}_{p,\sigma}(\Gamma)$. Therefore, λ^* is a set of optimal dual variables for $\text{Mod}_{p,\sigma}(\Gamma)$.

Moreover, $\lambda^*(\gamma) > 0$ if and only if $\gamma \in \tilde{\Gamma}$. To see this, assume that $\lambda^*(\gamma) = 0$ for some $\gamma \in \tilde{\Gamma}$. Then, γ can be removed from $\tilde{\Gamma}$ without affecting the modulus. But this contradicts the minimality of $\tilde{\Gamma}$.

Finally, we show that this choice of optimal dual variables is unique. Arguing by contradiction, assume that there are two distinct sets of optimal dual variables λ_1^* and λ_2^*, both supported and strictly positive on $\tilde{\Gamma}$. For $t \in \mathbb{R}$, define

$$\lambda_t^* := \frac{1+t}{2}\lambda_1^* + \frac{1-t}{2}\lambda_2^*$$

By linearity, $\mathcal{N}^T\lambda_t^* = \frac{1+t}{2}\mathcal{N}^T\lambda_1^* + \frac{1-t}{2}\mathcal{N}^T\lambda_2^*$. Also, by (12.9) and the uniqueness of ρ^*, we must have $\mathcal{N}^T\lambda_1^* = \mathcal{N}^T\lambda_2^*$. So, $\mathcal{N}^T\lambda_t^* = \mathcal{N}^T\lambda_1^* = \mathcal{N}^T\lambda_2^*$. Likewise, by linearity again, $\mathbf{1}^T\lambda_l^* = \frac{1+t}{2}\mathbf{1}^T\lambda_1^* + \frac{1-t}{2}\mathbf{1}^T\lambda_2^*$. And, since λ_1^* and λ_2^* are both optimal for (12.8) and since $\mathcal{N}^T\lambda_1^* = \mathcal{N}^T\lambda_2^*$, we also have that $\mathbf{1}^T\lambda_1^* = \mathbf{1}^T\lambda_2^* = \mathbf{1}^T\lambda_t^*$, for all $t \in \mathbb{R}$. Therefore, λ_t^* is a solution to the dual problem (12.8) as long as it is feasible, namely, as long as $\lambda_t^* \geq 0$. The idea then is to pick t so that λ_t^* falls on the boundary of the feasible set. More precisely, consider the set $\Gamma_0 := \{\gamma \in \Gamma : \lambda_1^*(\gamma) \neq \lambda_2^*(\gamma)\}$. Then $\Gamma_0 \subset \tilde{\Gamma}$ and by convexity λ_t^* is feasible for at least all $t \in [-1, 1]$. Fix $\gamma \in \Gamma_0$, and set $\lambda_t^*(\gamma) \geq 0$. Then,

$$(\lambda_1^*(\gamma) + \lambda_2^*(\gamma)) + t(\lambda_1^*(\gamma) - \lambda_2^*(\gamma)) \geq 0.$$

Hence, either

$$t \geq -\frac{\lambda_1^*(\gamma) + \lambda_2^*(\gamma)}{|\lambda_1^*(\gamma) - \lambda_2^*(\gamma)|} \quad \text{or} \quad t \leq \frac{\lambda_1^*(\gamma) + \lambda_2^*(\gamma)}{|\lambda_1^*(\gamma) - \lambda_2^*(\gamma)|},$$

depending on the sign of $\lambda_1^*(\gamma) - \lambda_2^*(\gamma)$. Set $t^* := \min_{\gamma \in \Gamma_0}(\lambda_1^*(\gamma) + \lambda_2^*(\gamma))/|\lambda_1^*(\gamma) - \lambda_2^*(\gamma)|$, and let γ^* be an object in Γ_0 that attains the minimum. Then, $\lambda_{t^*}^*(\gamma^*) = 0$ and $\lambda_{t^*}(\gamma) \geq 0$ for all $\gamma \in \Gamma_0$, and hence also for all $\gamma \in \Gamma$, since $\lambda^*_{t^*}(\gamma) = (\lambda^*_1(\gamma) + \lambda^*_2(\gamma))/2$, whenever $\gamma \notin \Gamma_0$. In other words, $\lambda^*_{t^*}$ is feasible and hence optimal for the dual problem; moreover, it is supported on $\tilde{\Gamma}$ and vanishes on $\gamma^* \in \tilde{\Gamma}$. This means that γ^* can be removed from $\tilde{\Gamma}$ without affecting the modulus. But this contradicts the minimality of $\tilde{\Gamma}$, so uniqueness follows, and (ii) is proved.

In particular, we can assume that $\tilde{\Gamma}$ is a Lagrangian subfamily for some (unique) choice of optimal λ^*. Let $\tilde{\Gamma} = \{\gamma_1, \gamma_2, \ldots, \gamma_r\}$ be an enumeration of $\tilde{\Gamma}$ and assume that there is a linear dependence relation in the rows of the usage matrix $\mathcal{N}(\tilde{\Gamma})$. Namely, assume that, after possibly reordering $\tilde{\Gamma}$, there exist real numbers $\{c_k\}_{k=1}^{r-1}$ such that

$$\mathcal{N}(\gamma_r, e) = \sum_{k=1}^{r-1} c_k \mathcal{N}(\gamma_k, e) \qquad \forall e \in E. \tag{12.21}$$

Then, for any density $\rho \in \mathbb{R}^E_{\geq 0}$, Equation (12.21) implies that

$$\begin{aligned} \ell_\rho(\gamma_r) &= \sum_{e \in E} \mathcal{N}(\gamma_r, e)\rho(e) = \sum_{e \in E} \sum_{k=1}^{r-1} c_k \mathcal{N}(\gamma_k, e)\rho(e) \\ &= \sum_{k=1}^{r-1} c_k \sum_{e \in E} \mathcal{N}(\gamma_k, e)\rho(e) = \sum_{k=1}^{r-1} c_k \ell_\rho(\gamma_k). \end{aligned} \tag{12.22}$$

This means that the first $r-1$ constraints, $\ell_\rho(\gamma_k) \geq 1$ for $k = 1, 2, \ldots, r-1$, imply the last constraint, $\ell_\rho(\gamma_r) \geq 1$, and, thus, that $\tilde{\Gamma}$ is not minimal for $\text{Mod}_{p,\sigma}(\Gamma)$. This is a contradiction. Hence, the rows of $\mathcal{N}(\tilde{\Gamma})$ must be linearly independent, (12.20) holds, and (iii) follows.

In particular, since $\mathcal{N}(\tilde{\Gamma})$ is a $\tilde{\Gamma} \times E$ matrix, $|\tilde{\Gamma}| = \operatorname{rank} \mathcal{N}(\tilde{\Gamma}) \leq |E|$, and (iv) follows as well.

Assertion (v) follows directly from (ii). Alternatively, note that by (12.9) any two optimal λ^* can only differ by an element of the kernel of $\mathcal{N}(\tilde{\Gamma})^T$. But (iii) states that the kernel of $\mathcal{N}(\tilde{\Gamma})^T$ is trivial. Also, positivity follows from minimality. □

12.10 THE OVERLAP MATRIX

Let $G = (V, E)$ be an unweighted graph and let $p = 2$. In this case, the modulus problem (12.1) simplifies and becomes the following quadratic problem:

$$\begin{aligned} \underset{\rho \in \mathbb{R}^E}{\text{minimize}} \quad & \rho^T \rho \\ \text{subject to} \quad & \mathcal{N}\rho \geq \mathbf{1}, \ \rho \geq 0. \end{aligned} \tag{12.23}$$

And the dual problem (12.8) in this case becomes:

$$\begin{aligned} \underset{\lambda\in\mathbb{R}^\Gamma}{\text{maximize}} \quad & \lambda^T\mathbf{1} - \frac{1}{4}\lambda^T\mathcal{N}\mathcal{N}^T\lambda \\ \text{subject to} \quad & \lambda \geq 0. \end{aligned} \tag{12.24}$$

Assume that Γ is itself a minimal family, so that, by Theorem 12.2, we have $\operatorname{rank}\mathcal{N} = |\Gamma|$ and the dual optimizer is unique. Let the *overlap matrix* for Γ be the matrix $C := \mathcal{N}\mathcal{N}^T$. The entries of C correspond to pairs of objects (γ_i, γ_j) and

$$C(\gamma_i, \gamma_j) = \sum_{e\in E} \mathcal{N}(\gamma_i, e)\mathcal{N}(\gamma_j, e). \tag{12.25}$$

In particular, if the objects γ are subsets of E, then

$$C(\gamma_i, \gamma_j) = |\gamma_i \cap \gamma_j|$$

counts the number of edges that γ_i and γ_j have in common.

Note that the overlap matrix C is symmetric and since $\operatorname{rank}\mathcal{N} = |\Gamma|$, we also have that $C = \mathcal{N}\mathcal{N}^T$ is invertible. So we get the following result.

Proposition 12.3. *Suppose $G = (V, E)$ is an unweighted graph and let Γ be a minimal family of objects on G. Then, the primal and dual problems have a unique solution given by*

$$\rho^* = \mathcal{N}^T C^{-1}\mathbf{1} \quad \textit{and} \quad \lambda^* = 2C^{-1}\mathbf{1}.$$

Moreover, the 2-Modulus of Γ is given by

$$\operatorname{Mod}_2(\Gamma) = \mathbf{1}^T C^{-1}\mathbf{1}.$$

Proof. Denote by λ^* the unique dual optimizer. Then, by part (v) of Theorem 12.2, $\lambda^*(\gamma) > 0$ for all $\gamma \in \Gamma$. Therefore, λ^* is a critical point of g. In other words, in order to maximize $g(\lambda)$ in (12.14) it's enough to compute the gradient of g and set it equal to zero. Recall that

$$g(\lambda) = \mathbf{1}^T\lambda - \frac{1}{4}\lambda^T C\lambda.$$

So we can compute the gradient of g to be:

$$\nabla g = \mathbf{1} - \frac{1}{2}C\lambda.$$

Since Γ is minimal, C is invertible and $\nabla g = 0$ yields $\lambda^* = 2C^{-1}\mathbf{1}$.

Moreover,

$$\operatorname{Mod}_2(\Gamma) = g(\lambda^*) = \mathbf{1}^T C^{-1}\mathbf{1}.$$

The extremal density is computed using (12.9). □

EXERCISES

12.1 *Weak duality.* The following is a very useful general fact. Suppose $F : X \times Y \to \mathbb{R}$. Show that

$$\inf_{x\in X}\sup_{y\in Y} F(x,y) \geq \sup_{y\in Y}\inf_{x\in X} F(x,y).$$

12.2 *Minimal subfamilies.* In this exercise we will lead you through the proof of Theorem 12.2. With all the notations in Theorem 12.2, we proceed point by point.

(i) To show this use Γ-monotonicity, see part 1. of Proposition 10.4.

(ii) Consider the extremal density $\rho^*(\tilde{\Gamma})$ and a set of optimal dual variables $\lambda^*(\tilde{\Gamma})$, for the modulus problem $\text{Mod}_{p,\sigma}(\tilde{\Gamma})$. Use the fact that $\text{Mod}_{p,\sigma}(\tilde{\Gamma}) = \text{Mod}_{p,\sigma}(\Gamma)$, to define new weights λ^* to give a solution to the dual problem (12.14) for Γ. Then, use (12.9) to show that $\rho^*(\tilde{\Gamma})$ is equal to the extremal density ρ^* for $\text{Mod}_{p,\sigma}(\Gamma)$. Finally, use the minimality of $\tilde{\Gamma}$, to conclude that $\lambda^*(\gamma) > 0$ for every $\gamma \in \tilde{\Gamma}$. To prove uniqueness, proceed by contradiction, noting that two distinct optimal λ^* would give rise to a line of optimal points, hence analyze where this line intersects the boundary of the feasible set. Finally, to show that this is a Beurling family use Proposition 12.2.

(iii) For this, show that any linear dependent relation in the rows of the usage matrix $\mathcal{N}$ implies that some constraints in the modulus problem are superfluous.

(iv) Use the properties of rank.

(v) Show that two optimal λ^* can only differ by an element of the kernel of $\mathcal{N}(\tilde{\Gamma})^T$.

12.3 *A Beurling subfamily that is not Lagrangian.* As we have seen in Proposition 12.2, every Lagrangian subfamily is also a Beurling subfamily. The goal of this exercise is to show that the converse does not necessarily hold.

Let G be the unweighted diamond graph with nodes and edges labeled as in Figure 1.1 (b), and let $\Gamma = \Gamma_{\text{path}}(x_1, x_3)$.

(i) Use Beurling's criterion for the subfamily $\Gamma^* = \{\gamma_1^*, \gamma_2^*\}$ where

$$\gamma_1^* = x_1\ x_0\ x_3 \qquad \text{and} \qquad \gamma_2^* = x_1\ x_2\ x_3$$

to show that the density

$$\rho(x_0, x_2) = 0 \qquad \text{and} \qquad \rho(e) \equiv \frac{1}{2} \;\text{ for } e \in E \setminus \{(x_0, x_2)\}$$

is extremal for $\text{Mod}_2(\Gamma)$.

Now, consider the subfamily $\tilde{\Gamma} = \{\tilde{\gamma}_1, \tilde{\gamma}_2\}$ where

$$\tilde{\gamma}_1 = x_1\ x_0\ x_2\ x_3 \qquad \text{and} \qquad \tilde{\gamma}_2 = x_1\ x_2\ x_0\ x_3$$

(ii) Show that $\tilde{\Gamma}$ is not a Beurling subfamily.

(iii) Note that $\Gamma = \Gamma^* \cup \tilde{\Gamma}$ and hence show that Γ is also a Beurling subfamily.

(iv) Finally, use (12.9) to show that Γ cannot be a Lagrangian subfamily.

12.4 *Modulus for single object families* Let $G = (V, E, \sigma)$ be a graph with edge weights $\sigma \in \mathbb{R}^E_{>0}$ and let $\Gamma = \{\gamma\}$ be a family of objects on G consisting of a single object with non-zero usage. To begin, fix $1 < p < \infty$ and let ρ_p denote the (unique) extremal density for $\text{Mod}_{p,\sigma}(\Gamma)$.

(i) Write down the Lagrangian for $\text{Mod}_{p,\sigma}(\Gamma)$, then use the KKT conditions and complementary slackness to get:

$$\text{Mod}_{p,\sigma}(\{\gamma\}) = \left(\sum_{e \in E} \sigma(e)^{-\frac{1}{p-1}} \mathcal{N}(\gamma, e)^{\frac{p}{p-1}}\right)^{1-p}$$

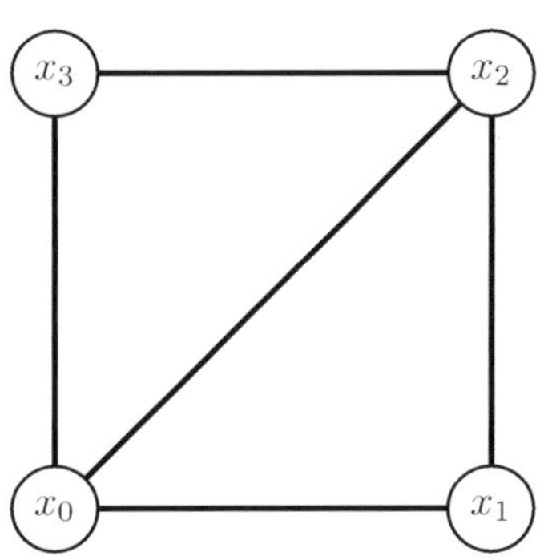

Figure 12.1 The diamond graph.

and

$$\rho_p(e) = \frac{\left(\sigma(e)^{-1}\mathcal{N}(\gamma,e)\right)^{\frac{1}{p-1}}}{\sum\limits_{e'\in E}\left(\sigma(e')^{-1}\mathcal{N}(\gamma,e')^p\right)^{\frac{1}{p-1}}}$$

(ii) To get a feeling for the formulas in (i), let $p=2$ and $\mathcal{N}(\gamma,\cdot)\in\{0,1\}$ and show that

$$\mathrm{Mod}_{2,\sigma}(\{\gamma\}) = \left(\sum_{\mathcal{N}(\gamma,e)=1}\sigma(e)^{-1}\right)^{-1}$$

and

$$\rho_2(e) = \frac{\sigma(e)^{-1}}{\sum\limits_{\mathcal{N}(\gamma,e')=1}\sigma(e')^{-1}}\mathcal{N}(\gamma,e)$$

(iii) Now let $p\downarrow 1$ and use the continuity of modulus in p to show that

$$\mathrm{Mod}_{1,\sigma}(\{\gamma\}) = \min_{e'\in E}\frac{\sigma(e')}{\mathcal{N}(\gamma,e')}.$$

Hence, show that when $\mathcal{N}(\gamma,\cdot)\in\{0,1\}$, we get

$$\mathrm{Mod}_{1,\sigma}(\{\gamma\}) = \min_{\mathcal{N}(\gamma,e')=1}\sigma(e')$$

(iv) Finally, show that the limit $\rho_1 := \lim\limits_{p\downarrow 1}\rho_p$ exists and when $\mathcal{N}(\gamma,e)\in\{0,1\}$ equals

$$\frac{\sigma(e)}{\sum_{j=1}^k\sigma(e_j)}\mathbb{1}_{\{e_1,\ldots,e_k\}}(e).$$

where $e_1,\ldots,e_k$ are all the edges that attain $\min\limits_{\mathcal{N}(\gamma,e')=1}\sigma(e')$.

(v) Check that ρ_1 is extremal for $\mathrm{Mod}_{1,\sigma}(\{\gamma\})$.

CHAPTER 13

The Probabilistic Interpretation and Blocking Duality

MODULUS has a probabilistic interpretation that is derived from its convex dual problem. Moreover, modulus duality is also related to combinatorial concepts such as blocking pairs.

13.1 THE PROBABILISTIC INTERPRETATION

For simplicity, we will always assume that Γ is a finite family of objects. In order to formulate the probabilistic interpretation of p-modulus, we let $\mathcal{P}(\Gamma)$ represent the set of probability mass functions (pmfs) on the set Γ. In other words, $\mathcal{P}(\Gamma)$ contains the set of vectors $\mu \in \mathbb{R}^{\Gamma}_{\geq 0}$ with the property that $\mu^T \mathbf{1} = 1$. Given a pmf μ, we can define a Γ-valued random variable $\underline{\gamma}$ with distribution:

$$\mathbb{P}_{\mu}\left(\underline{\gamma} = \gamma\right) = \mu(\gamma).$$

Then, for every edge $e \in E$, the usage $\mathcal{N}(\underline{\gamma}, e)$ is also a random variable, and we represent the expected edge usage (depending on the pmf μ) as $\mathbb{E}_{\mu}\left[\mathcal{N}(\underline{\gamma}, e)\right]$. With these notations, we will see that the Lagrangian dual problem of modulus (12.8) can be thought as the problem of minimizing the q-norm of the vector of edge usage expectations over all possible pmf's. Moreover, equation (13.2) below shows that the expected edge usage corresponding to optimal pmfs yields a probabilistic interpretation of the extremal density ρ^*.

DOI: 10.1201/9781003024866-16

Theorem 13.1. *Let $G = (V, E, \sigma)$ be a finite graph with edge weights $\sigma > 0$, and let Γ be a finite family of objects on G with usage matrix $\mathcal{N}$, as in Corollary 10.1. Then, for any $1 < p < \infty$, letting $q = p/(p-1)$ be the conjugate exponent to p, we have*

$$\mathrm{Mod}_{p,\sigma}(\Gamma)^{-\frac{1}{p}} = \left(\min_{\mu\in\mathcal{P}(\Gamma)} \sum_{e\in E} \sigma(e)^{-\frac{q}{p}} \mathbb{E}_\mu \left[\mathcal{N}(\underline{\gamma}, e) \right]^q \right)^{\frac{1}{q}}. \tag{13.1}$$

Moreover, $\mu \in \mathcal{P}(\Gamma)$ attains the minimum on the right-hand side of (13.1)*, if and only if*

$$\mathbb{E}_\mu \left[\mathcal{N}(\underline{\gamma}, e) \right] = \frac{\sigma(e)\rho^*(e)^{p-1}}{\mathrm{Mod}_{p,\sigma}(\Gamma)} \qquad \forall e \in E, \tag{13.2}$$

where ρ^* is the unique extremal density for $\mathrm{Mod}_{p,\sigma}(\Gamma)$.

Definition 13.1. If $\mu \in \mathcal{P}(\Gamma)$ attains the minimum on the right-hand side of (13.1), or equivalently, satisfies (13.2), we say that *μ is optimal for* $\mathrm{Mod}_{p,\sigma}(\Gamma)$. By a slight abuse of notation, we will use the expression "μ is optimal for $\mathrm{Mod}_{p,\sigma}(\Gamma)$" even in cases where $\mu \in \mathcal{P}(\tilde{\Gamma})$ for some subfamily $\tilde{\Gamma} \subset \Gamma$, since such pmfs embed into $\mathcal{P}(\Gamma)$ by setting them equal to zero outside of $\tilde{\Gamma}$.

Optimal pmfs for a given modulus problem always exist, but they might not be unique in general.

Proof of Theorem 13.1. The modulus primal problem (12.1) is a standard convex optimization problem. Its Lagrangian dual problem is (12.8). As we have seen, strong duality holds and when $1 < p < \infty$ both extrema are attained. Moreover, if ρ^* is the unique minimizer of the modulus problem and λ^* is any maximizer of the Lagrangian dual, then they must be related as in (12.9).

Since $\lambda \geq 0$ we can decompose it as $\lambda = \nu\mu$ with $\nu \geq 0$ and $\mu \in \mathcal{P}(\Gamma)$. Thus, we can rewrite the dual problem (12.8) as

$$\max_{\nu\geq 0} \left\{ \nu - (p-1)\left(\frac{\nu}{p}\right)^q \min_{\mu\in\mathcal{P}(\Gamma)} \sum_{e\in E} \sigma(e)^{-\frac{q}{p}} \left(\sum_{\gamma\in\Gamma} \mathcal{N}(\gamma, e)\mu(\gamma) \right)^q \right\}.$$

Let

$$\begin{aligned} \alpha &:= \min_{\mu\in\mathcal{P}(\Gamma)} \sum_{e\in E} \sigma(e)^{-\frac{q}{p}} \left(\sum_{\gamma\in\Gamma} \mathcal{N}(\gamma, e)\mu(\gamma) \right)^q \\ &= \min_{\mu\in\mathcal{P}(\Gamma)} \sum_{e\in E} \sigma(e)^{-\frac{q}{p}} \mathbb{E}_\mu \left[\mathcal{N}(\underline{\gamma}, e) \right]^q. \end{aligned}$$

Then, the maximum over $\nu \geq 0$ is attained at $\nu^* := p\alpha^{-\frac{p}{q}}$, and strong duality implies that

$$\mathrm{Mod}_{p,\sigma}(\Gamma) = \nu^* - (p-1)\left(\frac{\nu^*}{p}\right)^q \alpha = \alpha^{-\frac{p}{q}}.$$

Thus,

$$\min_{\mu \in \mathcal{P}(\Gamma)} \sum_{e \in E} \sigma(e)^{-\frac{q}{p}} \mathbb{E}_\mu \left[\mathcal{N}(\underline{\gamma}, e)\right]^q = \alpha = \mathrm{Mod}_{p,\sigma}(\Gamma)^{-\frac{q}{p}},$$

proving (13.1). The remainder of the theorem follows from (12.9):

$$\rho^*(e) = \left(\frac{\nu^*}{p\sigma(e)} \sum_{\gamma \in \Gamma} \mathcal{N}(\gamma, e)\mu^*(\gamma)\right)^{\frac{1}{p-1}} = \alpha^{-1}\sigma(e)^{-\frac{q}{p}} \mathbb{E}_{\mu^*}\left[\mathcal{N}(\underline{\gamma}, e)\right]^{\frac{q}{p}}$$

and the fact that, if $\mu \in \mathcal{P}(\Gamma)$ satisfies (13.2), then $\lambda := \nu^*\mu$ is admissible for (12.8) and has the same objective value as any optimal λ^*. □

In particular, equations (13.5) and (13.6) below show that computing the 2-modulus of a family Γ of edge subsets on an unweighted graph is equivalent to finding a pmf on Γ that *minimizes the expected overlap* of two iid Γ-valued random variables. Moreover, in this case, the extremal density ρ^* is proportional to the *edge usage probabilities* given by an optimal measure μ^*.

Corollary 13.1. *Let Γ be a finite family of objects on an unweighted graph $G = (V, E)$. Then*

$$\mathrm{Mod}_2(\Gamma)^{-1} = \min_{\mu \in \mathcal{P}(\Gamma)} \mu^T \mathcal{N}\mathcal{N}^T \mu. \tag{13.3}$$

Moreover, any pmf $\mu \in \mathcal{P}(\Gamma)$ is optimal if and only if,

$$\frac{\rho^*(e)}{\mathrm{Mod}_2(\Gamma)} = \left(\mathcal{N}^T\mu\right)(e) = \mathbb{E}_\mu\left[\mathcal{N}(\underline{\gamma}, e)\right], \qquad \forall e \in E, \tag{13.4}$$

where ρ^* is the extremal density for $\mathrm{Mod}_2(\Gamma)$.

Furthermore, if Γ is a collection of subsets of E with usage given by the indicator function. Then,

$$\mathrm{Mod}_2(\Gamma)^{-1} = \min_{\mu \in \mathcal{P}(\Gamma)} \mathbb{E}_\mu \left|\underline{\gamma} \cap \underline{\gamma}'\right|, \tag{13.5}$$

where $\underline{\gamma}$ *and* $\underline{\gamma}'$ *are two independent random variables chosen according to* μ, *and* $\left|\underline{\gamma}\cap\underline{\gamma}'\right|$ *is their* overlap. *While, any pmf* $\mu \in \mathcal{P}(\Gamma)$ *is optimal if and only if,*

$$\frac{\rho^*(e)}{\mathrm{Mod}_2(\Gamma)} = \mathbb{P}_\mu\left(e \in \underline{\gamma}\right) \qquad \forall e \in E. \tag{13.6}$$

Proof. Note that when $p = 2$ and $\sigma \equiv 1$, the quantity minimized in (13.1) becomes:

$$\begin{aligned}
\sum_{e\in E}\left(\mathbb{E}_\mu\left[\mathcal{N}(\underline{\gamma},e)\right]\right)^2 &= \sum_{e\in E}\left[\left(\mathcal{N}^T\mu\right)(e)\right]^2 \\
&= \sum_{e\in E}\left(\sum_{\gamma\in\Gamma}\mu(\gamma)\mathcal{N}(\gamma,e)\right)^2 \\
&= \sum_{\gamma,\gamma'\in\Gamma}\mu(\gamma)\mu(\gamma')\left(\sum_{e\in E}\mathcal{N}(\gamma,e)\mathcal{N}(\gamma',e)\right). \\
&= \mu^T\mathcal{N}\mathcal{N}^T\mu.
\end{aligned}$$

When the objects γ are subsets of E, then $\mathcal{N}(\gamma, e) = \mathbb{1}_\gamma(e)$ is $\{0,1\}$-valued and $C = \mathcal{N}\mathcal{N}^T$ is the overlap matrix defined in (12.25). □

An immediate consequence of Corollary 13.1 is the following, which can be useful for establishing the modulus.

Corollary 13.2. *Let* Γ *be a finite family of objects on a graph* G, *let* $\rho \in \mathrm{Adm}(\Gamma)$ *and let* $\mu \in \mathcal{P}(\Gamma)$, *then*

$$\rho^T\rho \geq \mathrm{Mod}_2(\Gamma) \geq \left(\mu^T\mathcal{N}\mathcal{N}^T\mu\right)^{-1},$$

with equalities if and only if ρ *and* μ *are optimal.*

Next, we shall focus on more combinatorial approach.

13.2 REVIEW OF CONVEX ANALYSIS CONCEPTS

First, we recall some general definitions. Let $\mathcal{K}$ be the set of all closed convex sets $K \subset \mathbb{R}^E_{\geq 0}$ that are *recessive*, in the sense that $K + \mathbb{R}^E_{\geq 0} = K$. To avoid trivial cases, we shall assume that $\varnothing \neq K \subsetneq \mathbb{R}^E_{\geq 0}$, for $K \in \mathcal{K}$. For each $K \in \mathcal{K}$, there is an associated *blocking polyhedron,*

$$\mathrm{BL}(K) := \left\{\eta \in \mathbb{R}^E_{\geq 0} : \eta^T\rho \geq 1, \;\; \forall \rho \in K\right\}. \tag{13.7}$$

The *dominant* of a set $P \subset \mathbb{R}^E_{\geq 0}$ is the recessive closed convex set

$$\mathrm{Dom}(P) := \overline{\mathrm{co}}(P) + \mathbb{R}^E_{\geq 0}, \tag{13.8}$$

where $\mathrm{co}(P)$ denotes the *convex hull* of P, and $\overline{\mathrm{co}}(P)$ the closure of the convex hull.

Given $K \in \mathcal{K}$ and a point $x \in K$, we say that x is an *extreme point* of K, if whenever $x = tx_1 + (1-t)x_2$ for some $x_1, x_2 \in K$ and some $t \in (0,1)$, then $x_1 = x_2 = x$. We write $\mathrm{Ext}(K)$ for the set of all extreme points of K.

Given $K \in \mathcal{K}$ and a *direction* y, i.e., a unit vector in $\mathbb{R}^E$, we say y is an *extreme direction* of K if y supports a half-line face of K. Recall that a convex subset K' of K is a *face* of K if every closed linear segment in K with a relative interior point in K' has both endpoints in K'. We write $\mathrm{Dir}(K)$ for the set of all extreme directions of K.

Recall that if S is a set of unit vectors in $\mathbb{R}^E$, then $\mathrm{cone}(S)$ is the convex cone generated by all vectors whose directions belong to S. Namely, S is the set of all positive linear combinations using vectors in S. Next, we quote [69, Theorem 18.5].

Theorem 13.2 ([69]). *Suppose C is a closed convex set containing no lines. Let $\mathrm{Ext}(C)$ be the set of extreme points of C and let $\mathrm{Dir}(C)$ be the set of extreme directions of C, then*

$$C = \mathrm{co}(\mathrm{Ext}(C)) + \mathrm{cone}(\mathrm{Dir}(C)).$$

Using Theorem 13.2 in our context, we obtain the following useful result.

Proposition 13.1. *Let $K \in \mathcal{K}$, i.e. K is a closed convex recessive set in $\mathbb{R}^E_{\geq 0}$, and let $\mathrm{Ext}(K)$ be the set of extreme points of K. Then,*

$$K = \mathrm{Dom}(\mathrm{Ext}(K)).$$

Proof. Suppose $x \in \mathrm{Dom}(\mathrm{Ext}(K))$. Then $x - \lambda \in \overline{\mathrm{co}}(\mathrm{Ext}(K))$, for some $\lambda \geq 0$. Since K is convex and $\mathrm{Ext}(K) \subset K$, we have $\mathrm{co}(\mathrm{Ext}(K)) \subset K$. Furthermore, since K is closed, we have $\overline{\mathrm{co}}(\mathrm{Ext}(K)) \subset K$. Therefore, $x - \lambda \in K$. Finally, since K is recessive, $x = (x - \lambda) + \lambda \in K$. So, $\mathrm{Dom}(\mathrm{Ext}(K)) \subset K$.

Now, suppose that $x \in K$. Note that, since $K \subset \mathbb{R}^E_{\geq 0}$, we have that K contains no lines. Hence, Theorem 13.2 applies, and $x \in \mathrm{co}(\mathrm{Ext}(K)) + \mathrm{cone}(\mathrm{Dir}(K))$. Note also that since $K \subset \mathbb{R}^E_{\geq 0}$, the extreme directions

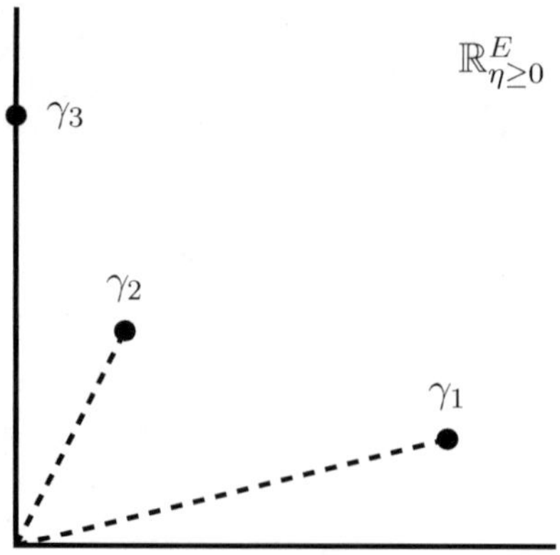

Figure 13.1 A family Γ in η-space.

of K also satisfy $\mathrm{Dir}(K) \subset \mathbb{R}^E_{\geq 0}$, and hence $\mathrm{cone}(\mathrm{Dir}(K)) \subset \mathbb{R}^E_{\geq 0}$. Therefore, $x \in \mathrm{Dom}(\mathrm{co}(\mathrm{Ext}(K))) = \mathrm{Dom}(\mathrm{Ext}(K))$. This shows that $K \subset \mathrm{Dom}(\mathrm{Ext}(K))$ and the result follows. $\square$

If Γ is a finite non-trivial family of objects on a graph G, the admissible set $\mathrm{Adm}(\Gamma)$ is determined by finitely many inequalities, in particular, $\mathrm{Adm}(\Gamma)$ has finitely many faces. However, $\mathrm{Adm}(\Gamma)$ is also determined by its finitely many extreme points, or "vertices". Indeed, it follows from Proposition 13.1 that $\mathrm{Adm}(\Gamma)$ is the dominant of its extreme points $\mathrm{ext}(\mathrm{Adm}(\Gamma))$.

Definition 13.2. Suppose $G = (V, E)$ is a finite graph and Γ is a finite non-trivial family of objects on G. We say that the family

$$\hat{\Gamma} := \mathrm{ext}(\mathrm{Adm}(\Gamma)) = \{\hat{\gamma}_1, \ldots, \hat{\gamma}_s\} \subset \mathbb{R}^E_{\geq 0},$$

consisting of the extreme points of $\mathrm{Adm}(\Gamma)$, is the *Fulkerson blocker* of Γ. Also, we define the usage matrix $\hat{\mathcal{N}} \in \mathbb{R}^{\hat{\Gamma}\times E}_{\geq 0}$ for $\hat{\Gamma}$ to be the matrix whose rows are the vectors $\hat{\gamma}^T$, for $\hat{\gamma} \in \hat{\Gamma} \subset \mathbb{R}^E_{\geq 0}$.

Fulkerson duality can best be illustrated using two copies of the non-negative orthant, which we call *ρ-space* $\mathbb{R}^E_{\rho\geq 0}$ and *η-space* $\mathbb{R}^E_{\eta\geq 0}$, respectively. It is useful to think of the usage vectors for objects in Γ as being in η-space, while the usage vectors for the Fulkerson dual family as being in ρ-space. See Figure 13.1 and Figure 13.2.

In the next few sections, we will loosely follow the presentation in [4].

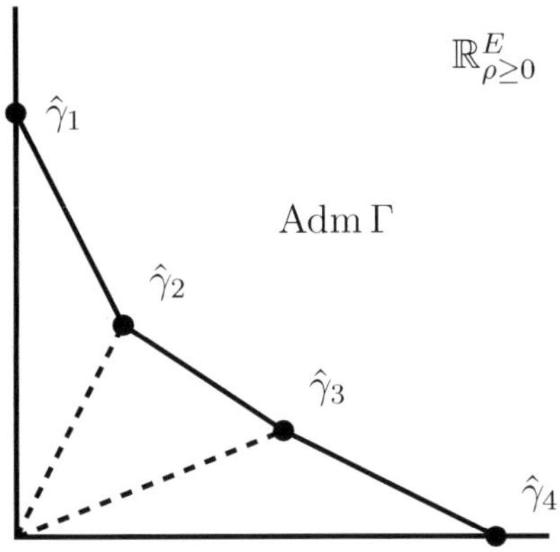

Figure 13.2 The Fulkerson dual $\hat{\Gamma}$ in ρ-space.

13.3 FULKERSON'S THEOREM

Here we rephrase some classical results of Fulkerson.

Theorem 13.3 (Fulkerson [41]). *Let $G = (V, E)$ be a graph and let Γ be a family of objects on G, as in Corollary 10.1. We identify Γ with the corresponding set of edge usage vectors, hence we think of Γ as a finite subset of $\mathbb{R}^E_{\eta\geq 0}$. Let $\hat{\Gamma}$ be the Fulkerson blocker of Γ in $\mathbb{R}^E_{\rho\geq 0}$. Then, with the notations in (10.3), (13.8), and (13.7):*

(1) $\mathrm{Adm}(\Gamma) = \mathrm{Dom}(\hat{\Gamma}) = \mathrm{BL}(\mathrm{Adm}(\hat{\Gamma}))$;

(2) $\mathrm{Adm}(\hat{\Gamma}) = \mathrm{Dom}(\Gamma) = \mathrm{BL}(\mathrm{Adm}(\Gamma))$;

(3) $\hat{\hat{\Gamma}} \subset \Gamma$.

In other words, (3) says that Γ always contains the extreme points of $\mathrm{Adm}(\hat{\Gamma})$. Combining (1) and (2) we get the following.

Corollary 13.3. *Let $G = (V, E)$ be a graph and let Γ be a family of objects on G, as in Corollary 10.1. We identify Γ with the corresponding set of edge usage vectors, hence we think of Γ as a finite subset of $\mathbb{R}^E_{\eta\geq 0}$. Then,*

$$\mathrm{BL}(\mathrm{BL}(\mathrm{Adm}(\Gamma))) = \mathrm{Adm}(\Gamma) \quad \textit{and} \quad \mathrm{BL}(\mathrm{BL}(\mathrm{Dom}(\Gamma))) = \mathrm{Dom}(\Gamma).$$

as well as

$$\mathrm{Adm}(\Gamma) = \mathrm{BL}\left(\mathrm{Dom}(\Gamma)\right) \quad \textit{and} \quad \mathrm{BL}(\mathrm{Adm}(\Gamma)) = \mathrm{Dom}(\Gamma).$$

We include a proof of Theorem 13.3 for the reader's convenience.

Proof of Theorem 13.3. We first prove (2). Suppose $\eta \in \mathrm{BL}(\mathrm{Adm}(\Gamma))$ in η-space. Then $\eta^T\rho \geq 1$, for every $\rho \in \mathrm{Adm}(\Gamma)$ in ρ-space. In particular,

$$\eta^T\hat{\gamma} \geq 1, \qquad \forall \hat{\gamma} \in \hat{\Gamma}. \tag{13.9}$$

In other words, $\eta \in \mathrm{Adm}(\hat{\Gamma})$. Hence, $\mathrm{BL}(\mathrm{Adm}(\Gamma)) \subset \mathrm{Adm}(\hat{\Gamma})$.

Conversely, suppose $\eta \in \mathrm{Adm}(\hat{\Gamma})$, that is, assume that (13.9) holds. By Proposition 13.1,

$$\mathrm{Adm}(\Gamma) = \mathrm{Dom}(\mathrm{Ext}(\mathrm{Adm}\,\Gamma)) = \mathrm{co}(\hat{\Gamma}) + \mathbb{R}^E_{\rho\geq 0}.$$

In other words, for every $\rho \in \mathrm{Adm}(\Gamma)$, there is a pmf $\mu \in \mathcal{P}(\hat{\Gamma})$ and a vector $z \geq 0$ such that

$$\rho = \sum_{\hat{\gamma}\in\hat{\Gamma}} \mu(\hat{\gamma})\hat{\gamma} + z.$$

And by (13.9),

$$\eta^T\rho = \sum_{\hat{\gamma}\in\hat{\Gamma}} \mu(\hat{\gamma})\eta^T\hat{\gamma} + \eta^T z \geq 1.$$

So $\eta \in \mathrm{BL}(\mathrm{Adm}(\Gamma))$, and $\mathrm{Adm}(\hat{\Gamma}) = \mathrm{BL}(\mathrm{Adm}(\Gamma))$.

Note that $\eta \in \mathrm{BL}(\mathrm{Adm}(\Gamma))$ if and only if the value of the following linear program is greater or equal 1.

$$\begin{array}{ll} \text{minimize} & \eta^T\rho \\ \text{subject to} & \mathcal{N}\rho \geq \mathbf{1},\ \rho \geq 0, \end{array} \tag{13.10}$$

where $\mathcal{N}$ is the usage matrix for Γ. The Lagrangian in this case is

$$\mathcal{L}(\rho,\lambda,\nu) := \eta^T\rho + \lambda^T(\mathbf{1} - \mathcal{N}\lambda) - \nu^T\rho = \lambda^T\mathbf{1} + \rho^T(\eta - \mathcal{N}^T\lambda - \nu),$$

with $\rho \in \mathbb{R}^E$, $\lambda \in \mathbb{R}^\Gamma_{\geq 0}$, and $\nu \in \mathbb{R}^E_{\geq 0}$. In particular, the dual problem of (13.10) is

$$\begin{array}{ll} \text{maximize} & \lambda^T\mathbf{1} \\ \text{subject to} & \mathcal{N}^T\lambda \leq \eta,\ \lambda \geq 0. \end{array} \tag{13.11}$$

Splitting $\lambda = \nu\mu$, with $\nu \geq 0$ and $\mu \in \mathcal{P}(\Gamma)$, we can rewrite this problem as

$$\begin{array}{ll} \text{maximize} & \nu \\ \text{subject to} & \nu\mathcal{N}^T\mu \leq \eta,\ \mu \in \mathcal{P}(\Gamma). \end{array} \tag{13.12}$$

By strong duality, $\eta \in \mathrm{BL}(\mathrm{Adm}(\Gamma))$ if and only if there is $\nu \geq 1$ and $\mu \in \mathcal{P}(\Gamma)$ so that

$$\eta \geq \nu \mathcal{N}^T \mu \tag{13.13}$$

Namely, $\eta \in \mathrm{BL}(\mathrm{Adm}(\Gamma))$ implies that $\eta \geq \mathcal{N}^T \mu$. Note that the columns of $\mathcal{N}^T$ are the usage vectors corresponding to the objects of Γ. Therefore, thinking of Γ as a subset of η-space, $\mathcal{N}^T \mu \in \mathrm{co}(\Gamma)$. In particular, $\eta \in \mathrm{Dom}(\Gamma)$. This shows that $\mathrm{BL}(\mathrm{Adm}(\Gamma)) \subset \mathrm{Dom}(\Gamma)$.

Conversely, if $\eta \in \mathrm{Dom}(\Gamma)$, then (13.13) holds with $\nu = 1$ and some $\mu \in \mathcal{P}(\Gamma)$. Therefore, $\eta \in \mathrm{BL}(\mathrm{Adm}(\Gamma))$ and $\mathrm{BL}(\mathrm{Adm}(\Gamma)) = \mathrm{Dom}(\Gamma)$. So (2) is proven.

In particular, since $\hat{\hat{\Gamma}}$ is the set of extreme points of $\mathrm{Adm}(\hat{\Gamma})$ and (2) implies that $\mathrm{Adm}(\hat{\Gamma}) = \mathrm{Dom}(\Gamma)$, we conclude that $\hat{\hat{\Gamma}} \subset \Gamma$, hence (3) is proved as well.

To prove (1), we apply (2) to $\mathrm{Adm}(\hat{\Gamma})$ and find that

$$\mathrm{BL}(\mathrm{Adm}(\hat{\Gamma})) = \mathrm{Adm}(\hat{\hat{\Gamma}}) \supset \mathrm{Adm}(\Gamma),$$

where the last inclusion follows from (3), since $\hat{\hat{\Gamma}} \subset \Gamma$. Also, by (3) applied to $\hat{\Gamma}$, the extreme points of $\mathrm{Adm}(\hat{\hat{\Gamma}})$ are a subset of $\hat{\Gamma}$ and therefore they are a subset of the vertices of $\mathrm{Adm}(\Gamma)$. This implies that $\mathrm{Adm}(\hat{\hat{\Gamma}}) \subset \mathrm{Adm}(\Gamma)$. So we have $\mathrm{BL}(\mathrm{Adm}(\hat{\Gamma})) = \mathrm{Adm}(\Gamma)$.

Moreover, by (2) applied to $\hat{\Gamma}$, we get that

$$\mathrm{BL}(\mathrm{Adm}(\hat{\Gamma})) = \mathrm{Dom}(\hat{\Gamma}).$$

So (1) is proved as well. □

13.4 FULKERSON DUALITY FOR p-MODULUS

As we have seen in (10.6), the admissible set is the feasible set for the convex minimization problem defining the modulus of a family. In particular, different families may have the same modulus if they happen to have the same admissible set. This led Adriana Ortiz-Aquino to develop the following concept in her thesis [67].

Definition 13.3. Two families of objects Γ and Γ' are said to be equivalent (in the sense of modulus) if $\mathrm{Adm}\,\Gamma = \mathrm{Adm}\,\Gamma'$. In this case, we write $\Gamma \simeq \Gamma'$.

One may notice, for instance, that for any family Γ, thinking of it as a set of usage vectors in $\mathbb{R}^E_{\geq 0}$, we always have $\Gamma \simeq \mathrm{Dom}\,\Gamma$.

Definition 13.4. Let $G = (V, E)$ be a graph. Consider two families of objects Γ and $\tilde{\Gamma}$, thought as sets of usage vectors in $\mathbb{R}^E_{\geq 0}$. We say that they form a *Fulkerson dual pair*, or $\tilde{\Gamma}$ is *a Fulkerson dual family* for Γ, if

$$\operatorname{Adm}\tilde{\Gamma} = \operatorname{Dom}\Gamma. \tag{13.14}$$

Proposition 13.2. *Consider two families of objects Γ and $\tilde{\Gamma}$, thought as sets of usage vectors in $\mathbb{R}^E_{\geq 0}$, and let $\hat{\Gamma}$ be the Fulkerson blocker dual family for Γ as in Definition 13.2. Then, $\tilde{\Gamma}$ is a Fulkerson dual family of Γ if and only if one of the following holds:*

(i) $\operatorname{Adm}\tilde{\Gamma} = \operatorname{BL}(\operatorname{Adm}\Gamma)$;

(ii) $\operatorname{Adm}\Gamma = \operatorname{Dom}\tilde{\Gamma}$;

(iii) $\operatorname{Adm}\tilde{\Gamma} = \operatorname{Adm}\hat{\Gamma}$;

(iv) $\operatorname{Dom}\tilde{\Gamma} = \operatorname{Dom}\hat{\Gamma}$;

Remark 13.1. In particular, any one of the equality statements (i), (ii), and (iii) in Proposition 13.2 can be substituted for (13.14) in the definition of the Fulkerson dual family. Also (ii) shows that $\tilde{\Gamma}$ is a Fulkerson dual family of Γ if and only if Γ is a Fulkerson dual family of $\tilde{\Gamma}$. That is the reason we may refer to Γ and $\tilde{\Gamma}$ as a dual pair. Finally, (iii) shows that a Fulkerson dual family is always equivalent to the Fulkerson blocker family $\hat{\Gamma}$.

Proof of Proposition 13.2. By Corollary 13.3, $\operatorname{Dom}\Gamma = \operatorname{BL}(\operatorname{Adm}\Gamma)$. So, (13.14) is equivalent to (i). Also by Corollary 13.3, $\operatorname{BL}(\operatorname{BL}(\operatorname{Adm}\Gamma) = \operatorname{Adm}\Gamma$. So (i) is equivalent to (ii) by taking BL of both sides. Moreover, (13.14) is equivalent to (iii) by Theorem 13.3 (2). Finally, (iv) follows from part (ii) and Theorem 13.3 (1). □

Theorem 13.4. *Let $G = (V, E)$ be a graph and let Γ be a family of objects on G, as in Corollary 10.1, with Fulkerson dual family $\tilde{\Gamma}$. Let the exponent $1 < p < \infty$ be given, with $q = (p-1)/p$ its conjugate exponent. For any set of weights $\sigma \in \mathbb{R}^E_{>0}$, define the dual set of weights $\tilde{\sigma}$ as $\tilde{\sigma}(e) := \sigma(e)^{1-q}$, for all $e \in E$. Then,*

$$\operatorname{Mod}_{p,\sigma}(\Gamma)^{\frac{1}{p}} \operatorname{Mod}_{q,\tilde{\sigma}}(\tilde{\Gamma})^{\frac{1}{q}} = 1. \tag{13.15}$$

Moreover, the optimal $\rho^* \in \mathrm{Adm}(\Gamma)$ *for* $\mathrm{Mod}_{p,\sigma}(\Gamma)$ *and the optimal* $\eta^* \in \mathrm{Adm}(\tilde{\Gamma})$ *for* $\mathrm{Mod}_{q,\tilde{\sigma}}(\tilde{\Gamma})$ *are unique and are related as follows:*

$$\eta^*(e) = \frac{\sigma(e)\rho^*(e)^{p-1}}{\mathrm{Mod}_{p,\sigma}(\Gamma)} \qquad \forall e \in E. \tag{13.16}$$

Also

$$\mathrm{Mod}_{1,\sigma}(\Gamma)\,\mathrm{Mod}_{\infty,\sigma^{-1}}(\tilde{\Gamma}) = 1. \tag{13.17}$$

Remark 13.2. The case for $p = 2$ is essentially contained in [59, Lemma 2], although not stated in terms of modulus and with a different proof. However, it is worth stating it separately. Namely,

$$\mathrm{Mod}_{2,\sigma}(\Gamma)\,\mathrm{Mod}_{2,\sigma^{-1}}(\tilde{\Gamma}) = 1. \tag{13.18}$$

And

$$\sigma(e)\rho^*(e) = \mathrm{Mod}_{2,\sigma}(\Gamma)\eta^*(e) \qquad \forall e \in E, \tag{13.19}$$

in this case. In particular, ρ^* and η^* are parallel when $\sigma \equiv 1$.

Proof of Theorem 13.4. Recall that by Proposition 13.2 (i), $\mathrm{Adm}(\tilde{\Gamma}) = \mathrm{BL}(\mathrm{Adm}(\Gamma))$. Hence, for all $\rho \in \mathrm{Adm}(\Gamma)$ and $\eta \in \mathrm{Adm}(\tilde{\Gamma})$,

$$\begin{aligned}
1 &\le \sum_{e\in E} \rho(e)\eta(e) && (\mathrm{Adm}(\tilde{\Gamma}) = \mathrm{BL}(\mathrm{Adm}(\Gamma)))\\
&= \sum_{e\in E} \left(\sigma(e)^{1/p}\rho(e)\right)\left(\sigma(e)^{-1/p}\eta(e)\right) && \text{(multiply and divide)}\\
&\le \left(\sum_{e\in E} \sigma(e)\rho(e)^p\right)^{1/p} \left(\sum_{e\in E} \tilde{\sigma}(e)\eta(e)^q\right)^{1/q} && \text{(by Hölder's inequality)},
\end{aligned}$$

since $-q/p = 1 - q$.

Taking the infimum of ρ and η, we get the lower bound

$$\mathrm{Mod}_{p,\sigma}(\Gamma)^{1/p}\,\mathrm{Mod}_{q,\tilde{\sigma}}(\tilde{\Gamma})^{1/q} \ge 1. \tag{13.20}$$

Conversely, let $\eta^* \in \mathrm{Adm}(\tilde{\Gamma})$ be the minimizer for $\mathrm{Mod}_{q,\tilde{\sigma}}(\tilde{\Gamma})$, and set

$$\alpha := \mathrm{Mod}_{q,\tilde{\sigma}}(\tilde{\Gamma}) = \sum_{e\in E} \tilde{\sigma}(e)\eta^*(e)^q.$$

Then, (13.20) implies that

$$\mathrm{Mod}_{p,\sigma}(\Gamma) \ge \alpha^{-\frac{p}{q}} = \alpha^{-\frac{1}{q-1}}. \tag{13.21}$$

Define a density $\tilde{\rho}$ as it would predicted by the relation in (13.16):

$$\tilde{\rho}(e) := \alpha^{-1}\left(\frac{\tilde{\sigma}(e)}{\sigma(e)}\eta^*(e)^q\right)^{1/p} = \alpha^{-1}\tilde{\sigma}(e)\eta^*(e)^{q/p} = \alpha^{-1}\tilde{\sigma}(e)\eta^*(e)^{q-1}. \tag{13.22}$$

Note that

$$\mathcal{E}_{p,\sigma}(\tilde{\rho}) = \sum_{e\in E}\sigma(e)\rho^*(e)^p = \alpha^{-p}\sum_{e\in E}\tilde{\sigma}(e)\eta^*(e)^q = \alpha^{1-p} = \alpha^{-\frac{1}{q-1}}.$$

Thus, if we can show that $\tilde{\rho}$ is admissible for Γ, then (13.21) is attained and $\tilde{\rho}$ must equal the extremal density ρ^* for $\text{Mod}_{p,\sigma}(\Gamma)$. In particular, (13.15) would follow. Moreover, (13.16) is another way of writing (13.19).

To see that $\tilde{\rho} \in \text{Adm}(\Gamma)$, we will use Theorem 13.3 (1) and Proposition 13.2 (iii), and verify that $\tilde{\rho} \in \text{BL}(\text{Adm}(\tilde{\Gamma}))$ instead, i.e., that $\sum_{e\in E}\tilde{\rho}(e)\eta(e) \geq 1$ for all $\eta \in \text{Adm}(\tilde{\Gamma})$. First, consider $\eta = \eta^*$. In this case

$$\sum_{e\in E}\tilde{\rho}(e)\eta^*(e) = \alpha^{-1}\sum_{e\in E}\tilde{\sigma}(e)\eta^*(e)^q = 1. \tag{13.23}$$

Now let $\eta \in \text{Adm}(\tilde{\Gamma})$ be arbitrary. Since $\text{Adm}(\tilde{\Gamma})$ is convex, we have that $\eta_t := (1-\theta)\eta^* + \theta\eta \in \text{Adm}(\tilde{\Gamma})$ for all $t \in [0,1]$. We will need the fact that the function $\phi(x) := x^q$ has a derivative

$$\phi'(x) = qx^{q-1} \qquad \text{whenever } x \geq 0 \tag{13.24}$$

Then,

$$\begin{aligned}
0 &\leq \lim_{t\downarrow 0}\frac{\mathcal{E}_{q,\tilde{\sigma}}(\eta_t) - \alpha}{t} && (\text{since } \alpha = \text{Mod}_{q,\tilde{\sigma}}(\tilde{\Gamma}))\\
&= \sum_{e\in E}\tilde{\sigma}(e)\lim_{t\downarrow 0}\frac{[(1-t)\eta^*(e) + t\eta(e)]^q - \eta^*(e)^q}{t}\\
&= q\sum_{e\in E}\tilde{\sigma}(e)\left[\eta(e) - \eta^*(e)\right]\eta^*(e)^{q-1} && ((13.24) \text{ and chain rule})\\
&= \alpha q\sum_{e\in E}\tilde{\rho}(e)\left[\eta(e) - \eta^*(e)\right] && (\text{by } (13.22))
\end{aligned}$$

Thus, by (13.23),

$$\sum_{e\in E}\tilde{\rho}(e)\eta(e) \geq \sum_{e\in E}\tilde{\rho}(e)\eta^*(e) = 1,$$

and the proof of (13.15) is complete. Moreover, (13.22) implies (13.16). We postpone the proof of (13.17) to Section 13.6. □

13.5 BLOCKING, THE PROBABILISTIC INTERPRETATION, AND MINIMAL SUBFAMILIES

Here, we reconcile the theory of Fulkerson duality with the probabilistic interpretation of modulus. Next, we will consider the special case of minimal subfamilies.

Proposition 13.3. *Let $G = (V, E, \sigma)$ be a graph and Γ a finite family of objects on G, as in Corollary 10.1. Let $\tilde{\Gamma}$ be the Fulkerson dual family for Γ. Fix $1 < p < \infty$, let $q = (p-1)/p$ its conjugate exponent, and set $\hat{\sigma}(e) := \sigma(e)^{1-q}$, for all $e \in E$. Assume that μ^* is an optimal pmf for the probabilistic interpretation of modulus in (13.1) and let η^* be the extremal density for $\mathrm{Mod}_{q,\hat{\sigma}}(\hat{\Gamma})$. Then,*

$$\eta^*(e) = \mathbb{E}_{\mu^*}\left[\mathcal{N}(\underline{\gamma}, e)\right], \qquad \forall e \in E. \tag{13.25}$$

Proof. Consider the extremal density η^* for $\mathrm{Mod}_{q,\hat{\sigma}}(\hat{\Gamma})$ and let μ^* be any optimal pmf for (13.1). Then, $\eta^* = \mathcal{N}^T\mu^* = \mathbb{E}_{\mu^*}\left[\mathcal{N}(\underline{\gamma}, \cdot)\right]$, by (13.2) and (13.16). □

Remark 13.3. (i) When $1 < p < \infty$, Proposition 13.3 shows that although the optimal pmfs μ^* for the probabilistic interpretation of modulus are not necessarily unique, for any given edge $e \in E$ the corresponding expected edge usage $\mathbb{E}_{\mu^*}\left[\mathcal{N}(\underline{\gamma}, \cdot)\right]$ is uniquely determined.

(ii) Also, if we combine the probabilistic interpretation (13.25) with the Fulkerson duality relation (13.16) between η^* and ρ^*, we see that an edge $e \in E$ satisfies

$$\rho^*(e) = 0 \qquad \text{if and only if} \qquad \mathcal{N}(\gamma, e) = 0, \ \forall \gamma \in \Gamma.$$

We conclude this section with the following very useful observation that a minimal subfamily always supports a unique optimal pmf for the probabilistic interpretation of modulus. See [10, Proposition 8] for the case $p = 2$.

Proposition 13.4. *Let $G = (V, E, \sigma)$ be a graph and Γ a finite family of objects on G, as in Corollary 10.1. Suppose $\tilde{\Gamma} \subset \Gamma$ is a minimal subfamily for Γ, then there is a unique pmf $\mu^* \in \mathcal{P}(\tilde{\Gamma})$ such that*

(a) μ^* is optimal for $\mathrm{Mod}_{p,\sigma}(\Gamma)$, in the sense of Definition 13.1;

(b) μ^* *is strictly supported on* $\tilde{\Gamma}$*, i.e.,* $\mu^*(\gamma) > 0$ *if and only if* $\gamma \in \tilde{\Gamma}$.

Proof. Let $\mathcal{M}$ be the set of pmfs in $\mathcal{P}(\tilde{\Gamma})$ that satisfy (a) and (b).

To simplify notation, define the energy for a pmf μ as

$$\mathcal{E}_{q,\tilde{\sigma}}(\mu) := \sum_{e \in E} \sigma(e)^{-\frac{q}{p}} \mathbb{E}_\mu \left[\mathcal{N}(\underline{\gamma}, e)\right]^q.$$

We want to show that, if $\tilde{\Gamma} \subset \Gamma$ is a minimal subfamily for Γ, then $|\mathcal{M}| = 1$. Since optimal pmfs always exist, there is at least one pmf $\mu^* \in \mathcal{P}(\tilde{\Gamma})$ that is optimal for $\mathrm{Mod}_{p,\sigma}(\tilde{\Gamma})$. Since $\mathrm{Mod}_{p,\sigma}(\tilde{\Gamma}) = \mathrm{Mod}_{p,\sigma}(\Gamma)$, Equation (13.1) implies that

$$\min_{\tilde{\mu} \in \mathcal{P}(\tilde{\Gamma})} \mathcal{E}_{q,\tilde{\sigma}}(\tilde{\mu}) = \min_{\mu \in \mathcal{P}(\Gamma)} \mathcal{E}_{q,\tilde{\sigma}}(\mu).$$

And, since $\tilde{\Gamma} \subset \Gamma$, the pmf μ^* is also in $\mathcal{P}(\Gamma)$ (extending it to be zero outside of $\tilde{\Gamma}$). Therefore, μ^* is also optimal for $\mathrm{Mod}_{p,\sigma}(\Gamma)$. Thus, we have shown that μ^* satisfies property (a).

Next, property (b) will follow from Claim 13.1 below, and that will show that $\mu^* \in \mathcal{M}$.

Claim 13.1. *If* $\tilde{\Gamma} \subset \Gamma$ *is a minimal subfamily for* Γ *and* μ_0 *is a pmf on* $\tilde{\Gamma}$ *that is optimal for* $\mathrm{Mod}_{p,\sigma}(\Gamma)$*, then* μ_0 *is strictly supported on* $\tilde{\Gamma}$.

Proof of Claim 13.1. Suppose that there is an object $\gamma \in \tilde{\Gamma}$ with $\mu_0(\gamma) = 0$. Define $\Gamma' := \tilde{\Gamma} \setminus \{\gamma\}$. Then, by minimality of $\tilde{\Gamma}$ and (13.1), we must have

$$\min_{\tilde{\mu} \in \mathcal{P}(\tilde{\Gamma})} \mathcal{E}_{q,\tilde{\sigma}}(\tilde{\mu}) < \min_{\mu' \in \mathcal{P}(\Gamma')} \mathcal{E}_{q,\tilde{\sigma}}(\mu'). \tag{13.26}$$

However, the restriction of μ_0 to $\Gamma' := \tilde{\Gamma} \setminus \{\gamma\}$ is also a pmf in $\mathcal{P}(\Gamma')$, which we again denote by μ_0. Hence,

$$\min_{\mu' \in \mathcal{P}(\Gamma')} \mathcal{E}_{q,\tilde{\sigma}}(\mu') \leq \mathcal{E}_{q,\tilde{\sigma}}(\mu_0) \tag{13.27}$$

Finally, we have

$$\begin{aligned}
\mathcal{E}_{q,\tilde{\sigma}}(\mu_0) &= \min_{\mu \in \mathcal{P}(\Gamma)} \mathcal{E}_{q,\tilde{\sigma}}(\mu) && \text{(since } \mu_0 \text{ is optimal for } \mathrm{Mod}_{p,\sigma}(\Gamma)) \\
&\leq \min_{\tilde{\mu} \in \mathcal{P}(\tilde{\Gamma})} \mathcal{E}_{q,\tilde{\sigma}}(\tilde{\mu}) && \text{(since } \tilde{\Gamma} \subset \Gamma) \\
&< \min_{\mu' \in \mathcal{P}(\Gamma')} \mathcal{E}_{q,\tilde{\sigma}}(\mu') && \text{(by (13.26))} \\
&\leq \mathcal{E}_{q,\tilde{\sigma}}(\mu_0), && \text{(by (13.27))}
\end{aligned}$$

which is a contradiction. Therefore, μ_0 is strictly supported on $\tilde{\Gamma}$. □

We have shown the existence of a pmf μ^* in $\mathcal{M}$. To prove uniqueness, suppose $\mu_0, \mu_1 \in \mathcal{M}$ are two distinct pmfs. Define the linear combination $\mu_t := t\mu_1 + (1-t)\mu_0$, for $t \in \mathbb{R}$. We claim that there is a maximal strictly larger interval $(t_0, t_1) \supset [0,1]$ so that $\mu_t \in \mathcal{P}(\tilde{\Gamma})$ for $t \in [t_0, t_1]$. Indeed, by linearity, $\mathbf{1}^T\mu_t = 1$, for every $t \in \mathbb{R}$. Also, since μ_0 and μ_1 are strictly supported on $\tilde{\Gamma}$, we have $\min_{j=0,1}\mu_j(\gamma) > 0$, for every $\gamma \in \tilde{\Gamma}$. In particular, if $\gamma \in \tilde{\Gamma}$ is such that $\mu_0(\gamma) = \mu_1(\gamma)$, then $\mu_t(\gamma) > 0$ for all $t \in \mathbb{R}$. Moreover, since $\mu_0 \neq \mu_1$ and $\mathbf{1}^T(\mu_0 - \mu_1) = 0$, there is at least one pair (γ, γ') of objects in $\tilde{\Gamma}$ such that $\mu_0(\gamma) > \mu_1(\gamma)$ and $\mu_0(\gamma') < \mu_1(\gamma')$. Then, $\mu_t(\gamma) > 0$ if and only if $t < t_1(\gamma) := \frac{\mu_0(\gamma)}{\mu_0(\gamma)-\mu_1(\gamma)}$, and $\mu_t(\gamma') > 0$ if and only if $t > t_0(\gamma') := -\frac{\mu_0(\gamma')}{\mu_1(\gamma')-\mu_0(\gamma')}$. Note that $(t_0(\gamma'), t_1(\gamma)) \supset [0,1]$. If we take the (finite) intersection of all such intervals $(t_0(\gamma'), t_1(\gamma))$, over all possible pairs (γ, γ') as above, we find an interval $(t_0, t_1) \supset [0,1]$, such that $\mu_t(\gamma) \geq 0$ for all $t \in [t_0, t_1]$ and for all $\gamma \in \tilde{\Gamma}$, with the further property that when $t = t_0$ or $t = t_1$, then μ_t must vanish on some object in $\tilde{\Gamma}$. In particular, we have shown that $\mu_t \in \mathcal{P}(\Gamma)$ for all $t \in [t_0, t_1]$.

Finally, since μ_0 and μ_1 are optimal pmfs for $\mathrm{Mod}_{p,\sigma}(\Gamma)$, they both satisfy Equation (13.2). By linearity, μ_t will also satisfy Equation (13.2), for $t \in [t_0, t_1]$, and therefore, by Theorem 13.1, the pmf μ_t will also be optimal for $\mathrm{Mod}_{p,\sigma}\,\Gamma$ for $t \in [t_0, t_1]$. Hence, by construction, μ_{t_0} and μ_{t_1} are pmfs in $\mathcal{P}(\tilde{\Gamma})$ that are optimal for $\mathrm{Mod}_{p,\sigma}\,\Gamma$, and that vanish on some object $\gamma \in \tilde{\Gamma}$. This contradicts Claim 13.1. Hence we have shown that the minimality of $\tilde{\Gamma}$ implies $|\mathcal{M}| = 1$. □

13.6 THE CASES $p = 1$ AND $p = \infty$

Now we turn our attention to establishing the duality relationship in the cases $p = 1$ and $p = \infty$. Recall that by Proposition 11.1,

$$\lim_{p\to\infty} \mathrm{Mod}_{p,\sigma}(\Gamma)^{\frac{1}{p}} = \mathrm{Mod}_{\infty,1}(\Gamma) = \frac{1}{\ell(\Gamma)},$$

where $\ell(\Gamma)$ is defined to be the smallest element of the vector $\mathcal{N}\mathbf{1}$.

In order to pass to the limit in (13.15), we need to establish the limits for the second term in the left-hand side product.

Lemma 13.1. *Under the assumptions of Theorem 13.4,*

$$\begin{aligned} &\lim_{q\to 1} \mathrm{Mod}_{q,\hat{\sigma}}(\hat{\Gamma})^{\frac{1}{q}} = \mathrm{Mod}_{1,1}(\hat{\Gamma}) \quad \textit{and} \\ &\lim_{q\to\infty} \mathrm{Mod}_{q,\hat{\sigma}}(\hat{\Gamma})^{\frac{1}{q}} = \mathrm{Mod}_{\infty,\sigma^{-1}}(\hat{\Gamma}), \end{aligned} \tag{13.28}$$

where $\sigma^{-1}(e) := \sigma(e)^{-1}$ *and* $\hat{\sigma}(e) := \sigma(e)^{1-q}$, *for every* $e \in E$.

Proof. Let $\mathcal{N}, \hat{\mathcal{N}}$ be the usage matrices for Γ and $\hat{\Gamma}$ respectively. Let $\boldsymbol{\sigma} \in \mathbb{R}^{E\times E}$ be the diagonal matrix with entries $\boldsymbol{\sigma}(e,e) = \sigma(e)$, and define $\tilde{\mathcal{N}} = \hat{\mathcal{N}}\boldsymbol{\sigma}$, with $\tilde{\Gamma}$ its associated family in $\mathbb{R}^E_{\eta\geq 0}$. Note that $\eta \in \mathrm{Adm}(\hat{\Gamma})$, if and only if $\hat{\mathcal{N}}\eta \geq 1$, if and only if $\tilde{\mathcal{N}}\boldsymbol{\sigma}^{-1}\eta \geq 1$, if and only if $\boldsymbol{\sigma}^{-1}\eta \in \mathrm{Adm}(\tilde{\Gamma})$. Moreover, for every $\eta \in \mathrm{Adm}(\hat{\Gamma})$,

$$\mathcal{E}_{q,\hat{\sigma}}(\eta) = \sum_{e\in E} \hat{\sigma}(e)\eta(e)^q = \sum_{e\in E} \sigma(e)\left(\frac{\eta(e)}{\sigma(e)}\right)^q = \mathcal{E}_{q,\sigma}(\boldsymbol{\sigma}^{-1}\eta).$$

This implies that

$$\mathrm{Mod}_{q,\hat{\sigma}}(\hat{\Gamma}) = \mathrm{Mod}_{q,\sigma}(\tilde{\Gamma}).$$

Taking the limit as $q \to 1$, and using the continuity statement in Theorem 11.1, then gives (13.17).

$$\begin{aligned}\lim_{q\to 1} \mathrm{Mod}_{q,\hat{\sigma}}(\hat{\Gamma})^{\frac{1}{q}} &= \lim_{q\to 1} \mathrm{Mod}_{q,\sigma}(\tilde{\Gamma})^{\frac{1}{q}} = \mathrm{Mod}_{1,\sigma}(\tilde{\Gamma}) \\ &= \min_{\eta\in\hat{\mathcal{A}}} \sum_{e\in E} \sigma(e)\left(\frac{\eta(e)}{\sigma(e)}\right) = \mathrm{Mod}_{1,1}(\hat{\Gamma}).\end{aligned}$$

Taking the limit as $q \to \infty$ and using Proposition 11.1 shows that

$$\begin{aligned}\lim_{q\to\infty} \mathrm{Mod}_{q,\hat{\sigma}}(\hat{\Gamma})^{\frac{1}{q}} &= \lim_{q\to\infty} \mathrm{Mod}_{q,\sigma}(\tilde{\Gamma})^{\frac{1}{q}} = \mathrm{Mod}_{\infty,1}(\tilde{\Gamma}) \\ &= \min_{\eta\in\hat{\mathcal{A}}} \max_{e\in E} \left(\frac{\eta(e)}{\sigma(e)}\right) = \mathrm{Mod}_{\infty,\sigma^{-1}}(\hat{\Gamma}).\end{aligned}$$

□

Taking the limit as $p \to 1$ in (13.15), then gives (13.17). Note that taking the limit as $p \to \infty$ in (13.15) simply yields the same result for the unweighted case. Equation (13.17) can be thought of as a generalized max-flow min-cut statement, as will be explained next in Section 13.7

13.7 MAX-FLOW MIN-CUT REVISITED

Suppose that $G = (V, E, \sigma)$ is a weighted graph, with weights $\sigma \in \mathbb{R}^E_{>0}$, and Γ is a non-empty, finite family of subsets of E, where the usage of each object $\gamma \in \Gamma$ is given by the indicator function $\mathbb{1}_\gamma \in \mathbb{R}^E_{\geq 0}$. Then, we can identify each object with its usage vector and think of Γ as a subset of $\mathbb{R}^E_{\geq 0}$. Recall from (10.6) that $\mathrm{Mod}_{1,\sigma}(\Gamma)$ is the value of the linear program:

$$\begin{aligned} \underset{\rho\in\mathbb{R}^E}{\text{minimize}} \quad & \sigma^T\rho \\ \text{subject to} \quad & \rho \geq 0, \quad \mathcal{N}\rho \geq \mathbf{1} \end{aligned} \tag{13.29}$$

Since this is a linear program, strong duality holds, and the dual problem is

$$\begin{aligned} \underset{\lambda\in\mathbb{R}^\Gamma}{\text{maximize}} \quad & \lambda^T\mathbf{1} \\ \text{subject to} \quad & \lambda \geq 0, \quad \mathcal{N}^T\lambda \leq \sigma. \end{aligned} \tag{13.30}$$

If we consider the max-flow problem in its path formulation (1.13), then the dual of 1-modulus (13.30) can be thought as a *generalized max-flow* problem. Indeed, the goal in (13.30) is to assign some non-negative weights $\lambda(\gamma)$ for each object γ in Γ, so as to maximize the total weight, given the condition $\mathcal{N}^T\lambda \leq \sigma$. Namely, the constraint in this case says that for every $e \in E$

$$\sum_{\substack{\gamma\in\Gamma \\ e\in\gamma}} \lambda(\gamma) \leq \sigma(e),$$

which is reminiscent of the capacity condition in the classical max-flow problem, see Definition 1.9.

To think of 1-modulus (13.29) as a (generalized) min-cut problem, one needs to be able to restrict the densities ρ to be indicator functions of some subsets of E. That's exactly what the Fulkerson dual family tries to do.

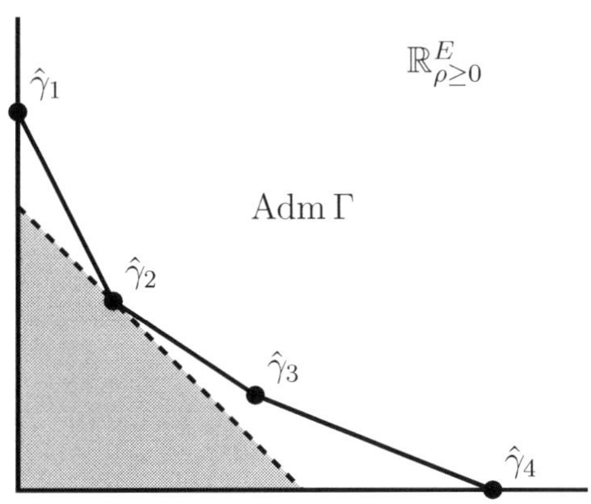

Figure 13.3 The unit ball in L^1.

Proposition 13.5. *Suppose $G = (V, E)$ is a finite graph and Γ is a finite family of objects on G, as in Corollary 10.1. Let $\hat{\Gamma}$ be the Fulkerson blocker family for Γ (as in Definition 13.2). Then, for any set of weights $\sigma \in \mathbb{R}^E_{\geq 0}$,*

$$\mathrm{Mod}_{1,\sigma}(\Gamma) = \min_{\hat{\gamma}\in\hat{\Gamma}} \sigma^T \hat{\gamma}. \tag{13.31}$$

Moreover, for every $\hat{\gamma} \in \hat{\Gamma}$ there is a choice of $\sigma \in \mathbb{R}^E_{\geq 0}$ such that $\hat{\gamma}$ is the unique solution for both problems in (13.31).

Proof. This proof is due to Huy Truong [80]. Since $\mathrm{Adm}(\Gamma)$ is a closed convex recessive set, by Proposition 13.1, we have $\mathrm{Adm}(\Gamma) = \mathrm{Dom}(\hat{\Gamma})$. Therefore, there always is an extremal density ρ^* for $\mathrm{Mod}_{1,\sigma}(\Gamma)$ that belongs to the set of extreme points $\mathrm{Ext}(\mathrm{Adm}(\Gamma)) = \hat{\Gamma}$. Hence, the optimization problem (13.29) can be restricted to densities of the form $\hat{\gamma}$, for $\hat{\gamma} \in \hat{\Gamma}$. This proves (13.31).

For the "Moreover" part recall that the polyhedron $\mathrm{Adm}(\Gamma)$ is defined as

$$\mathrm{Adm}(\Gamma) = \left\{\rho \in \mathbb{R}^E : \mathcal{N}(\gamma,\cdot)\rho \geq 1,\ \forall\gamma \in \Gamma \text{ and } \mathbb{1}^T_{\{e\}}\rho \geq 0,\ \forall e \in E\right\}.$$

Let $\widehat{\gamma}$ be an extreme point of $\mathrm{Adm}(\Gamma)$. Then, by [18, Theorem 2.3], there exists $|E|$ constraints of $\mathrm{Adm}(\Gamma)$, which we denote by $\{a_i^T\rho \geq b_i : i = 1, 2, \ldots, |E|\}$, where $a_i^T\widehat{\gamma} = b_i$, for $i = 1, 2, \ldots, |E|$, and the vectors a_i, for $i = 1, 2, \ldots, |E|$, are linearly independent.

Let $\sigma := \sum\limits_{i=1}^{|E|} a_i$. For any $\rho \in \mathrm{Adm}(\Gamma)$, we have $a_i^T\rho \geq b_i$, for $i = 1, 2, \ldots, |E|$. Hence,

$$\sigma^T\rho = \left(\sum_{i=1}^{|E|} a_i^T\right)\rho \geq \sum_{i=1}^{|E|} b_i = \left(\sum_{i=1}^{|E|} a_i^T\right)\widehat{\gamma} = \sigma^T\widehat{\gamma}. \tag{13.32}$$

That means $\widehat{\gamma}$ is an optimal solution for $\mathrm{Mod}_{1,\sigma}(\Gamma)$. Furthermore, equality holds in (13.32) if and only if $a_i^T\rho = b_i$ for $i = 1, 2, \ldots, |E|$. Since the vectors a_i, for $i = 1, 2, \ldots, |E|$, are linearly independent, $\widehat{\gamma}$ is the unique solution of the system $\{a_i^T\rho = b_i : i = 1, 2, \ldots, |E|\}$. Therefore, $\widehat{\gamma}$ is the unique optimal solution for $\mathrm{Mod}_{1,\sigma}(\Gamma)$.

Notice that, since $a_i \in \mathbb{R}^E_{\geq 0}$, for $i = 1, 2, \ldots, |E|$, we have $\sigma \in \mathbb{R}^E_{\geq 0}$. Moreover, since the determinant of the matrix $|E| \times |E|$ formed by $|E|$ columns a_i, for $i = 1, 2, \ldots, |E|$, is nonzero (since all $|E|$ columns are linearly independent), there is no row in this matrix containing all zeros. Thus, the row sums satisfy $\sigma(e) > 0$ for all $e \in E$. Therefore, $\sigma \in \mathbb{R}^E_{>0}$. □

13.8 EXAMPLES OF FULKERSON DUALITY

In this section we consider, a couple of instances where Γ is a non-empty finite family of subsets γ of E with edge usage given by the indicator function $\mathbb{1}_\gamma \in \mathbb{R}^E_{\geq 0}$. Therefore, we can think of Γ as a subset of $\{0,1\}^E \subset \mathbb{R}^E_{\geq 0}$.

Example 13.1 (Connecting families). Let $\Gamma = \Gamma_{\text{path}}(s,t)$ be the family of all simple paths connecting node s to node t in G. Then, Proposition 13.6 below shows that, by the Max-flow Min-cut Theorem 1.4, the Fulkerson dual $\hat{\Gamma}$ is the family $\Gamma_{\text{cut}}(s,t)$ of all minimal st-cuts, see Definition 1.8. Namely, recall that an st-cut is a set $C \subset V$ such that $s \in C$ and $t \notin C$. Moreover, its usage is given by the indicator function on its edge-boundary $\partial C = \{\{x,y\} \in E : x \in C, y \notin C\}$. Finally, an st-cut is said to be *minimal*, if there is no other st-cut with strictly smaller usage.

Proposition 13.6. *Let $G = (V,E)$ be a graph. Fix nodes $s \neq t \in V$. Then, the Fulkerson blocker for the family of simple paths from s to t, $\Gamma_{\text{path}}(s,t)$, is the family of all minimal st-cuts, $\Gamma_{\text{cut}}(s,t)$.*

Proof. Recall that $\text{Mod}_{1,\sigma}(\Gamma) = \min_{\rho\in\text{Adm}\,\Gamma} \sigma^T\rho$ and when $\Gamma = \Gamma_{\text{path}}(s,t)$, the dual problem (13.30) is exactly the max-flow problem in its path formulation, see Section 1.5. Therefore, by the Max-flow Min-cut Theorem 1.4, $\text{Mod}_{1,\sigma}(\Gamma)$ coincides with the value of the min-cut problem $\min_{C\in\Gamma_{\text{cut}}(s,t)} \sigma^T \mathbb{1}_{\partial C}$. Hence, by Proposition 13.5, for any given edge weights $\sigma \in \mathbb{R}^E_{>0}$, we have

$$\text{Mod}_{1,\sigma}(\Gamma) = \min_{C\in\Gamma_{\text{cut}}(s,t)} \sigma^T \mathbb{1}_{\partial C} = \min_{\hat{\gamma}\in\hat{\Gamma}} \sigma^T \hat{\gamma}.$$

By the "Moreover" part of Proposition 13.5, we can always choose σ appropriately, so that any given element of $\hat{\Gamma}$ becomes the unique solution to $\text{Mod}_{1,\sigma}(\Gamma)$. In particular, since for every $C \in \Gamma_{\text{cut}}(s,t)$, we have $\mathbb{1}_{\partial C} \in \text{Adm}\,\Gamma$, we deduce that each $\hat{\gamma} \in \hat{\Gamma}$ is some minimal st-cut. This shows that $\hat{\Gamma} \subset \Gamma_{\text{cut}}(s,t)$.

Conversely, let C be a minimal st-cut. We will show that $\mathbb{1}_{\partial C} \in \text{Ext}(\text{Adm}(\Gamma))$. Assume that $\rho_0, \rho_1 \in \text{Adm}(\Gamma)$ and

$$\mathbb{1}_{\partial C} = (1-t)\rho_0 + t\rho_1, \qquad \text{for some } 0 < t < 1.$$

If $e \notin \partial C$, then

$$0 = (1-t)\rho_0(e) + t\rho_1(e),$$

and since $\rho_0(e), \rho_1(e) \geq 0$, we must have $\rho_0(e) = \rho_1(e) = 0$. Therefore, ρ_0 and ρ_1 are supported on ∂C. Now, we claim that $\rho_0 \geq 1$ and $\rho_1 \geq 1$ on ∂C. Indeed, assume that $e \in \partial C$ and (without loss of generality) $\rho_0(e) < 1$. then, by Menger's Theorem, see Exercise 1.15, there is a path $\gamma \in \Gamma_{\text{path}}(s,t)$, such that $\gamma \cap \partial C = \{e\}$. In particular, we have $\ell_{\rho_0}(\gamma) = \rho_0(e) < 1$. But this contradicts the admissibility of ρ_0. Therefore, $\rho_0, \rho_1 \geq 1$ on ∂C. So, for $e \in \partial C$,

$$1 = (1-t)\rho_0(e) + t\rho_1(e),$$

hence, $\rho_0(e) = \rho_1(e) = 1$. This shows that $\rho_0 = \rho_1 = \mathbb{1}_{\partial C}$. So, $\Gamma_{\text{cut}}(s,t) \subset \hat{\Gamma}$, and we have shown that $\hat{\Gamma} = \Gamma_{\text{cut}}(s,t)$. □

We can now revisit the Fulkerson duality given by (13.17) in the case when $\Gamma = \Gamma_{\text{path}}(s,t)$ and hence $\hat{\Gamma} = \Gamma_{\text{cut}}(s,t)$. Recall that,

$$\text{Mod}_{1,\sigma}(\Gamma)\,\text{Mod}_{\infty,\sigma^{-1}}(\hat{\Gamma}) = 1. \tag{13.33}$$

As discussed above, $\text{Mod}_{1,\sigma}(\Gamma)$ takes the value of the minimum st-cut, given the edge weights σ. As we have seen in (12.18), the infinity modulus in (13.33), can be transformed into the linear program

$$\begin{aligned} \underset{t\in\mathbb{R},\eta\in\mathbb{R}^E}{\text{minimize}} \quad & t \\ \text{subject to} \quad & \sigma(e)^{-1}\eta(e) \leq t \;\forall e \in E \\ & \eta \geq 0, \quad \hat{\mathcal{N}}\eta \geq \mathbf{1} \end{aligned}$$

The minimum must occur in $\text{Adm}(\hat{\Gamma})$. By Theorem 13.3(2), $\text{Adm}(\hat{\Gamma}) = \text{Dom}(\Gamma)$, so the minimum must be attained at some η of the form

$$\eta = \sum_{\gamma\in\Gamma} \lambda(\gamma)\mathbb{1}_\gamma + z,$$

for some $\lambda(\gamma) \geq 0$, such that $\sum_{\gamma\in\Gamma}\lambda(\gamma) = 1$, and some $z \in \mathbb{R}^E_{\geq 0}$.

Note that z can be set to zero since that does not invalidate the constraints on η, so we can think of η as an st-flow in path formulation.

In other words, the minimum occurs at a unit st-flow η, and the problem can be restated as

$$\begin{aligned} \underset{t\in\mathbb{R},f\in\mathbb{R}^{\hat{E}}}{\text{minimize}} \quad & t \\ \text{subject to} \quad & \frac{1}{t}|f(e)| \leq \sigma(e), \quad \forall e \in E \\ & f \text{ is a unit } st\text{-flow} \end{aligned}$$

The minimum is attained when $\frac{1}{t}f$ is a maximum st-flow respecting edge capacities σ. The value of such a flow is $1/t$.

Example 13.2 (Planar connecting families). Suppose $G = (V, E)$ is a planar graph embedded in the plane with edge weights $\sigma \in \mathbb{R}^E_{>0}$. Adding the bounded faces, we obtained a domain $\hat{G}$ that can be decomposed as $\text{int}(G) \cup \partial G$, into an interior domain $\text{int}(G)$ and a boundary curve ∂G. For simplicity, assume that ∂G is a simple Jordan curve. Now suppose that A and B are two disjoint arcs on ∂G. Let $\Gamma = \Gamma_{\text{path}}(A, B)$ be the family of all simple paths connecting A to B in G. As in Example 13.1, the Fulkerson dual family $\hat{\Gamma} = \Gamma_{\text{cut}}(A, B)$ is the family of all minimal AB-cuts. Recall that an AB-cut is a subset of nodes $C \subset V$ such that $A \cap V \subset C$ and $B \cap C = \emptyset$. The edge usage associated with such a cut is the indicator vector for the set $\partial C = \{\{x, y\} \in E : x \in C, y \notin C\}$.

In this case, however, the dual family can be identified with a path family, albeit on a "dual" graph. The idea is to place a dual node in each bounded face of G. Then, form dual edges crossing edges shared by neighboring faces. Care must be shown when an edge is shared by the unbounded face (this is where we use the fact that ∂G is a simple curve). We say that an edge belongs to the arc A (resp. B), if both its end nodes belong to A (resp. B). As one can see in Figure 13.4, this creates two complementary arcs E and F on ∂G. For each edge $e \in E \cup F$, we simply create an extra node in the unbounded face of G, so as to make a dual edge that crosses e.

Remark 13.4. More generally, when Γ is a family of subsets of E, it is customary to say that Γ *has the max-flow min-cut property* if its Fulkerson blocker $\hat{\Gamma}$ is also a family of subsets of E. For more details, we refer to the discussion in [58, Chapter 3].

In the next example, Γ is the family of spanning trees of G. In this case, the Fulkerson dual family is known and can have fractional edge usages. Therefore, this is an example of a family that does not have the max-flow min-cut property.

Example 13.3 (The spanning trees family). Assume $G = (V, E)$ is a finite unweighted multigraph, without self-loops. More specifically, we assume that the vertex set V and the edge set E are finite, but we allow parallel edges connecting two nodes while discarding all *self-loops*, namely edges that connect a node to itself. Let Γ_G be the set of spanning trees of G. More specifically, each $\gamma \in \Gamma_G$ is a tree, as in Definition 1.3, with $V(\gamma) = V(G)$. It is useful to also think of γ as a subset of E, where the associated usage vector is the indicator $\mathcal{N}(\gamma, \cdot) = \mathbb{1}_\gamma(\cdot)$. In this case,

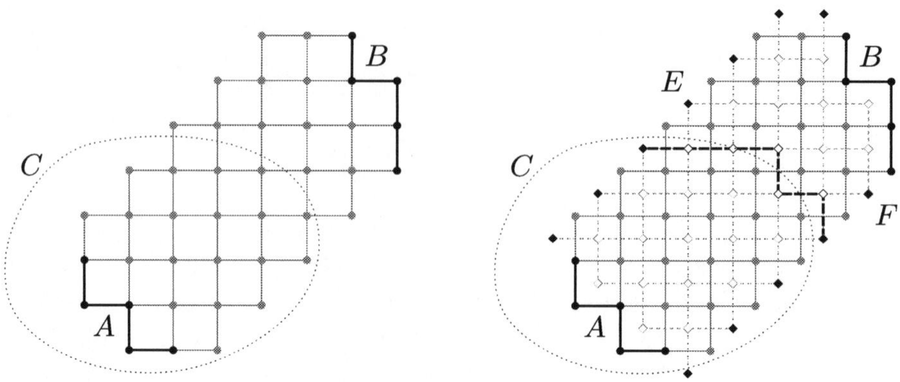

Figure 13.4 The figure on the left shows a planar domain whose boundary, ∂G, is a simple curve. The solid black edges mark two disjoint arcs, labeled A and B. The dotted curve indicates one possible AB-cut. On the right we have added the dual graph, using diamonds to mark the vertices and dashed lines to mark the edges. The vertices on the complementary arcs, E and F, are shaded black. Shaded black diamonds with an apparent degree of two (e.g., the vertex near the F label) should be thought of as two overlapping vertices of degree one. The path from E to F corresponding to the cut C is indicated by a black dashed line.

the Fulkerson dual family $\hat{\Gamma}_G$ can be interpreted as the set of *feasible partitions*, with a particular (fractional) usage vector.

Definition 13.5. A feasible partition P of a graph $G = (V, E)$ is a partition of the vertex set V into two or more subsets, $\{V_1, \ldots, V_{k_P}\}$, such that the induced subgraphs $G(V_i)$ are connected. The corresponding edge set, E_P, is defined to be the set of edges in G that connect vertices belonging to different parts V_i's.

The results of [27] imply the following theorem. See also [81].

Theorem 13.5 ([27]). *Let $G = (V, E)$ be a connected, unweighted, undirected multigraph with no self-loops, and let Γ_G be the family of spanning trees on G. Let Φ be the set of all feasible partitions. If $P \in \Phi$ is a feasible partition with k_P parts and edge cut E_P, we set the usage vector for P to be the weighted indicator*

$$\frac{1}{k_P - 1}\mathbb{1}_{E_P}.$$

Then, the set $\tilde{\Gamma}$ of all such usage vectors for $P \in \Phi$ is a Fulkerson dual family for Γ.

13.9 THE THEORY OF CLUTTERS

In the last four sections of this chapter we will review the theory of clutters in combinatorics, by following an article by Gérard Cornuéjols [30]. The goal is to translate some of the concepts in the theory of clutters to our language of modulus and Fulkerson duality, and highlight the main differences.

A *clutter* $\mathcal{C}$ is a pair $(\Omega, \mathcal{P})$ where the *ground set* Ω is finite, and $\mathcal{P} \subset 2^{\Omega}$, is a family of subsets of Ω with $P \not\subset Q$ when P and Q are distinct in $\mathcal{P}$. In a clutter, a *matching* is a set of pairwise disjoint subsets $\{P_1, \ldots, P_k\} \subset \mathcal{P}$. A *transversal* is a subset of Ω that intersects all $P \in \mathcal{P}$.

Definition 13.6. A clutter is said to *pack* if the maximum cardinality of a matching equals the minimum cardinality of a transversal.

Remark 13.5. The cardinality of a transversal is always greater than the cardinality of a matching. Indeed, suppose $\mathcal{M} = (P_1, \ldots, P_k) \subset \mathcal{P} \subset 2^{\Omega}$ is a matching and suppose S is a transversal. By definition, S intersects every $P \in \mathcal{P}$. Therefore, $|S \cap P_j| \neq \emptyset$, for all $j = 1, \ldots, k$. In particular, there exists $e_j \in S \cap P_j$ for each $j = 1, \ldots, k$. Since the P_j are mutually disjoint, the edges e_j are also mutually disjoint. Thus $|S| \geq k = |\mathcal{M}|$.

Example 13.4. Let s and t be distinct nodes of a graph G. Then, Menger's theorem, see Exercise 1.15, states that the maximum number of pairwise edge-disjoint st-paths in G equals the minimum number of edges in an st-cut. Let $\mathcal{C}$ be the clutter where the ground set is Ω is the edge set E, and $\mathcal{P} = \Gamma_{\text{path}}(s,t)$. This is sometimes referred to as the *clutter of st-paths* of G. The transversals of $\mathcal{C}$ are all the st-cuts. Therefore, Menger's theorem states that the clutter of st-paths packs. Similarly, we can define the clutter of minimal st-cuts $\hat{\mathcal{C}}$. Its transversals are all connecting st-subgraphs, and $\hat{\mathcal{C}}$ packs, namely, the maximum number of pairwise disjoint minimal st-cuts equals the length of the shortest st-path.

To make a connection with the theory of modulus, we assume from now on that $\Omega = E(G)$ and instead of the letter $\mathcal{P}$, we use Γ to denote a clutter. We say that a family of object Γ is a *graph clutter* if each $\gamma \in \Gamma$ is a subset of E with $0/1$ usage, and $\gamma \not\subset \gamma'$, for all distinct $\gamma, \gamma' \in \Gamma$.

Recall that given a weight $\sigma \in \mathbb{R}^E_{>0}$ on the edges of G, $\mathrm{Mod}_{1,\sigma}(\Gamma)$ is the value of the linear program in (13.29) and its dual problem is (13.30). As mentioned before, we think of the dual problem to 1-modulus as a generalized max-flow problem, given the edge capacities σ. However, to think of 1-modulus as a generalized min-cut problem, we would need to be able to restrict the densities ρ to some given subsets of E. This is not always possible.

Proposition 13.7. *Assume $G = (V, E)$ is a graph and Γ is a graph clutter on G.*

(i) If Γ is packing, then there are optimal $\rho^* \in \{0,1\}^E$ and $\lambda^* \in \{0,1\}^\Gamma$ for the unweighted ($\sigma \equiv 1$) problems (13.29) and (13.30) respectively.

(ii) Conversely, if there are optimal $\rho^* \in \{0,1\}^E$ and $\lambda^* \in \mathbb{Z}^\Gamma_{\geq 0}$ for the unweighted ($\sigma \equiv 1$) problems (13.29) and (13.30) respectively, then Γ is packing.

Remark 13.6. When Γ is a graph clutter, we have $\mathcal{N}_{\min} = 1$, where $\mathcal{N}_{\min}$ is defined in (10.9). Therefore, by Proposition 10.2 (2) any optimal ρ^* satisfies $0 \leq \rho^* \leq 1$. In particular,

$$\rho^* \in \mathbb{Z}^E_{\geq 0} \quad \Longleftrightarrow \quad \rho^* \in \{0,1\}^E.$$

Proof of Proposition 13.7. Assume that Γ is packing as in Definition 13.6. Let $\tilde{\Gamma} := \{\gamma_1, \ldots, \gamma_k\}$ be a maximal matching in Γ, and let $S = \{e_1, \ldots, e_j\} \subset E$ be transversal, so that, for every $\gamma \in \Gamma$,

$$|\gamma \cap S| \geq 1.$$

Let $\rho := \mathbb{1}_S \in \mathbb{R}^E_{\geq 0}$ and let $\lambda = \mathbb{1}_{\tilde{\Gamma}} \in \mathbb{R}^\Gamma_{\geq 0}$. Then, λ is a feasible for (13.30), since every edge belongs to at most one object in $\tilde{\Gamma}$. Also, ρ is an admissible density for (13.29), since for every $\gamma \in \Gamma$:

$$\ell_\rho(\gamma) = \sum_{e \in E} \mathcal{N}(\gamma, e)\rho(e) = \sum_{e \in E} \mathcal{N}(\gamma, e)\mathbb{1}_S(e) = |\gamma \cap S| \geq 1.$$

Evaluating the primal and dual objectives in this case we get:

$$\mathbf{1}^T\rho = |S| \qquad \text{and} \qquad \mathbf{1}^T\lambda = |\tilde{\Gamma}|.$$

Since Γ is packing, we have $|S| = |\tilde{\Gamma}|$. Hence, λ and ρ are optimal.

Therefore, in this case, we have found optimal $\rho^* \in \{0,1\}^E$ and $\lambda^* \in \{0,1\}^\Gamma$.

Conversely, assume that there are optimal $\rho^* \in \{0,1\}^E$ and $\lambda^* \in \mathbb{Z}_{\geq 0}^\Gamma$ for (13.29) and (13.30) respectively. Then, dual feasibility implies that, for every $e \in E$:

$$\sum_{\gamma \in \Gamma} \mathcal{N}(\gamma, e)\lambda^*(\gamma) \leq 1.$$

Since λ is integer-valued, this sum is either 0 or 1. When it equals 0, the edge e does not belong to any $\gamma \in \Gamma$. When it equals 1, then e belongs to exactly one γ, and $\lambda^*(\gamma) = 1$ while $\lambda^*(\gamma') = 0$, for $\gamma' \neq \gamma$. Let $\tilde{\Gamma} = \{\gamma_1, \ldots, \gamma_m\} \subset \Gamma$, be the set of objects with $\lambda^*(\gamma) = 1$. Then, as we have seen, the objects in $\tilde{\Gamma}$ must be mutually disjoint. So $\tilde{\Gamma}$ is a matching for Γ.

Moreover, since ρ^* is an admissible density, for every $i = 1, \ldots, m$, we have

$$\ell_{\rho^*}(\gamma_i) = \sum_{e \in E} \mathcal{N}(\gamma_i, e)\rho^*(e) \geq 1.$$

Hence, for each $i = 1, \ldots, m$, there is at least one edge $e_i \in E$, such that

$$\mathcal{N}(\gamma_i, e_i)\rho^*(e_i) > 0.$$

Since $\rho^* \in \{0,1\}^E$ and $\mathcal{N}(\gamma_i, e_i) = 1$, this means that, for each $i = 1, \ldots, m$, there is an edge $e_i \in \gamma_i$ such that $\rho^*(e_i) = 1$. Since the objects in $\tilde{\Gamma}$ are pairwise disjoint, the edges $\{e_i\}_{i=1}^m$ are also pairwise disjoint. Define the set $S = \{e_i\}_{i=1}^m$. By strong duality, then

$$|\tilde{\Gamma}| = \sum_{i=1}^m \lambda^*(\gamma_i) = \sum_{e \in E} \rho^*(e) \geq \sum_{i=1}^m \rho^*(e_i) \geq |\tilde{\Gamma}|.$$

In particular, all inequalities must be an equalities, and

$$\rho^*(e_i) = 1 \qquad \forall i = 1, \ldots, m,$$

while

$$\rho^*(e) = 0 \qquad \forall e \in E \setminus S.$$

This implies that S is a transversal for Γ.

In conclusion, since $|\tilde{\Gamma}| = |S| = m$, Remark 13.5 implies that $\tilde{\Gamma}$ is a maximal matching and S is a minimal transversal. Thus, Γ is packing. □

13.10 THE PACKING, MFMC, AND IDEAL PROPERTIES

Definition 13.7. Let $G = (V, E)$ be a graph and let Γ be a graph clutter on G. We say that:

- Γ has the *packing property* if and only if, for all weights $\sigma \in \{0, 1, \infty\}^E$, there are optimal $\rho^* \in \mathbb{Z}^E_{\geq 0}$ and $\lambda^* \in \mathbb{Z}^\Gamma_{\geq 0}$ for the problems (13.29) and (13.30) respectively.

- Γ has the *max-flow min-cut* property (MFMC) if and only if, for all weights $\sigma \in \mathbb{Z}^E_{\geq 0}$, there are optimal $\rho^* \in \mathbb{Z}^E_{\geq 0}$ and $\lambda^* \in \mathbb{Z}^\Gamma_{\geq 0}$ for the problems (13.29) and (13.30) respectively.

- Γ is *ideal* if for every $\sigma \in \mathbb{R}^E_{\geq 0}$, there is an optimal $\rho^* \in \mathbb{Z}^E_{\geq 0}$ for (13.29).

Remark 13.7. The packing property asks for the edge weights to be either $0, 1$, or ∞. When $\sigma(e) = 0$, this corresponds to removing e from the graph. And, when $\sigma(e) = \infty$, this corresponds to contracting the edge.

Definition 13.8 (Semi-ideal and semi-MFMC property). A clutter Γ has the $1/k$*-MFMC property* for a positive integer k, when it is ideal and, for all $\sigma \in \mathbb{Z}^E_{>0}$, the dual modulus problem (13.30) has an optimal solution vector λ^* such that $k\lambda^*$ is integral. If the clutter has the $1/k$-MFMC property for some k, we say that it has the semi-MFMC property. When $k = 1$, this definition reduces to the MFMC property. If Γ has the $1/k$-MFMC property, then it also has the $1/q$-MFMC property, for every integer q that is a multiple of k.

The definition of the semi-idealness is similar. Let k be a positive integer. A clutter Γ has the $1/k$*-ideal property* if and only if, $k\hat{\Gamma}$ is integral, where $\hat{\Gamma}$ is the Fulkerson blocker of Γ from Definition 13.2. If the clutter is $1/k$-ideal for some k, we say that it has the semi-ideal property.

For instance, it follows from Chopra's Theorem 13.5, that the family of spanning trees is semi-ideal.

Remark 13.8. We have the following implications, see [30, Figure 1.2].

- MFMC implies the packing property. The converse is a conjecture of Conforti and Cornuejol.

- Also MFMC implies ideal, because if $\text{Mod}_{1,\sigma}(\Gamma)$ admits an extremal density $\rho^* \in \mathbb{Z}^E_{\geq 0}$ whenever $\sigma \in \mathbb{Z}^E_{\geq 0}$, then by taking common denominators in the energy, the same holds whenever

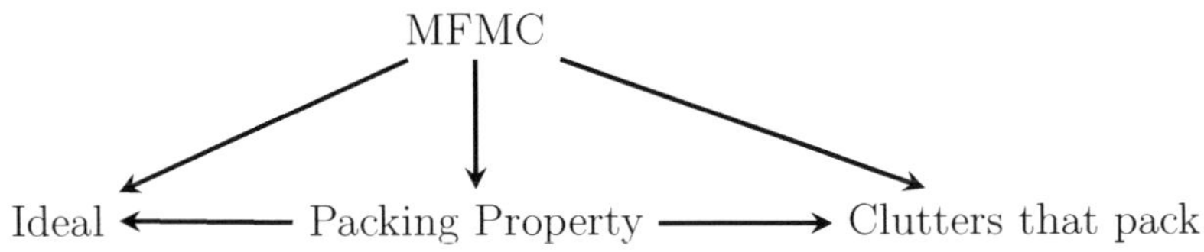

Figure 13.5 Summary of the implications.

$\sigma \in \mathbb{Q}^E_{\geq 0}$. And, by continuity of $\text{Mod}_{1,\sigma}(\Gamma)$ and ρ^*_σ in σ, it follows that the same holds for $\sigma \in \mathbb{R}^E_{\geq 0}$.

- By a result of Lehman, the packing property implies idealness, see [30, Theorem 4.1].
- The fact that a clutter packs if it has the packing property follows from Proposition 13.7.

Figure 13.5 below summarizes the above remark.

13.11 THE BLOCKER OF A CLUTTER

Definition 13.9. The *blocker* $b(\Gamma)$ of a graph clutter Γ is the clutter $(E, b(\Gamma))$, with the same edge set E and whose objects are the minimal transversals of Γ. That is, $b(\Gamma)$ consists of the minimal members of the family

$$\tilde{\Gamma} := \{B \subseteq E : |B \cap A| \geq 1, \quad \text{for all } A \in \Gamma\}.$$

In other words, the rows of $\mathcal{N}(b(\Gamma))$ are the minimal 0,1 vectors x^T such that x belongs to the polyhedron $\text{Adm}\,\Gamma = \{x \geq 0 : \mathcal{N}(\Gamma)x \geq \mathbf{1}\}$.

Example 13.5. Let $G = (V, E)$ be a graph and s and t be distinct nodes in V. If Γ is the clutter of st-paths, then $b(\Gamma)$ is the clutter of minimal st-cuts.

Theorem 13.6 (Edmonds and Fulkerson). *If* Γ *is a clutter, then* $b(b(\Gamma)) = \Gamma$.

Proof. Let A be an object of Γ. The definition of $b(\Gamma)$ implies that $|A \cap B| \geq 1$, for all objects $B \in b(\Gamma)$. Thus A is a transversal of $b(\Gamma)$. This implies that A contains an object of $b(b(\Gamma))$.

Conversely, let A be an object of $b(b(\Gamma))$. We claim that A contains an object of Γ. If not, then the complement $E \setminus A$ must intersect each object in Γ. Thus $E \setminus A$ is a transversal of Γ. Thus, $E \setminus A$ contains an object $B \in b(\Gamma)$, and B has the property that $B \cap A = \emptyset$. However, since A is in $b(b(\Gamma))$, we must also have $|B \cap A| \geq 1$. This is a contradiction. Therefore, A contains an object of Γ.

The conclusion follows as an exercise for the reader. □

The next result gives a relationship between the blocker $b(\Gamma)$ of a graph clutter Γ and the Fulkerson blocker of Γ.

Proposition 13.8. *Consider a graph clutter Γ on a graph G. Let $b(\Gamma)$ be the blocker of Γ and let $\hat{\Gamma}$ be the Fulkerson blocker of Γ from Definition 13.2. Then,*

(i) $b(\Gamma) = \hat{\Gamma} \cap \{0,1\}^E$.

(ii) $\hat{\Gamma} = b(\Gamma)$ *if and only if* Γ *is ideal.*

Proof. We begin with the proof of (i). Let $T \in b(\Gamma)$ be a minimal transversal for Γ. We identify T with the indicator $\rho := \mathbb{1}_T \in \{0,1\}^E$. Since $|T \cap \gamma| \geq 1$ for all $\gamma \in \Gamma$, the density ρ is admissible for Γ. Assume, by contradiction, that $\rho = t\rho_1 + (1-t)\rho_0$, with $t \in (0,1)$ and $\rho_0 \neq \rho_1 \in \text{Adm}(\Gamma)$. Since admissible densities are non-negative, this implies that $\rho_0(e) = \rho_1(e) = 0$, for every $e \in E \setminus T$. In particular, since $\rho_0 \neq \rho_1$, there must be an edge $e' \in T$ such that $0 \leq \rho_1(e') < 1$. Hence, if $\gamma \in \Gamma$ is an object with $T \cap \gamma = \{e'\}$, then $\rho_1^T \gamma = \rho_1(e') < 1$, and this contradicts the admissibility of ρ_1. This means that for all objects $\gamma \in \Gamma$, if $e' \in \gamma$, then there must be another edge $e'' \neq e'$ such that $e'' \in \gamma \cap T$. In other words, the edge e' can be removed from T without affecting the property of transversality of T with respect to Γ. This contradicts the fact that T is a minimal transversal. Therefore, ρ is an extremal point for $\text{Adm}(\Gamma)$, and $T \in \hat{\Gamma}$. This shows that $b(\Gamma) \subset \hat{\Gamma} \cap \{0,1\}^E$.

Conversely, suppose $x \in \hat{\Gamma} \cap \{0,1\}^E$. Then, $x = \mathbb{1}_S$ for some $S \in E$, and $\mathbb{1}_S \in \hat{\Gamma} \subset \text{Adm}(\Gamma)$. Therefore,

$$|S \cap \gamma| \geq 1, \qquad \text{for all } \gamma \in \Gamma.$$

This shows that S is a transversal for Γ. Now, assume that S is not minimal. This means that there is an edge $e \in E$ that can be safely removed from S so as to get a new transversal $T := S \setminus \{e\}$. Moreover, $\mathbb{1}_T \in \text{Adm}(\Gamma)$ and $\mathbb{1}_T < \mathbb{1}_S$. This contradicts the fact that $\mathbb{1}_S \in \hat{\Gamma} = \text{Ext}(\text{Adm}(\Gamma))$. Therefore, we have shown that $S \in b(\Gamma)$ and the proof of (i) is done.

Next, we prove (ii). Assume first that Γ is ideal. Fix $\hat{\gamma} \in \hat{\Gamma}$, an extreme point of $\mathrm{Adm}(\Gamma)$. By the "Moreover" part of Proposition 13.5, there is a set of weights $\sigma \in \mathbb{R}^E_{>0}$, such that $\rho^*(\cdot) := \mathcal{N}(\hat{\gamma}, \cdot)$ is the unique optimal solution to (13.29). By idealness, ρ^* must be in $\mathbb{Z}^E_{\geq 0}$. Therefore, $\hat{\gamma}$ has integer-valued edge usages. By contradiction, assume that $\hat{\gamma} \in \mathbb{Z}^E_{\geq 0} \setminus \{0,1\}^E$. Let T be the support of $\hat{\gamma}$, meaning that $T := \{e \in E : \mathcal{N}(\hat{\gamma}, e) \geq 1\}$. Since $\hat{\gamma} \in \mathrm{Adm}(\Gamma)$, we have $\sum_{e\in\gamma} \mathcal{N}(\hat{\gamma}, e) \geq 1$. In particular, since $\mathcal{N}(\hat{\gamma}, \cdot) \in \mathbb{Z}^E_{\geq 0}$, there must exist an edge $e \in \gamma$ such that $\mathcal{N}(\hat{\gamma}, e) \geq 1$. In other words, we have $|\gamma \cap T| \geq 1$, for every $\gamma \in \Gamma$. Hence, T is a transversal for Γ. We may assume that T is minimal. Then, $\mathbb{1}_T \in \mathrm{Adm}(\Gamma)$ and $\mathbb{1}_T < \hat{\gamma}$. This contradicts the fact that $\hat{\gamma} \in \hat{\Gamma} = \mathrm{Ext}(\mathrm{Adm}(\Gamma))$. Therefore, $\hat{\gamma} = \mathbb{1}_T \in \{0,1\}^E$. Namely, $\hat{\gamma} \in b(\Gamma)$. So we have shown that $\hat{\Gamma} \subset b(\Gamma)$. On the other hand, if $b(\Gamma) \subset \hat{\Gamma}$, by part (i). Therefore, we have shown that when Γ is ideal, $\hat{\Gamma} = b(\Gamma)$.

Conversely, suppose that $\hat{\Gamma} = b(\Gamma)$. Then, by Proposition 13.5, for any $\sigma \in \mathbb{R}^E_{>0}$, the set of optimal solutions for (13.29) always includes at least one extreme point of $\mathrm{Adm}(\Gamma)$, which we are now assuming is integer-valued. So Γ is ideal. □

Corollary 13.4 (Lehman). *A clutter Γ on a graph G is ideal if and only if its blocker $b(\Gamma)$ is also ideal. And, in this case,*

$$\Gamma = b(b(\Gamma)) = \widehat{b(\Gamma)} = b(\hat{\Gamma}) = \hat{\hat{\Gamma}}.$$

Proof. By Theorem 13.6, it is enough to show that $b(\Gamma)$ is ideal if Γ is. By Proposition 13.8, it is enough to show that if $\hat{\Gamma} = b(\Gamma)$, then the Fulkerson dual family of $b(\Gamma)$ is also integral, and in fact, equal to Γ.

Since $b(\Gamma) = \hat{\Gamma}$, by Theorem 13.3 (3), we have $\widehat{b(\Gamma)} = \hat{\hat{\Gamma}} \subset \Gamma$. So $b(\Gamma)$ is ideal. In particular, $\widehat{b(\Gamma)} = b(b(\Gamma)) = \Gamma$, by Theorem 13.6. □

13.12 THE WIDTH-LENGTH INEQUALITY

The following width-length inequality can be used as a characterization of idealness.

Theorem 13.7 (Lehman). *For a clutter Γ and its blocker $b(\Gamma)$, the following statements are equivalent.*

(i) Γ and $b(\Gamma)$ are ideal, i.e., $\hat{\Gamma} = b(\Gamma)$.

(ii) For all $w \in \mathrm{Adm}(\Gamma)$, and $z \in \mathrm{Adm}(b(\Gamma))$:

$$\ell_w(\Gamma)\ell_z(b(\Gamma)) \leq w^T z, \tag{13.34}$$

where $\ell_\rho(\Gamma) := \min_{\gamma\in\Gamma} \ell_\rho(\gamma) = \min_{\gamma\in\Gamma} \sum_{e\in E} \mathcal{N}(\gamma, e)\rho(e)$.

Proof. Let $\mathcal{N}$ and $b\mathcal{N}$ be the usage matrices for Γ and $b(\Gamma)$, respectively. First, assume that Γ and $b(\Gamma)$ are ideal, i.e., assume that $\hat{\Gamma} = b(\Gamma)$. Observe that (13.34) is invariant under independent scaling of w and z. Therefore, we may assume that $\ell_w(\Gamma) = \ell_z(b(\Gamma)) = 1$. Let J denote the $|b(\Gamma)| \times |\Gamma|$ matrix of all ones. Note that $(b\mathcal{N})\mathcal{N}^T \geq J$ since the rows of $b\mathcal{N}$ represent the minimal transversals for the columns in $\mathcal{N}^T$. Therefore, by Theorem 13.3, there are pmfs $\mu \in \mathcal{P}(\Gamma)$ and $\nu \in \mathcal{P}(b(\Gamma))$ such that

$$w^T z \geq \nu^T (b\mathcal{N})\mathcal{N}^T \mu \geq \nu^T J \mu = 1 = \ell_w(\Gamma)\ell_z(b(\Gamma)).$$

Conversely, assume (ii) holds. Let $w \in \hat{\Gamma}$. Since $w \in \hat{\Gamma} \subset \mathrm{Adm}(\Gamma)$, by (13.34), we have that,

$$z^T w \geq \ell_w(\Gamma)\ell_z(b(\Gamma)) \geq 1, \qquad \forall z \in \mathrm{Adm}(b(\Gamma)).$$

This means that $w \in \mathrm{BL}(\mathrm{Adm}(b(\Gamma)))$. By Theorem 13.3, $\mathrm{BL}(\mathrm{Adm}(b(\Gamma))) = \mathrm{Dom}(b(\Gamma))$, and $\mathrm{Dom}(b(\Gamma)) \subset \mathrm{Dom}(\hat{\Gamma}) = \mathrm{Adm}(\Gamma)$. And, since w is an extreme point of $\mathrm{Adm}(\Gamma)$, it must also be an extreme point of $\mathrm{Dom}(b(\Gamma))$. Therefore, $w \in b(\Gamma)$ and we have shown that $\hat{\Gamma} = b(\Gamma)$ in this case. So the converse follows from Proposition 13.8 (ii). □

EXERCISES

13.1 *Spanning trees of the Paw graph.* Let G be the graph of Figure 13.6 and let Γ be the set of spanning trees, with usage function $\mathcal{N}(\gamma, e)$ equaling 1 if e is included in the tree γ and 0 otherwise.

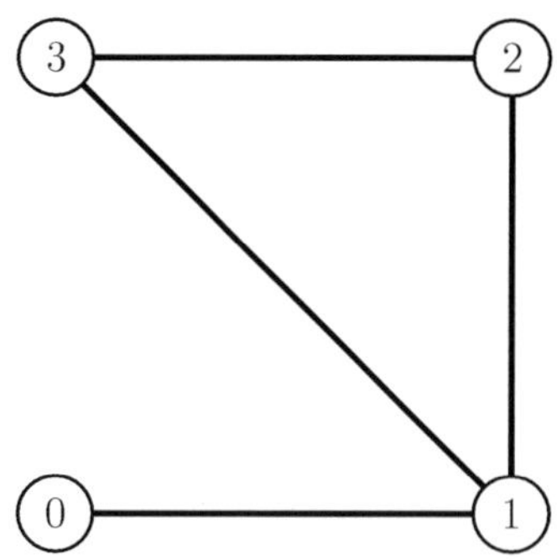

Figure 13.6 The Paw graph.

(i) Draw the 3 spanning trees for G and write down the overlap matrix C.

(ii) Let μ be the uniform pmf on Γ and compute the expected overlap using this law.

(iii) Assume ρ is an admissible density that it is constant on the triangle (1 2 3) and then minimize its 2-energy.

(iv) Finally, use Corollary 13.2 to compute the modulus.

13.2 *A connecting family of the Paw graph.* Let G be again the graph of Figure 13.6 and let Γ be the connecting family of simple paths from 0 to 3, with usage function $\mathcal{N}(\gamma, e)$ equaling 1 if e is included in the path γ and 0 otherwise.

(i) Draw the 2 simple paths connecting 0 and 3 and write down the overlap matrix C.

(ii) Let μ be an arbitrary pmf on Γ and minimize the expected overlap using this law.

(iii) Define an admissible density ρ from (ii) using the expected edge usage formula (13.6), replacing the modulus with the reciprocal of the minimum in (ii).

(iv) Finally, use Corollary 13.2 to verify that the density in (iii) is extremal.

13.3 *Spanning trees for the house graph.* Let G be the unweighted house graph with nodes and edges labeled as in Figure 1.1 (c). Let Γ be the family of spanning trees of G. A tree $\gamma \in \Gamma$ is obtained by removing two edges from G in such a way as to break the two cycles $x_0\ x_1\ x_2\ x_3$ and $x_2\ x_3\ x_4$. To this end, partition the edges into $E_1 = \{e_1, e_2, e_3\}$ and $E_2 = \{e_4, e_5, e_6\}$.

(i) Show that there are 9 ways to remove an edge from E_1 and edge from E_2, giving rise to a subfamily $\Gamma^* \subset \Gamma$. Deduce that $|\Gamma| = 11$.
Hint: See Figure 18.5.

(ii) Use Corollary 13.2 and the constant density $\rho \equiv 1/4$ to show that the uniform pmf restricted to the subfamily Γ^* from part (i) is optimal for the minimum expected overlap problem (13.5).

Thus, deduce $\mathrm{Mod}_2\,\Gamma$ and η^* as well.
Hint: Use Figure 18.5 to compute the overlap of a given tree with all the trees in Γ^*.

(iii) Show that the trees in $\Gamma \setminus \Gamma^*$ cannot be in the support of an optimal pmf μ^*.
Hint: consider the sum $\sum_{e\in E_2} \eta^*(e)$ *and show that*

$$\sum_{e\in E_2} \mathcal{N}(\gamma, e) < 2 \quad \textit{if and only if} \quad \gamma \in \Gamma \setminus \Gamma^*.$$

(iv) Show that both Γ^* and Γ itself are Beurling subfamilies, but Γ cannot be a Lagrangian subfamily. Hence, this gives another example as in Exercise 12.3.

13.4 *Graph clutters with two objects.* Assume that $\Gamma = \{\gamma_0, \gamma_1\}$ is a graph clutter on an unweighted graph G.

(i) Compute the overlap matrix C and solve the problem

$$\begin{aligned} \underset{\mu\in\mathbb{R}^\Gamma}{\text{minimize}} \quad & \mu^T C \mu \\ \text{subject to} \quad & \mathbf{1}^T \mu = 1 \\ & \mu \geq 0. \end{aligned} \tag{P13.1}$$

(ii) Deduce that the optimal pmf μ^* is unique and positive in this case and provide a formula.

(iii) Finally, provide a formula for $\mathrm{Mod}_2(\Gamma)$.

13.5 *All-star modulus.* Let $G = (V, E)$ be a finite simple graph. For $x \in V$, let the *star* at x be the set of edges

$$E_x := \{\{x, y\} \in E : y \sim x\}.$$

Consider the family $\Gamma = \Gamma_{\text{star}} = \{E_x\}_{x\in V}$ consisting of all the possible stars in G. The usage matrix $\mathcal{N}$ is known in other contexts as the *unoriented gradient matrix* B of the graph G. Namely, for $x \in V$ and $e \in E$:

$$\mathcal{N}(E_x, e) = B(x, e) := \mathbb{1}_{\{x\in e\}}.$$

(i) Show that, in this case, the overlap matrix is

$$C = \mathcal{N}\mathcal{N}^T = D + A,$$

which is known as the *unoriented Laplacian.* Here, D and A are the degree and adjacency matrices respectively defined in Section 3.2.

(ii) Write

$$C = D(I + D^{-1}A) = 2D\tilde{P}$$

where $\tilde{P}$ is the transition matrix for the lazy random walk. Use Proposition 7.3 and Exercise 5.6 to show that C is always invertible for non-bipartite graphs. By Proposition 12.3, we deduce that, if we also know that Γ is minimal, then

$$\mathrm{Mod}_{2,\sigma}(\Gamma) = \mathbf{1}^T C^{-1}\mathbf{1}, \qquad \text{and} \qquad \rho^* = \mathcal{N}^T C^{-1}\mathbf{1}.$$

For more information about all-star modulus and its connections to the modulus of edge covers and fractional edge covers, see [67].

13.6 *All-star families examples.* Let $\Gamma_{\text{star}}(G)$ be the family of all stars in a graph G, as defined in Exercise 13.5. Show the following.

- The family $\Gamma_{\text{star}}(K_4)$ does not pack but is ideal.
- The family $\Gamma_{\text{star}}(K_3)$ does not pack and is not ideal.
- The star family $\Gamma_{\text{star}}(C_4)$ does not pack, is ideal and has the MFMC property.

Here K_N is the complete graph on N nodes, and C_N is the cycle graph on N nodes. These facts can be found in [39, Chapter 5].

IV

Families of Paths

CHAPTER 14

Connecting Families of Walks

SHORTEST-PATHS, max-flow min-cut, and effective resistance are three of the most important tools for graph and network theory, as we saw in Chapters 2 and 3. Here we will see that all three of these concepts are instances of modulus for connecting families of walks. This topic has a long history, see for example [36, 66].

14.1 CONNECTING MODULUS GENERALIZES CLASSICAL GRAPH QUANTITIES

It will be useful in this chapter to refer to the given data (G, σ, s, t) as a *network*, where $G = (V, E)$ is a connected finite graph with edge weights $\sigma \in \mathbb{R}^E_{>0}$, and $s \neq t$ are two chosen nodes in V.

The concept of p-modulus of connecting families generalizes three classical ways of measuring the richness of a family of walks:

- Min-cut and max-flow;
- Effective conductance and resistance;
- Shortest paths and shortest path distance.

We summarize this in the following theorem.

Theorem 14.1. *Let (G, σ, s, t) be a network. Let $\Gamma = \Gamma_{\text{path}}(s, t)$ be the family of all paths in G that start at s and end at t. Computing p-modulus $\text{Mod}_{p,\sigma}(\Gamma)$, we obtain the following:*

DOI: 10.1201/9781003024866-18

(i) When $p = \infty$, we recover the shortest path distance (see Remark 1.2):

$$\mathrm{Mod}_{\infty,\sigma}(\Gamma) = \frac{1}{d_r(s,t)},$$

where $r = 1/\sigma$.

(ii) When $p = 1$, we recover the min-cut (see Definition 1.8):

$$\mathrm{Mod}_{1,\sigma}(\Gamma) = \min\{\sigma(C) : \ C \textit{ a cut of } \Gamma\}.$$

(iii) When $p = 2$, we recover effective conductance (see Theorem 2.3):

$$\mathrm{Mod}_{2,\sigma}(\Gamma) = \mathrm{eff}\mathcal{C}_\sigma(s,t) = \frac{1}{\mathrm{eff}\mathcal{R}_\sigma(s,t)}.$$

Remark 14.1. An early version of the case $p = 2$ is due to Duffin [36]. The proof we give below in Theorem 14.4, which is from [3], was guided by a very general result in metric spaces, see [49, Theorem 7.31].

For the proof of Theorem 14.1, note that claim (i) follows from Proposition 11.1; claim (ii) follows from Propositions 13.6 and 13.5; and in (iii) the result follows from Theorem 14.4 below.

In order to apply the tools from convex optimization, it is more convenient to deal with finite families of objects. Unfortunately, connecting families of walks are usually infinite families. Here we will show that any family of walks Γ on a finite graph can be replaced with a finite subfamily with the same set of admissible densities and hence the same modulus. We call such finite subfamily $\Gamma^* \subset \Gamma$ an *essential subfamily.*

Theorem 14.2. *Any family of walks Γ admits an essential subfamily, that is, a finite subfamily $\Gamma^* \subseteq \Gamma$ such that $\mathrm{Adm}(\Gamma^*) = \mathrm{Adm}(\Gamma)$.*

Theorem 14.2 follows from the following general result about sets of vectors in $\mathbb{Z}^n_{\geq 0}$.

Theorem 14.3. *Let $X \subseteq \mathbb{Z}^n_{\geq 0}$. Then there exists a finite subset $X^* \subseteq X$ such that*

$$X = \bigcup_{x^* \in X^*} \{x \in X : x^* \preceq x\},$$

where $x^* \preceq x$ is the partial order induced by the coordinate-wise orders.

The proof of Theorem 14.3 is given in Exercise 14.3.

Proof of Theorem 14.2. Let $X = \{x_\gamma : \gamma \in \Gamma\}$ and let $X^* \subseteq X$ be the finite essential subset guaranteed by Theorem 14.3. We may construct Γ^* by choosing, for each $x^* \in X^*$, a corresponding $\gamma \in \Gamma$ such that $x_\gamma = x^*$. Then Γ^* is finite, and every $\gamma \in \Gamma$ dominates some element of Γ^*. Since $\Gamma^* \subseteq \Gamma$, we have that $\mathrm{Adm}(\Gamma^*) \supseteq \mathrm{Adm}(\Gamma)$. Moreover, if $\rho \in \mathrm{Adm}(\Gamma^*)$ and $\gamma \in \Gamma$ then there must exist a $\gamma^* \in \Gamma^*$ with $\gamma^* \preceq \gamma$. In particular, since $\rho \geq 0$, we have

$$\ell_\rho(\gamma) = \sum_{e\in E} \mathcal{N}(\gamma, e)\rho(e) \geq \sum_{e\in E} \mathcal{N}(\gamma^*, e)\rho(e) = \ell_\rho(\gamma^*) \geq 1.$$

Thus, we also have the reverse inequality $\mathrm{Adm}(\Gamma^*) \subseteq \mathrm{Adm}(\Gamma)$, and equality is proven. □

Example 14.1. Let $G = (V, E)$ be a simple connected finite graph and let $s \neq t \in V$ be two nodes. Then the connecting family of walks $\Gamma(s,t)$ is infinite. However, the subfamily of paths $\Gamma_{\text{path}}(s,t)$ is finite and is an essential subfamily, see Exercise 14.2 for the details.

The next result is an expanded version of [3, Theorem 4.2].

Theorem 14.4. *Let (G, σ, s, t) be a network. Let $\Gamma = \Gamma_{\text{path}}(s,t)$ be the connecting family of paths from s to t. Let ρ^* be the extremal density for $\mathrm{Mod}_{2,\sigma}(\Gamma)$. Then, there exists a vertex potential $\phi^* : V \to \mathbb{R}$ such that $\phi^*(s) = 0$, $\phi^*(t) = 1$, and*

$$\rho^*(x,y) = |\phi^*(x) - \phi^*(y)| \qquad \forall (x,y) \in E. \tag{14.1}$$

Moreover, this ϕ^* solves the optimization problem

$$\begin{aligned} \textit{minimize} \quad & \sum_{\{x,y\}\in E} \sigma(x,y)|\phi(x) - \phi(y)|^2 \\ \textit{subject to} \quad & \phi(s) = 0, \quad \phi(t) = 1. \end{aligned}$$

In addition, ϕ^* is harmonic at every node $x \in V \setminus \{s,t\}$:

$$\sum_{y\sim x} \sigma(x,y)(\phi^*(x) - \phi^*(y)) = 0 \tag{14.2}$$

In particular, it follows that $\mathrm{Mod}_2(\Gamma(s,t)) = \mathrm{eff}\mathcal{C}(s,t)$.

Later in Theorem 14.5 we will generalize Theorem 14.4 to the case $1 < p < \infty$ and also expand the list of properties.

Proof of Theorem 14.4. Let $\rho^* \in \mathrm{Adm}(\Gamma)$ be the extremal density for $\mathrm{Mod}_{2,\sigma}\,\Gamma$. Note that $\ell_{\rho^*}(\Gamma) = 1$, for, if not, the density $\rho' = \rho^*/\ell_{\rho^*}(\Gamma)$ is also admissible and has lower p-energy than ρ^*. Define $\phi^*(s) = 0$ and for $x \in V \setminus \{s\}$

$$\phi^*(x) = \min_{\gamma \in \Gamma(s,x)} \ell_{\rho^*}(\gamma),$$

where $\Gamma(s,x) = \Gamma_{\text{path}}(s,x)$. In particular, since $\Gamma = \Gamma(s,t)$, we have $\phi^*(t) = 1$.

Lemma 14.1. *Let (G, σ, s, t) be a network. Let $\Gamma = \Gamma_{\text{path}}(s,t)$ be the connecting family of paths from s to t. Suppose $h \in \mathbb{R}^V$ satisfies $h(s) = 0$ and $h(t) = 1$. Let*

$$\rho(e) = |h(y) - h(x)|, \qquad \forall e = \{x, y\} \in E.$$

Then, $\rho \in \mathrm{Adm}\,\Gamma$.

Proof of Lemma 14.1. Consider a path $\gamma \in \Gamma$. Write $\gamma = x_0 x_1 v_2 \dots x_{k-1} x_k$, where $x_0 = s$ and $x_k = t$. We have,

$$\begin{aligned}
&\ell_\rho(\gamma) \\
&= |h(x_1) - h(x_0)| + |h(x_2) - h(x_1)| + \\
&\cdots + |h(x_k) - h(x_{k-1})| \\
&\geq \left| \sum_{j=1}^{k} (h(x_j) - h(x_{j-1})) \right| && \text{(Triangle inequality)} \\
&\geq |h(t) - h(s)| && \text{(Telescoping sums)} \\
&= 1.
\end{aligned}$$

So $\rho \in \mathrm{Adm}(\Gamma)$. □

To see that (14.1) holds, define $\rho'(x,y) = |\phi^*(x) - \phi^*(y)|$ for $(x,y) \in E$. Then, by Lemma 14.1, $\rho' \in \mathrm{Adm}(\Gamma)$. Let $(x,y) \in E$ be an edge. Without loss of generality, $\phi^*(x) \leq \phi^*(y)$ and $y \neq s$. We claim that

$$0 \leq \rho'(x,y) = \phi^*(y) - \phi^*(x) \leq \rho^*(x,y). \tag{14.3}$$

The case $x = s$ is trivial. Suppose that $x \neq s$ and let $\gamma \in \Gamma(s,x)$ be a walk such that $\ell_{\rho^*}(\gamma) = \phi^*(x)$. Letting $\gamma' \in \Gamma(s,y)$ be the walk obtained by first traversing γ and then traversing the edge (x,y), we have

$$\phi^*(y) \leq \ell_{\rho^*}(\gamma') = \ell_{\rho^*}(\gamma) + \rho^*(x,y),$$

which implies (14.3).

Furthermore, (14.3) implies that $\mathcal{E}_2(\rho') \leq \mathcal{E}_2(\rho^*)$ and, by uniqueness of the extremal density, $\rho^* = \rho'$. To see that ϕ^* solves the optimization problem, let $\phi' : V \to \mathbb{R}$ with $\phi'(s) = 0$ and $\phi'(t) = 1$ and define $\rho'(x,y) = |\phi'(x) - \phi'(y)|$. As before, $\rho' \in \mathrm{Adm}(\Gamma)$ and so

$$\sum_{(x,y)\in E} \sigma(x,y)|\phi^*(x) - \phi^*(y)|^2 = \mathcal{E}_2(\rho^*)$$

$$\leq \mathcal{E}_2(\rho') = \sum_{(x,y)\in E} \sigma(x,y)|\phi'(x) - \phi'(y)|^2.$$

Finally, to show that ϕ^* satisfies (14.2), fix $x \in V\setminus\{s,t\}$ and let $\delta_x \in \mathbb{R}^V$ be the indicator function at x. Then, if $t \in (-\epsilon, \epsilon)$ for some $\epsilon > 0$ small,

$$\begin{aligned}
&\mathcal{E}_{2,\sigma}(\phi^* + t\delta_x) \\
&= \mathcal{E}_{2,\sigma}(\phi^*) + \sum_{y\sim x} \sigma(x,y)\left\{|\phi^*(x) - \phi^*(y) + t|^2 - |\phi^*(x) - \phi^*(y)|^2\right\}, \\
&= \mathcal{E}_{2,\sigma}(\phi^*) + \sum_{y\sim x} \sigma(x,y)\left\{2t\left(\phi^*(x) - \phi^*(y)\right) + t^2\right\},
\end{aligned}$$

Hence, by the minimizing property,

$$0 = \lim_{t\downarrow 0} \frac{\mathcal{E}_{p,\sigma}(\phi^* + t\delta_x) - \mathcal{E}_{2,\sigma}(\phi^*)}{t} = 2\sum_{y\sim x} \sigma(x,y)\left(\phi^*(x) - \phi^*(y)\right)$$

Therefore, (14.2) holds. In particular, 2-modulus equals effective conductance, see the discussion in Section 2.6, namely,

$$\mathrm{Mod}_2(\Gamma(s,t)) = \mathrm{eff}\mathcal{C}(s,t).$$

□

Next, we show that the case $p = 2$ generalizes to $1 < p < \infty$.

14.2 THE p-DIRICHLET PRINCIPLE

Set $1 < p, q < \infty$ so that $(p-1)(q-1) = 1$. Let $G = (V, E, \sigma)$ be a simple finite connected graph with edge conductances $\sigma \in \mathbb{R}^E_{>0}$. Fix a source and a target: $s \neq t \in V$. Let

$$\mathcal{H} := \{h \in \mathbb{R}^V : h(s) = 0, h(t) = 1\}.$$

The (p, σ)-*capacity problem* for the network (G, σ, s, t), also known as the *p-Dirichlet principle* is the following optimization problem

$$\begin{aligned} &\text{minimize} \qquad \mathcal{E}_{p,\sigma}(h) := \sum_{e=\{x,y\}\in E} \sigma(x,y)\,|h(x)-h(y)|^p \\ &\text{subject to} \qquad h \in \mathcal{H}. \end{aligned} \tag{14.4}$$

Optimal solutions for (14.4) will be called *minimizers.*

Remark 14.2. At times, we will be interested in solving capacity problems where the source node s and the target node t are replaced by subsets of nodes A and B respectively. The theory in this case is not substantially different. One can either consider a modified space $\mathcal{H}$ with $h_{|A} \equiv 0$ and $h_{|B} \equiv 1$. Or, alternatively, one can identify all the nodes in A into a single node a and all the nodes in B to a single node b, remove all the self-loops that might be created when doing this identification and then work with this new network.

The (p, σ)-*Dirichlet problem* for the network (G, σ, s, t), is to find a function $u \in \mathbb{R}^V$ that solves

$$\begin{cases} (L_{p,\sigma}u)(x) = 0, & \text{for all } x \in V \setminus \{s,t\}, \\ u(s) = 0 \quad \text{and} \quad u(t) = 1, \end{cases} \tag{14.5}$$

where the operator

$$(L_{p,\sigma}u)(x) := \sum_{y\sim x} \sigma(x,y)\,|u(y)-u(x)|^{p-2}\,(u(y)-u(x)) \tag{14.6}$$

is called *the weighted* (p, σ)-*Laplacian.* Note that, when $p = 2$, we recover the usual Laplacian defined in (3.14).

Remark 14.3. Let $m > 0$. If h^* is a minimizer for (14.4), then mh^* also minimizes the energy but with boundary values 0 at s and m at t. Likewise, if u^* solves (14.5), then mu^* also solves the Laplacian equation but with boundary values 0 at s and m at t.

Next, we use the extremal density for the p-modulus of the path family connecting s and t, to establish existence and uniqueness properties for the capacity problem (14.4) and the Dirichlet problem (14.5). Existence and uniqueness for (14.4) can also be deduced directly with convex optimization methods. See for instance Exercise 14.4 for an alternative approach to the uniqueness.

Theorem 14.5. *Let (G, σ, s, t) be a network as described above. Consider the family $\Gamma := \Gamma_{\text{path}}(s,t)$ of all paths connecting s and t. Let $\rho^* \in \operatorname{Adm}\Gamma$ be the extremal density for $\operatorname{Mod}_{p,\sigma}\Gamma$. Set*

$$h^*(x) := \min_{\gamma \in \Gamma_{\text{path}}(s,x)} \ell_{\rho^*}(\gamma) \tag{14.7}$$

for $x \neq s$, and $h^*(s) := 0$. Then,

(i) *The function h^* is a minimizer for the (p, σ)-capacity problem (14.4). Moreover,*

   (a) *the value of (14.4) equals the modulus:*

$$\mathcal{E}_{p,\sigma}(h^*) = \operatorname{Mod}_{p,\sigma}\Gamma;$$

   (b) *on each edge, the potential drops equal the extremal density for modulus:*

$$|h^*(x) - h^*(y)| = \rho^*(e), \tag{14.8}$$

   for all $e = \{x, y\} \in E$.

(ii) *If u^* is a minimizer for (14.4). Then, u^* solves the (p, σ)-Dirichlet problem (14.5).*

(iii) *If u solves the (p, σ)-Dirichlet problem (14.5), set*

$$F(x,y) := \sigma(x,y)\,|u(y) - u(x)|^{p-2}\,(u(y) - u(x)), \tag{14.9}$$

for $\{x, y\} \in E$, and $F(x, y) := 0$, if $\{x, y\} \notin E$. Then, F satisfies the Node Law, meaning that $\operatorname{Div}_F(x) = 0$ for every $x \in V \setminus \{s, t\}$, i.e., F is an st-flow as per Definition 1.6.

Moreover,

$$\operatorname{Val} F = \mathcal{E}_{p,\sigma}(u).$$

(iv) *Solutions to the (p, σ)-Dirichlet problem (14.5) are unique. Thus, h^* as defined in (14.7) is the unique solution for both the capacity problem (14.4) and the Dirichlet problem (14.5).*

In particular, the value of the flow F defined in part (iii) is

$$\operatorname{Val} F = \operatorname{Mod}_{p,\sigma}\Gamma(s,t) = \mathcal{E}_{p,\sigma}(h^*). \tag{14.10}$$

The unique minimizer $h^* \in \mathbb{R}^V$ is called the (p,σ)-*harmonic potential*, or equivalently, the (p,σ)-*capacitary* function for the network (G,σ,s,t). For $p \neq 2$, the p-Laplacian is a non-linear generalization of the combinatorial Laplacian. Yet, as we will see, many of the "electric network" properties that hold for the case $p = 2$ survive.

For instance, in view of the flow constructed in Theorem 14.5 (iv), we may define the notion of p-*effective conductance*, $\text{eff}\mathcal{C}_{p,\sigma}(s,t)$, to be the constant so that

$$\text{Val}\, F = \text{eff}\mathcal{C}_{p,\sigma}(s,t)\, |h^*(t) - h^*(s)|^{p-2}\, (h^*(t) - h^*(s)).$$

In particular, since $h^*(t) - h^*(s) = 1$, we have

$$\text{Val}\, F = \text{eff}\mathcal{C}_{p,\sigma}(s,t).$$

Thus, by (14.10),

$$\text{Mod}_{p,\sigma}\,\Gamma = \text{eff}\mathcal{C}_{p,\sigma}(s,t) = \mathcal{E}_{p,\sigma}(h^*) = \text{Val}\, F.$$

In Section 14.3 on the Thomson principle, we will show that the flow F defined in (14.9) is also an energy minimizer as current is when $p = 2$.

Part of the proof of Theorem 14.5 is contained in [3, Theorem 4.2]. We record here a complete proof for the reader's convenience.

Proof of Theorem 14.5. We begin by showing (i)(a) and (i)(b). Since $\Gamma = \Gamma_{\text{path}}(s,t)$, we have $h^*(t) = 1$. Indeed, note that

$$\ell_{\rho^*}(\Gamma) := \min_{\gamma\in\Gamma(s,t)} \ell_{\rho^*}(\gamma) = 1. \tag{14.11}$$

If not, we would have $\ell_{\rho^*}(\Gamma) > 1$, because ρ^* is admissible. Hence, the density $\rho := \rho^*/\ell_{\rho^*}(\Gamma)$ would also be admissible and would have a lower p-energy than ρ^*, which would contradict the optimality of ρ^*.

Now, define $\rho'(x,y) := |h^*(x) - h^*(y)|$ for $\{x,y\} \in E$. Then, by Lemma 14.1, $\rho' \in \text{Adm}(\Gamma)$. Next, fix an edge $\{x,y\} \in E$. Without loss of generality, assume that $h^*(x) \leq h^*(y)$. We claim that

$$0 \leq \rho'(x,y) = h^*(y) - h^*(x) \leq \rho^*(x,y). \tag{14.12}$$

The only thing to prove is the right-most inequality. The case when $x = s$ follows directly from the definition of $h^*(y)$. Suppose that $x \neq s$ and let $\gamma \in \Gamma_{\text{path}}(s,x)$ be a path such that $\ell_{\rho^*}(\gamma) = h^*(x)$. Let $\gamma' \in \Gamma_{\text{path}}(s,y)$

be the path obtained by first traversing γ and then traversing the edge (x, y). This gives

$$h^*(y) \leq \ell_{\rho^*}(\gamma') = \ell_{\rho^*}(\gamma) + \rho^*(x, y) = h^*(x) + \rho^*(x, y),$$

which implies (14.12). In particular, (14.12) establishes the '$\geq$' part of (14.8). In turn, (14.12) implies that $\mathcal{E}_{p,\sigma}(\rho') \leq \mathcal{E}_{p,\sigma}(\rho^*)$. Hence, by uniqueness of the extremal density, we must have $\rho^* = \rho'$. In other words, (14.8) holds. In particular, we have shown that $\mathcal{E}_{p,\sigma}(\rho^*) = \mathcal{E}_{p,\sigma}(h^*)$. So both (i)(a) and (i)(b) hold.

To finish the proof of (i) we must show that h^* solves the capacity problem (14.4). Consider an arbitrary candidate function $\phi : V \to \mathbb{R}$ with $\phi(s) = 0$ and $\phi(t) = 1$ and define $\tilde{\rho}(x, y) := |\phi(x) - \phi(y)|$, for every $\{x, y\} \in E$. As before, by Lemma 14.1, $\tilde{\rho} \in \mathrm{Adm}(\Gamma)$. So

$$\begin{aligned}\sum_{\{x,y\}\in E} \sigma(x, y)|h^*(x) - h^*(y)|^p &= \mathcal{E}_{p,\sigma}(\rho^*) \\ &\leq \mathcal{E}_{p,\sigma}(\tilde{\rho}) = \sum_{\{x,y\}\in E} \sigma(x, y)|\phi(x) - \phi(y)|^p.\end{aligned}$$

This shows that h is a minimizer for (14.4). Therefore, we have established (i).

Next we show (ii). Let u^* be a minimizer for the capacity problem. Fix $x \in V \setminus \{s, t\}$ and let $\delta_x \in \mathbb{R}^V$ be the indicator function at x. If $t \in (-\epsilon, \epsilon)$ for some $\epsilon > 0$ small, we have

$$\begin{aligned}&\mathcal{E}_{p,\sigma}(u^* + t\delta_x) \\ &= \mathcal{E}_{p,\sigma}(u^*) + \sum_{y\sim x} \sigma(x, y) \left\{|u^*(x) - u^*(y) + t|^p - |u^*(x) - u^*(y)|^p\right\}.\end{aligned}$$

Hence, by the minimizing property of u^*, and using the fact that $(|x|^p)' = p|x|^{p-2}x$,

$$\begin{aligned}0 &= \lim_{t\downarrow 0} \frac{\mathcal{E}_{p,\sigma}(u^* + t\delta_x) - \mathcal{E}_{p,\sigma}(u^*)}{t} \\ &= p\sum_{y\sim x} \sigma(x, y) \left|u^*(x) - u^*(y)\right|^{p-2} \left(u^*(x) - u^*(y)\right).\end{aligned}$$

This means that u^* satisfies the discrete (p, σ)-harmonic equation (14.6). Hence (ii) is proved.

Next, for part (iii), note that the Node Law for F follows directly from (14.6).

The 'Moreover part' follows from the "integration by parts" Lemma 2.2. We need the version of this lemma with $r(e) \equiv 1$, which we restate here for convenience. However, we skip the proof, since it is the same.

Lemma 14.2. *If $f = dh$ for some potential h, and k is an arbitrary st-flow, then*

$$\langle f, k\rangle := \sum_{e\in E} f(e)k(e) = \mathrm{Val}(k)(h(t) - h(s)). \tag{14.13}$$

Next, we apply Lemma 14.2 with $f := du$ and $k := F$. Then

$$\mathcal{E}_{p,\sigma}(u) = \langle F, du\rangle = \mathrm{Val}\, F(u(t) - u(s)) = \mathrm{Val}\, F.$$

This ends the proof of part (iii).

Finally, to prove part (iv), we use the following version of the maximum principle for p-harmonic functions, inspired by [57, Theorem 2.15].

Lemma 14.3 (Comparison principle). *Suppose $1 < p < \infty$. Let (G, σ, s, t) be a network and let h_0, h_1 be two solutions for the Dirichlet problem* (14.5). *Then,*

$$h_0 = h_1.$$

Lemma 14.3 says that uniqueness holds for the Dirichlet problem, hence the proof of Theorem 14.5 is concluded. □

Proof of Lemma 14.3. Let F_0 and F_1 be the st-flows associated to h_0 and h_1, respectively, as in (14.9). Then, $k := F_1 - F_0$ is an st-flow of value zero. Also, consider the gradient flow $f := d(h_1 - h_0)$. By Lemma 14.2, we have

$$\begin{aligned} 0 &= \langle f, k\rangle = \langle dh_1 - dh_0, \sigma|dh_1|^{p-2}dh_1 - \sigma|dh_0|^{p-2}dh_0\rangle \\ &= \sum_{e\in E} \sigma(e)(dh_1(e) - dh_0(e))(|dh_1(e)|^{p-2}dh_1(e) - |dh_0(e)|^{p-2}dh_0(e)). \end{aligned} \tag{14.14}$$

To conclude, we will need the following estimate valid for all real numbers.

Claim 14.1. *For any two numbers $a, b \in \mathbb{R}$ and any $1 < p < \infty$, the following identity always holds:*

$$(b - a)(|b|^{p-2}b - |a|^{p-2}a) \geq 0,$$

and equality holds if and only if $a = b$.

Assuming Claim 14.1 for the moment, we get that

$$\sigma(e)(dh_1(e) - dh_0(e))(|dh_1(e)|^{p-2}dh_1(e) - |dh_0(e)|^{p-2}dh_0(e)) \geq 0,$$

for all $e \in E$. Hence, by (14.14),

$$\sigma(e)(dh_1(e) - dh_0(e))(|dh_1(e)|^{p-2}dh_1(e) - |dh_0(e)|^{p-2}dh_0(e)) = 0,$$

for all $e \in E$. Therefore, by the equality case in Claim 14.1, we must have $dh_1(e) = dh_0(e)$ for every $e \in E$, namely,

$$h_1(y) - h_1(x) = h_0(y) - h_0(x) \qquad \text{for all } \{x, y\} \in E.$$

Thus, since G is connected, $h_1 = h_0 + C$ for some constant C. But, $h_0, h_1 \in \mathcal{H}$ means that they agree at s, hence must agree on all of V. This ends the proof of Lemma 14.3. □

Proof of Claim 14.1.

$$\begin{aligned}(b-a)(|b|^{p-2}b - |a|^{p-2}a) &= |b|^p + |a|^p - ab(|b|^{p-2} + |a|^{p-2}) \\ &\geq |b|^p + |a|^p - |a||b|(|b|^{p-2} + |a|^{p-2}) \\ &= (|b| - |a|)(|b|^{p-1} - |a|^{p-1})\end{aligned}$$

and the last expression is non-negative because the function $x \mapsto |x|^{p-1}$ is increasing for $x \geq 0$. Also, the case of equality follows from the fact that the function $x \mapsto x|x|^{p-2}$ is strictly increasing on all of $\mathbb{R}$. □

Note that the Dirichlet problem tells us that any choice of $h \in \mathcal{H}$ yields an upperbound for $\text{Mod}_{p,\sigma} \Gamma$, since $\text{Mod}_{p,\sigma} \Gamma \leq \mathcal{E}_{p,\sigma}(h)$. Is it possible to establish some lowerbounds for $\text{Mod}_{p,\sigma} \Gamma$ in a similar way? The answer will arise once we understand the duality between the Dirichlet problem and the Thomson problem, which we discuss next. In particular, we will show that the flow F defined in (14.9) solves the Thomson problem.

14.3 THE p-THOMSON PROBLEM

For $\{x, y\} \in E$, we write $r(x, y) := \sigma(x, y)^{-1}$, representing the edge resistance. Consider the affine subspace

$$\mathcal{F} := \{\phi \in \mathbb{R}^{V \times V} : \phi \text{ is an } st\text{-flow with value one}\},$$

meaning that for $\phi \in \mathcal{F}$, then $\text{Div}_\phi(x) = 0$ for every $x \in V \setminus \{s, t\}$ and $\text{Div}_\phi(s) = -\text{Div}_\phi(t) = 1$. Again, fix $1 < p, q < \infty$ with $(p-1)(q-$

$1) = 1$. Then, the (p,r)-*Thomson problem*, or *principle*, is the following minimization problem:

$$\begin{aligned} \text{minimize} \quad & \mathcal{E}_{p,r}(f) := \sum_{e=\{x,y\}\in E} r(x,y)|f(x,y)|^p \\ \text{subject to} \quad & f \in \mathcal{F}. \end{aligned} \tag{14.15}$$

The following is similar to Remark 14.3.

Remark 14.4. Let $m > 0$. If f^* is a minimizer for (14.15), then mf^* also minimizes the flow energy $\mathcal{E}_{r,p}$ among all st-flows with a value equal to m.

Theorem 14.6. *Let (G,σ,s,t) be a network. Let $1 < p,q < \infty$ with $pq = p+q$.*

(i) *Suppose f^* is a minimizer for the (p,r)-Thomson problem (14.15). Consider the flow*

$$\phi^* := |f^*|^{p-2} f^*. \tag{14.16}$$

Then, ϕ^* satisfies the Cycle Law, meaning that whenever $W = x_0e_1x_1\cdots e_kx_k$ is a cycle with $e_j = \{x_{j-1},x_j\} \in E$, for $j = 1,\ldots,k$, $x_0,\ldots,x_{k-1}$ distinct and $x_k = x_0$, then

$$\sum_{j=1}^{k} r(e_j)\phi^*(x_{j-1},x_j) = 0. \tag{14.17}$$

(ii) *The flow ϕ^* defined in (14.16) admits a potential $h \in \mathbb{R}^V$ such that*

$$h(y) - h(x) = r(x,y)\phi^*(x,y), \qquad \text{for all } \{x,y\} \in E, \tag{14.18}$$

i.e., $r\phi^* = dh$, and we can assume that $h(s) = 0$.

Moreover, h is $(q,\tilde{\sigma})$-harmonic in $V \setminus \{s,t\}$, in the sense that

$$(L_{q,\tilde{\sigma}}h)(x) = \sum_{y\sim x} \tilde{\sigma}(x,y)\,|h(x)-h(y)|^{q-2}\,(h(x)-h(y)) = 0. \tag{14.19}$$

for every $x \in V \setminus \{s,t\}$, where $\tilde{\sigma} := r^{1-q}$.

(iii) *The potential h defined in (ii) satisfies*

$$h(t) = \mathcal{E}_{p,r}(f^*) = \mathcal{E}_{q,r}(\phi^*). \tag{14.20}$$

(iv) *The Thomson problem* (14.15) *has the existence and uniqueness property.*

Remark 14.5. In analogy with the $p = 2$ case, we may use (14.20) to define the *p-effective resistance* $\text{eff}\mathcal{R}_{p,r}(s,t)$ by setting

$$\text{eff}\mathcal{R}_{p,r}(s,t) := \mathcal{E}_{p,r}(f^*) = h(t).$$

Proof of Theorem 14.6. For (i), define ϕ^* as in (14.16), and let k_0 be a st-flow with $\text{Val}(k_0) = 0$. Let $g_t := f^* + tk_0 \in \mathcal{F}$. If $t \in (-\epsilon, \epsilon)$ for some $\epsilon > 0$ small, we have

$$\begin{aligned}
&\mathcal{E}_{p,r}(f^* + tk_0) - \mathcal{E}_{p,r}(f^*) \\
&= \sum_{\{x,y\}\in E} r(x,y) \left\{ |f^*(x,y) + tk_0(x,y)|^p - |f^*(x,y)|^p \right\}.
\end{aligned}$$

Hence, by the minimizing property of f^*, and using the fact that $(|x|^p)' = p|x|^{p-2}x$,

$$\begin{aligned}
0 &= \lim_{t\downarrow 0} \frac{\mathcal{E}_{p,r}(f^* + tk_0) - \mathcal{E}_{p,r}(f^*)}{t} \\
&= p \sum_{\{x,y\}\in E} r(x,y) \left| f^*(x,y) \right|^{p-2} f^*(x,y) k_0(x,y) \\
&= p \sum_{\{x,y\}\in E} r(x,y) \phi^*(x,y) k_0(x,y).
\end{aligned}$$

In particular, given a cycle W, the corresponding cycle flow k_W is always an st-flow of value zero. This means that ϕ^* satisfies the Cycle Law (14.17).

For (ii), recall that the Cycle Law implies that there is a potential h with $h(s) = 0$ such that $r\phi^* = dh$. Indeed, given $x \in V$, find a path $\gamma_x = x_0 x_1 \cdots x_n$, with $x_0 = s$ and $x_n = x$, then set

$$h(x) := \sum_{j=1}^{n} r(x_{j-1}, x_j) \phi^*(x_{j-1}, x_j).$$

Then, the Cycle Law implies that h is well defined and by setting γ_y to be γ_x followed by y, whenever $y \sim x$, we see that (14.18) holds. So, $r\phi^* = dh$.

Next, we establish the following useful facts.

Lemma 14.4. *Let $1 < p, q < \infty$ with $pq = p + q$. Given two numbers $f, \phi \in \mathbb{R}$, we have*

$$\phi = |f|^{p-2}f \qquad \textit{if and only if} \qquad f = |\phi|^{q-2}\phi. \tag{14.21}$$

and, in this case,

$$|f|^p = |\phi|^q. \tag{14.22}$$

Proof. We will check only one direction of (14.21). The other one follows by symmetry. So suppose $\phi = |f|^{p-2}f$. Then, we just compute:

$$\begin{aligned} |\phi|^{q-2}\phi &= |f|^{(p-1)(q-2)}|f|^{p-2}f && (\text{since } |\phi| = |f|^{p-1}) \\ &= |f|^{(1-(p-1))}|f|^{p-2}f = f && (\text{since } (p-1)(q-1) = 1) \end{aligned}$$

Finally, assuming $f = |\phi|^{q-2}\phi$, we have

$$|f|^p = \left||\phi|^{q-2}\phi\right|^p = |\phi|^{pq-2p+p} = |\phi|^q.$$

□

To finish the proof of (iii) we have to check that h is $(q, \tilde{\sigma})$-harmonic. Fix, $x \neq s, t$. We have

$$\begin{aligned} &(L_{q,\tilde{\sigma}}h)(x) \\ &= \sum_{y\sim x} r(x,y)^{1-q}\,|h(x) - h(y)|^{q-2}\,(h(x) - h(y)) && (\text{since } \tilde{\sigma} := r^{1-q}) \\ &= -\sum_{y\sim x} |\phi^*(x,y)|^{q-2}\,\phi^*(x,y) && (\text{by (14.18)}) \\ &= -\sum_{y\sim x} f^*(x,y) = 0, && (\text{by (14.16) and (14.21)}) \end{aligned}$$

where the last equality follows from the fact that $\mathrm{Div}_{f^*}(x) = 0$, since $f^* \in \mathcal{F}$.

For (iii), we apply Lemma 14.2 to the gradient flow $r\phi^* = dh$ and the st-flow f^*, and get

$$\langle f^*, r\phi^*\rangle = \mathrm{Val}(f^*)(h(t) - h(s)) = h(t),$$

since $h(s) = 0$ and $f^* \in \mathcal{F}$. On the other hand, using (14.16), we get

$$\begin{aligned}
\langle f^*, r\phi^* \rangle &= \sum_{\{x,y\}\in E} r(x,y) f^*(x,y)\phi^*(x,y) \\
&= \sum_{\{x,y\}\in E} r(x,y)|f^*(x,y)|^p = \mathcal{E}_{p,r}(f^*) \qquad \text{(by (14.16))} \\
&= \sum_{\{x,y\}\in E} r(x,y)|\phi^*(x,y)|^q = \mathcal{E}_{q,r}(\phi^*) \qquad \text{(by (14.22))}
\end{aligned}$$

Finally for (iv), we begin by showing uniqueness. Suppose that f_i^*, $i = 0, 1$, are two minimizers for (14.15). Define the flows $\phi_i^* = |f_i^*|^{p-2} f_i^*$ as in part (i) and let h_i be the corresponding potentials from part (ii). Then, by part (iii),

$$h_0(t) = \mathcal{E}_{p,r}(f_0^*) = \mathcal{E}_{p,r}(f_1^*) = h_1(t).$$

And, since both h_0 and h_1 are $(q, \tilde{\sigma})$-harmonic in $V \setminus \{s,t\}$, with $h_0(s) = h_1(s) = 0$, by the uniqueness of the Dirichlet problem, Theorem 14.5 (iv), we have $h_0 \equiv h_1$. In particular, $\phi_0^* \equiv \phi_1^*$, and therefore $f_0^* \equiv f_1^*$ by (14.21).

Next, we show existence. Let $h \in \mathbb{R}^V$ be the unique solution to the $(q, \tilde{\sigma})$-Dirichlet problem with $h(s) = 0$ and $h(t) = 1$ we constructed in Theorem 14.5. Let

$$F := \tilde{\sigma}|dh|^{q-2} dh \tag{14.23}$$

be the corresponding st-flow. By Theorem 14.5 (iii), we have

$$\mathrm{Val}(F) = \mathcal{E}_{q,\tilde{\sigma}}(h). \tag{14.24}$$

Define

$$f = \frac{F}{\mathrm{Val}(F)}.$$

Then, f is an st-flow with value 1. Also we compute

$$\begin{aligned}
\mathcal{E}_{p,r}(f) &= \sum_{e\in E} r(e)|f(e)|^p \\
&= \mathcal{E}_{q,\tilde{\sigma}}(h)^{-p} \sum_{e\in E} r(e)|F(e)|^p \qquad \text{(by (14.24))} \\
&= \mathcal{E}_{q,\tilde{\sigma}}(h)^{-p} \sum_{e\in E} r(e)\tilde{\sigma}(e)^p |dh(e)|^{pq-2p+p} \qquad \text{(by (14.23))} \\
&= \mathcal{E}_{q,\tilde{\sigma}}(h)^{1-p}.
\end{aligned}$$

To show that f is a minimizer for (14.15) we argue by contradiction. Suppose f^* is an st-flow with value 1 and

$$\mathcal{E}_{p,r}(f^*) < \mathcal{E}_{p,r}(f) = \mathcal{E}_{q,\tilde{\sigma}}(h)^{1-p}.$$

Then

$$\mathcal{E}_{p,r}(f^*)^{1/p}\mathcal{E}_{q,\tilde{\sigma}}(h)^{1/q} < 1. \tag{14.25}$$

However, by Lemma 14.2,

$$\langle f^*, dh\rangle = \mathrm{Val}(f^*)(h(t) - h(s)) = 1.$$

So,

$$\begin{aligned}
1 = \langle f^*, dh\rangle &= \sum_{e\in E} f^*(e)dh(e) \\
&= \sum_{e\in E} r(e)^{1/p} f^*(e)\sigma(e)^{1/p} dh(e) \\
&\le \left(\sum_{e\in E} r(e)|f^*(e)|^p\right)^{1/p} \left(\sum_{e\in E} \sigma(e)^{q/p}|dh(e)|^q\right)^{1/q} && \text{(Holder's Inequality)} \\
&= \mathcal{E}_{p,r}(f^*)^{1/p}\mathcal{E}_{q,\tilde{\sigma}}(h)^{1/q} \\
&< 1. && \text{(by (14.25))}
\end{aligned}$$

This is a contradiction. So f is a minimizer.

This ends the proof of Theorem 14.6. □

Now, we put the Dirichlet and Thomson principles together.

Theorem 14.7 (Electric duality). *Let (G, σ, s, t) be a network. Let $1 < p, q < \infty$ with $pq = p + q$. Let h^* solve the (p, σ)-capacity problem (14.4) and let f^* solve the $(q, \hat{r})$-Thomson problem, where $\hat{r} := r^{q-1}$ and $r = \sigma^{-1}$. Then,*

$$\mathcal{E}_{p,\sigma}(h^*)\mathcal{E}_{q,\hat{r}}(f^*)^{p-1} = 1 \tag{14.26}$$

Moreover, if $F = \sigma|dh^*|^{p-2}dh^*$ is the st-flow corresponding to h^* as defined in (14.9), and H is the harmonic potential such that $dH = \hat{r}|f^*|^{q-2}f^*$, as defined in (14.18), then

$$\frac{F}{\mathrm{Val}\,F} = f^* \qquad \textit{and} \qquad \frac{H}{H(t)} = h^*.$$

Remark 14.6. Note that the energy duality in (14.26) implies that any choice of unit st-flow $\phi \in \mathcal{F}$ yields a lowerbound for modulus, since

$$\operatorname{Mod}_{p,\sigma} \Gamma \geq \frac{1}{\mathcal{E}_{q,\hat{r}}(\phi)^{p-1}}.$$

Also, in the language of effective conductance and effective resistance, the duality (14.26) can be rephrased as saying that

$$\operatorname{eff}\mathcal{C}_{p,\sigma}(s,t)^{\frac{1}{p}} \cdot \operatorname{eff}\mathcal{R}_{q,\hat{r}}(s,t)^{\frac{1}{q}} = 1. \tag{14.27}$$

Proof of Theorem 14.7. By definition, f^* solves the $(q, \hat{r})$-Thomson problem, where $\hat{r} := r^{q-1}$. By Theorem 14.6, the corresponding potential H, which satisfies $dH = \hat{r}|f^*|^{q-2} f^*$, is $(p, \tilde{\sigma})$-harmonic in $V \setminus \{s,t\}$, with

$$\tilde{\sigma} = \hat{r}^{1-p} = r^{-(p-1)(q-1)} = r^{-1} = \sigma.$$

Hence, H is (p,σ)-harmonic in $V \setminus \{s,t\}$. In particular, $H/H(t) \in \mathcal{H}$, and hence, by uniqueness, Theorem 14.5 (iv), we have $h^* = H/H(t)$. Moreover, by (14.20),

$$H(t) = \mathcal{E}_{q,\hat{r}}(f^*).$$

Thus, computing the (p,σ)-energy of h^*, since $|dH| = \hat{r}|f^*|^{q-1}$, we get

$$\begin{aligned}
\mathcal{E}_{p,\sigma}(h^*) &= H(t)^{-p}\mathcal{E}_{p,\sigma}(H) \\
&= \mathcal{E}_{q,\hat{r}}(f^*)^{-p} \sum_{e=\{x,y\}\in E} \sigma(x,y)\,|H(x) - H(y)|^p \\
&= \mathcal{E}_{q,\hat{r}}(f^*)^{-p} \sum_{e=\{x,y\}\in E} \hat{r}(x,y)^{1-p}\hat{r}(x,y)^p\,|f^*(x,y)|^q \\
&= \mathcal{E}_{q,\hat{r}}(f^*)^{1-p}
\end{aligned}$$

Therefore, the duality (14.26) holds.

Finally, let F be the st-flow corresponding to h^* as defined in (14.9). Recall that, $\operatorname{Val} F = \mathcal{E}_{p,\sigma}(h^*)$, by (14.10). Define the flow $G := F/\operatorname{Val} F$, so that $G \in \mathcal{F}$. Thus, computing the $(q,\hat{r})$-energy of G, since

$|F| = \sigma|dh^*|^{p-1}$, we get

$$\begin{aligned}
\mathcal{E}_{q,\hat{r}}(G) &= \mathcal{E}_{p,\sigma}(h^*)^{-q}\mathcal{E}_{q,\hat{r}}(F) \\
&= \mathcal{E}_{p,\sigma}(h^*)^{-q} \sum_{e=\{x,y\}\in E} \hat{r}(x,y)|F(x,y)|^q \\
&= \mathcal{E}_{p,\sigma}(h^*)^{-q} \sum_{e=\{x,y\}\in E} \hat{r}(x,y)\sigma(x,y)^q\,|h^*(x) - h^*(y)|^p \\
&= \mathcal{E}_{p,\sigma}(h^*)^{-q} \sum_{e=\{x,y\}\in E} \sigma(x,y)\,|h^*(x) - h^*(y)|^p \\
&= \mathcal{E}_{p,\sigma}(h^*)^{1-q}
\end{aligned}$$

Hence, by the duality (14.26),

$$\mathcal{E}_{q,\hat{r}}(G) = \mathcal{E}_{p,\sigma}(h^*)^{1-q} = \mathcal{E}_{q,\hat{r}}(f^*)^{(p-1)(q-1)} = \mathcal{E}_{q,\hat{r}}(f^*).$$

By uniqueness in the Thomson problem, we see that $G = f^*$. □

14.4 MODULUS AND THE THOMSON PROBLEM

Next, we address the following question: can the probabilistic interpretation of modulus (13.1) and Thomson's principle be reconciled? As we will see, the key is to formulate a path flow decomposition for the flow that arises in the Thomson problem.

Theorem 14.8. *Let (G, σ, s, t) be a network. Let $p, q > 1$ with $(p-1)(q-1) = 1$. Set $\Gamma = \Gamma_{\text{path}}(s,t)$. Let F^* solve the $(q, \hat{r})$-Thomson problem as in Theorem 14.7. Then, F^* admits a path flow decomposition*

$$F^* = \sum_{\gamma\in\tilde{\Gamma}\subset\Gamma} a(\gamma) f_\gamma, \tag{14.28}$$

where

(i) $\tilde{\Gamma}$ is a subfamily of Γ such that one every edge $e \in E$ the orientation induced on e by the paths in $\tilde{\Gamma}$ coincides with the one induced by F^*;

(ii) f_γ is the unit flow along the path γ;

(iii) The coefficients form a pmf $a \in \mathcal{P}(\Gamma)$, that is strictly supported on $\tilde{\Gamma}$, and that is optimal for $\mathrm{Mod}_{p,\sigma}(\Gamma)$, as in Definition 13.1.

The proof of Theorem 14.8 is left for the reader in Exercise 14.5. A possible algorithm for producing the path flow decomposition of F^* goes like this:

- Consider the modulus problem $\mathrm{Mod}_{p,\sigma}\Gamma$ and let ρ^* be the corresponding extremal density.
- Do a gradient ascent excursion from s to t by always choosing the edge with the largest value of $\rho^*(e)$, and get a path γ_0.
- Then, let $R := \min_{e\in\gamma_0}\rho^*(e)$, and subtract R from each value of ρ^* along the path γ_0.
- Repeat with the new values of ρ^*.

This is formalized in Algorithm 2. We would like to thank N. Shanmugalingam for a helpful discussion regarding this method.

Algorithm 2 The steepest ascent path decomposition

Input: A weighted network with boundary $G = (V, E, \sigma, A, B)$ and the extremal density ρ^* for the family of paths from A to B in G.

Initialization: set $R(e) = \sigma(e)\rho^*(e)^{p-1}$ for each edge $e \in E$.

Iteration:

```
1: Remove any edges e for which R(e) = 0 from G (call such edges zero edges).
2: Run the Greatest Ascent subroutine.
3: if there is a steepest ascent path γ in G then
4:    add it to a growing family Γ̃
5: else
6:    terminate.
7: end if
8: For the path γ identified in step 2, set m(γ) = min_{e∈γ} R(e).
9: for each edge e in the path γ identified in step 2 do
10:    update R(e) to be R(e) − m(γ).
11: end for
```

Output: curve family $\tilde{\Gamma}$ and the weights $m : \Gamma \to (0, \infty)$.

Greatest ascent subroutine: Find an edge $e_1 = ax_1$ so that e_1 maximizes R among all edges for which $a \in A$. Next, find an edge $e_2 = x_1x_2$ so that e_2 maximizes R among all edges with an endpoint at x_1 and which are not the previous edge (no backtracking). Continue this way until one hopefully reaches B.

EXERCISES

14.1 *Asymptotics of modulus of connecting path families.* Let $G = (V, E, \sigma)$ be a graph with edge weights $\sigma \in \mathbb{R}^E_{>0}$. Fix $a \neq b \in V$ and let $\Gamma = \Gamma_{\text{path}}(a, b)$ be the family of all paths from a to b.

(i) Show that

$$\lim_{p\to\infty} \operatorname{Mod}_{p,\sigma} \Gamma = \begin{cases} \sigma(a,b) & \text{if } a \sim b, \\ 0 & \text{if } a \not\sim b. \end{cases}$$

(ii) Consider the *Erdős-Rényi random graph* model $G(N, \pi)$. Recall this means that we build a graph G on N nodes $V = \{1, \ldots, N\}$, by sampling $m := \binom{N}{2}$ independent π-Bernoulli variables X_e, for $e \in \binom{V}{2}$. Namely, $X_e \in \{0, 1\}$ and $\mathbb{P}(X_e = 1) = \pi$, where $\pi \in [0, 1]$ is given. Then, $G = (V, E)$ where $E := \left\{e \in \binom{V}{2} : X_e = 1\right\}$, and we write $G \sim G(N, \pi)$, meaning that G was sampled from this random model. We will assume $N \geq 2$ and consider the family $\Gamma(1, 2)$ of all paths connecting nodes 1 and 2 in G. With this set up, $\operatorname{Mod}_p \Gamma(1, 2)$ is a random variable. Show that

$$\lim_{p\to\infty} \mathbb{E}\left[\operatorname{Mod}_p \Gamma(1, 2)\right] = \pi.$$

(We thank Colin Williams for this exercise).

14.2 *Essential subfamilies of connecting families of walks.* Let $G = (V, E)$ be a simple connected finite graph and let $s \neq t \in V$ be two nodes.

(i) Show that the connecting family of walks $\Gamma(s, t)$ is infinite.

(ii) Show that the subfamily of paths $\Gamma_{\text{path}}(s, t)$ is finite.

(iii) Show that every walk γ from s to t dominates a path γ^* from s to t, which is obtained from γ by removing all the loops. Deduce that $\operatorname{Adm}(\Gamma(s, t)) = \operatorname{Adm}(\Gamma_{\text{path}}(s, t))$, and hence that $\Gamma_{\text{path}}(s, t)$ is an essential subfamily for $\Gamma(s, t)$.

14.3 *Essential subfamilies.* Here we give a proof of Theorem 14.3.

(i) Use the mapping $h : \mathbb{Z}^n_{\geq 0} \to \mathbb{Z}_{\geq 0}$ defined by $h(x) = \sum_{i=1}^n x_i$, to show that every non-empty set $X \subseteq \mathbb{Z}^n_{\geq 0}$ has a minimal element (not necessarily unique) with respect to the partial ordering defined in Theorem 14.3.

(ii) The rest of the proof proceeds by induction on n. First, establish the base case, when $n = 1$.

(iii) Now, suppose the result holds for dimensions $k = 1, 2, \ldots, n-1$. We want to show that it also holds for dimension n. Let $X \subseteq \mathbb{Z}^n_{\geq 0}$. If X is empty, we are finished. Otherwise, let x^0 be a minimal element of X, whose existence is guaranteed by part (i). Let X^0 be the set of elements in X that do not dominate x^0:

$$X^0 = \{x \in X : x^0 \not\preceq x\}.$$

For any subset I of $\mathcal{I} := \{1, 2, \ldots, n\}$, define

$$X^0_I = \{x \in X^0 : x_i < x^0_i, \forall i \in I, \text{ and } x_i \geq x^0_i, \forall i \notin I\}.$$

Show that X^0_I is empty, if I is the empty set or, if $I = \mathcal{I}$.

(iv) Now, let I be a subset of $\mathcal{I}$ such that X^0_I is nonempty, and let $k = |I|$. We have seen that necessarily $1 \leq k \leq n - 1$. After reordering, we can write any $x \in X^0_I$ as $x = (x', x'')$ with the property that $x' \prec (x^0)'$ and $x'' \succeq (x^0)''$, where x' is made up of the first k entries of x and x'' of the remaining $n - k$ entries. To simplify the notation, consider the projections $p(x) = x'$ and $q(x) = x''$. Since $p(x) \prec (x^0)'$, there can only be finitely many such projections. So let $\{z_i\}_{i=1}^{m_I} \subset \mathbb{Z}^k_{\geq 0}$ be an enumeration of the nonnegative integer vectors satisfying $z_i = p(x)$ for some $x \in X^0_I$ and define for each $i = 1, 2, \ldots, m_I$ the set

$$X^0_{I,i} := \{q(x) \in \mathbb{Z}^{n-k}_{\geq 0} : (z_i, q(x)) \in X^0_I\}.$$

14.4 *Uniqueness of minimizers for the capacity problem.* In Theorem 14.5, we deduced the uniqueness of minimizers of the capacity problem (14.4) from the fact that they must also solve the Dirichlet problem and thus the maximum principle can be used. Here we will give an alternative, more direct proof that uses strict convexity.

(i) Assume $h_0, h_1 \in \mathcal{H}$ are two distinct minimizers for (14.4). For $t \in (0, 1)$, define $h_t := th_1 + (1 - t)h_0$. Show that

$$\mathcal{E}_{p,\sigma}(h_t) \leq t\mathcal{E}_{p,\sigma}(h_1) + (1 - t)\mathcal{E}_{p,\sigma}(h_0) = \mathrm{Mod}_{p,\sigma}\,\Gamma. \quad \text{(P14.1)}$$

(ii) Next, show that

$$t|dh_1(e)|^p + (1-t)|dh_0(e)|^p - |tdh_1(e) + (1-t)dh_0(e)|^p = 0,$$

for all $e \in E$.

(iii) Finally, use the function $x \mapsto |x|^p$ to reach a contradiction.

14.5 *Path flow decomposition.* Let (G, σ, s, t) be a network. Let $1 < p, q < \infty$ with $(p-1)(q-1) = 1$. Let ρ^* be the unique extremal density for $\text{Mod}_{p,\sigma}\,\Gamma$ where $\Gamma = \Gamma_{\text{path}}(s,t)$. The goal is to show that Algorithm 2 outputs a minimal subfamily $\tilde{\Gamma}$ such that its corresponding optimal pmf is proportional to m. So, let $E_0 = \{e \in E : \rho^*(e) = 0\}$.

(i) Show that there exists an orientation of all edges in $E \setminus E_0$ so that the following hold:

(a) At any vertex $x \in V \setminus \{s,t\}$, the sum of $\sigma(e)\rho^*(e)^{p-1}$ over the edges leading into x equals the sum of $\sigma(e)\rho^*(e)^{p-1}$ over the edges leading out of x.

(b) All edges in $E \setminus E_0$ with a vertex at s (resp. t) are oriented away from s (resp. towards t). *Hint: Show that the solution h to the (p,σ)-Dirichlet problem for the network (G, σ, s, t), satisfies $0 \le h \le 1$ by the maximum principle for the p-Laplacian.*

(c) Let γ be a path in $G \setminus E_0$ from s to t. Then γ is a directed path from s to t if and only if γ has ρ^*-length 1, i.e. $\ell_{\rho^*}(\gamma) = 1$.

(d) $G \setminus E_0$ is a directed acyclic graph.

(ii) Let $E \setminus E_0$ be oriented as in (i). After each completion of step 1 in the Greatest Ascent Path Decomposition algorithm 2, the connected components of the resulting graph are isolated vertices or subgraphs with all their source vertices in s and all their sink vertices in t.

(iii) We want to use algorithm 2 to decompose F^*. The algorithm must terminate because we remove at least one edge from the graph during every iteration. When the algorithm terminates, there are no edges remaining in the graph by (ii). Thus for each e edge in $E \setminus E_0$, the algorithm's output path family $\tilde{\Gamma}$ must include a path containing e.

Apply Beurling's Criterion to show that $\tilde{\Gamma}$ is a Beurling subfamily of Γ.

(iv) Show that $\tilde{\Gamma}$ is also a minimal subfamily for $\text{Mod}_{p,\sigma}(\Gamma)$.

(v) Set $a(\gamma) := \mu'(\gamma)$ for $\gamma \in \tilde{\Gamma}$ and set $a(\gamma) = 0$ otherwise. Consider the flow

$$G := \sum_{\gamma \in \Gamma} a(\gamma) f_\gamma$$

where f_γ is the unit flow along γ. Show that $G = F^*$.

CHAPTER 15

Modulus on Planar Maps

IN THIS CHAPTER we will specialize to the case of planar graphs and more specifically to orthodiagonal maps.

15.1 ORTHODIAGONAL MAPS

A finite *orthodiagonal map* is a finite connected plane graph $G = (V, E)$ that satisfies the following conditions:

(i) Each edge is a straight line segment (unless it borders a non-convex face, see below).

(ii) Each inner face is a quadrilateral with orthogonal diagonals.

(iii) The boundary of the outer face is a closed Jordan curve.

Since G is a bipartite graph, see Exercise 15.1, we obtain a bipartition of the vertices $V = V^\bullet \sqcup V^\circ$. Each inner face Q is determined by the four vertices on its boundary, which must alternate between $V^\bullet$ and V°. We will write $Q = [z, w, z', w']$ with the convention that $z, z' \in V^\bullet$, and $w, w' \in V^\circ$. Also, the vertices will be ordered counterclockwise around Q. We can create two additional graphs associated with G, the *primal graph* $G^\bullet = (V^\bullet, E^\bullet)$ and the *dual graph* $G^\circ = (V^\circ, E^\circ)$. There is one primal edge $e_Q^\bullet$ and one dual edge e_Q° for each inner face of G. More specifically, let $e_Q^\bullet \in E^\bullet$ be an edge connecting z and z' in Q; it is defined to be the line segment zz', if this segment is contained in Q, and otherwise it is the union of the line segments zp and pz', where $p = (w + w')/2$ is the midpoint of ww'. The edge $e_Q^\circ \in E^\circ$ between w and w' in Q is

DOI: 10.1201/9781003024866-19

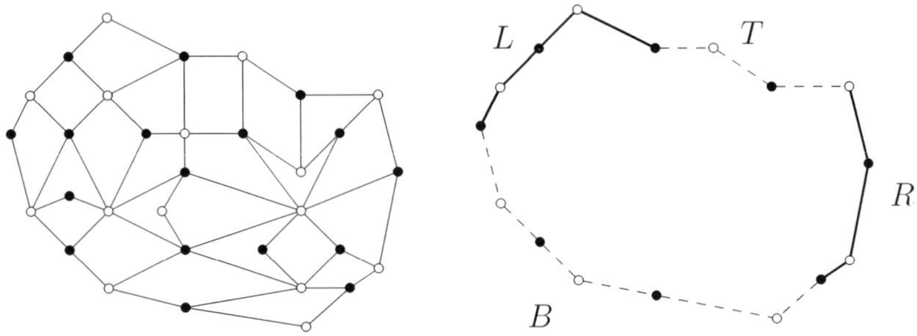

Figure 15.1 Left: An example of an orthodiagonal map with the bipartition of the vertices. Right: The topological boundary of the map is decomposed into four paths: L, R (solid) and B, T (dashed).

defined similarly. For an edge $e \in E^\bullet$, the *dual edge to* e is the unique edge $e^\circ \in E^\circ$ so that there is an inner face Q with $e = e_Q^\bullet$ and $e^\circ = e_Q^\circ$.

Furthermore, we let ∂G be the topological boundary of the outer face of G, which we assumed is a closed Jordan curve; we let the interior $\text{int}(G)$ be the open subset enclosed by ∂G; and we set $\hat{G}$ to be $\text{int}(G) \cup \partial G$; finally, we set $\tilde{G}$ to be the union of the convex hulls of the closures of inner faces of G.

The graph $G = (V^\bullet \sqcup V^\circ, E)$ is a finite *orthodiagonal map with boundary arcs* L, B, R, T, if G is a finite orthodiagonal map with the following boundary condition:

(iv) (Boundary condition) The topological boundary of the outer face of G is subdivided into four nonempty paths L, B, R, T, listed counterclockwise, with $L \cap B$, $B \cap R$, $R \cap T$, and $T \cap L$ each equal to a single vertex, yielding four corners. See Figure 15.1.

Orthodiagonal maps arise naturally in numerical analysis in the context of Delaunay triangulations and Voronoi graphs, see [74, Section 5.1], and also in the context of circle packing, see [47, Section 2].

15.2 ORTHODIAGONAL TILINGS OF THE SQUARE

We will begin by describing a remarkable result for modulus on orthodiagonal tilings of the square. This was discovered in collaboration with J. Lind and P. Pankka, see [9]. In personal communication, we learned that Asaf Nachmias found a similar result, in the slightly different language of electrical networks and circle packings, and with a different proof.

Let $S = [0,1] \times [0,1]$ be the unit square in the complex plane. It is well-known that the (continuous) modulus for the family of all curves in S connecting the two vertical sides of S is equal to one. Here we will show that the same continues to be true for discrete modulus that arises from orthodiagonal tilings of the square.

We begin by describing a concrete example. Assume that S is tiled arbitrarily by an orthodiagonal map, as in Figure 15.2. Here S is tiled by five quadrilaterals $ABDJ$, CDB, $DEFJ$, $FGHJ$, and $AJHI$. The points have coordinates

$$\begin{aligned} &A = (0,0),\ B = (1-x,0),\ C = (1,0),\ D = (1,1/2), \\ &E = (1,1), F = (1-2x,1),\ G = (1/2,1),\ H = (4x,1), \\ &I = (0,1),\ J = (1/2,1-2x) \end{aligned}$$

where x is a real number with $0 < x < 1/8$.

To be an orthodiagonal map all the quadrilaterals in the tiling must have orthogonal diagonals. The quadrilaterals don't have to be convex, i.e., the diagonals may not meet in the interior of the quadrilateral. As a union of quadrilaterals that can only meet along an edge, the corresponding graph is bipartite, so we can color the vertices black and white in a unique way (once the color of one node is determined). However, in the tiling we allow the quadrilaterals to degenerate into triangles as depicted in Figure 15.3. Another way to think about this is that some edges connecting a black to a white vertex may have a length zero.

As can be seen in Figure 15.2, the diagonals connecting the black vertices form a graph called the *primal graph*, which we denote by $G^\bullet = (V^\bullet, E^\bullet)$. Likewise, the diagonals connecting the white vertices form a graph called the *dual graph*, which we denote by $G^\circ = (V^\circ, E^\circ)$. Note that both graphs are connected and touch all four sides of S.

Furthermore, both graphs are weighted in a canonical way. Suppose $Q = [z, w, z', w']$ is a face where $z, z' \in V^\bullet$ and $w, w' \in V^\circ$. Then

$$\sigma(zz') := \frac{|w - w'|}{|z - z'|} \qquad \text{and} \qquad \sigma(ww') := \frac{|z - z'|}{|w - w'|}. \tag{15.1}$$

We use the letter σ in both cases and we interpret these weights as electric conductances. In Figure 15.2, each canonical weight is marked next to the corresponding edge.

We now define four families of paths $\Gamma^\bullet_{\mathrm{LR}}, \Gamma^\circ_{\mathrm{LR}}, \Gamma^\bullet_{\mathrm{BT}}$, and $\Gamma^\circ_{\mathrm{BT}}$, where $\Gamma^\bullet_{\mathrm{LR}}$ consists of all the paths in the primal graph $G^\bullet$ that connect the left vertical side to the right vertical side of the square S. Likewise $\Gamma^\circ_{\mathrm{LR}}$

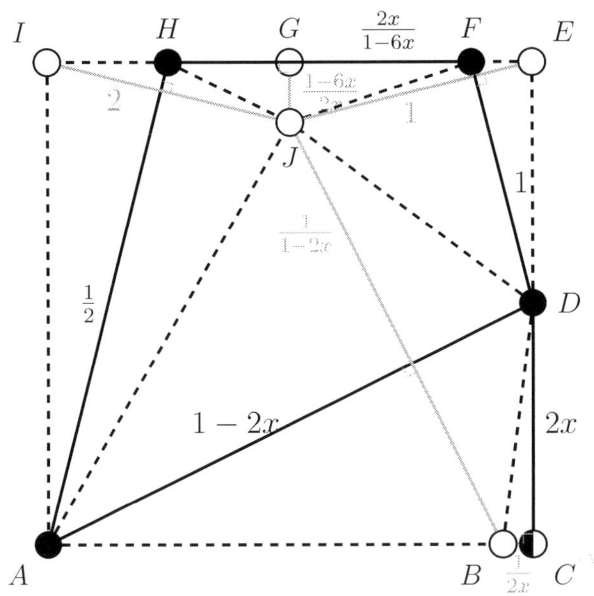

Figure 15.2 An orthodiagonal map tiling of the unit square.

consists of all the paths in the dual graph G° that connect the left vertical side to the right vertical side of S. The subscript BT indicates paths that connect the bottom to the top of S in the relevant graph. We want to

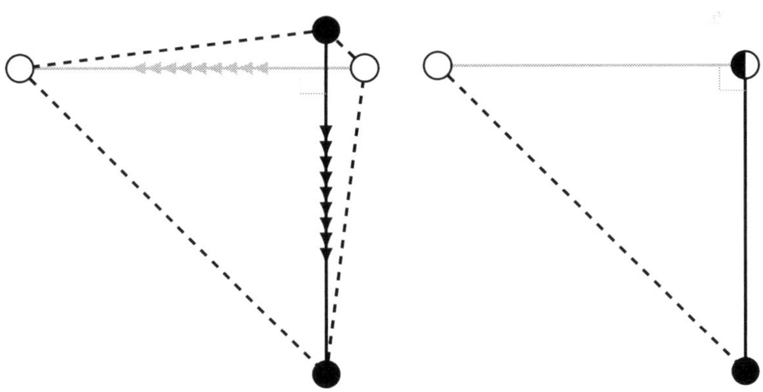

Figure 15.3 Orthodiagonal quadrilaterals may degenerate into a triangle. One of the four sides may shrink until a black and white vertex superimposes, in which case we denote this by a single node half white and half black.

compute the weighted $\mathrm{Mod}_{2,\sigma}$ modulus of these families. For simplicity, in this example, we will use the fact the $\mathrm{Mod}_{2,\sigma}$ modulus coincides with the notion of effective conductance, see Theorem 14.1 (iii). The fact that the path families here connect two sets of nodes rather than two single nodes is not a problem because one can always identify all the vertices in a set into a single node and then work in the shrunk graph.

Recall that the serial and parallel rules for conductances work as follows:

$$\begin{aligned} &\sigma_1 + \sigma_2 && \text{for conductances in parallel} \\ &\frac{1}{\frac{1}{\sigma_1} + \frac{1}{\sigma_2}} && \text{for conductances in series} \end{aligned}$$

With this at hand, one can verify that

$$\mathrm{Mod}\,\Gamma^{\bullet}_{\mathrm{LR}} = (1-2x) + \frac{1}{2 + \frac{1-6x}{2x} + 1} = 1.$$

Likewise,

$$\mathrm{Mod}\,\Gamma^{\circ}_{\mathrm{LR}} = \frac{1}{\frac{1}{2} + \frac{1}{1+\frac{1}{(1-2x)+2x}}} = 1.$$

Similarly,

$$\mathrm{Mod}\,\Gamma^{\bullet}_{\mathrm{BT}} = \frac{1}{2} + \frac{1}{1 + \frac{1}{(1-2x)+2x}} = 1.$$

and,

$$\mathrm{Mod}\,\Gamma^{\circ}_{\mathrm{BT}} = \frac{1}{(1-2x) + \frac{1}{1+2+\frac{1-6x}{2x}}} = 1.$$

Namely, we see that the discrete modulus is always equal to one, which is also the value of the continuous modulus. We claim that this is always the case, no matter what orthogonal map one uses to tile the square.

Definition 15.1. We say that G is a *tiling* of the square S if G is an orthodiagonal map such that the boundary of the outer face of G coincides with ∂S, and where some of the faces may degenerate as in Figure 15.3.

Theorem 15.1. *When a square $S \subset \mathbb{C}$ is tiled by an orthodiagonal map as described in Definition 15.1, the discrete potential for the Dirichlet problem with boundary values 0 on the left side of S and 1 on the right side of S, coincides exactly with the continuous potential, for both the primal and the dual graph. In particular, the discrete modulus of the family of paths connecting the two vertical sides of S is always equal to one for both primal and dual graphs.*

Before we can prove Theorem 15.1, we will adapt Fulkerson duality to orthodiagonal tilings of the square.

15.3 FULKERSON DUALITY FOR ORTHODIAGONAL MAPS

We investigate Fulkerson duality in the specific case of orthodiagonal maps and tilings of the square.

Proposition 15.1. *Suppose that G is an orthodiagonal map tiling of the square S as in Definition 15.1. Let A be the left vertical side of S and B be the right vertical side of S. Likewise, let C be the bottom side and D the top side. Write $\Gamma^{\bullet}_{\mathrm{LR}}$ for the family of paths on the primal graph $G^{\bullet}$ that connect A and B, and write $\Gamma^{\circ}_{\mathrm{BT}}$ for the family of paths on the dual graph G° that connect C and D. Then,*

$$\operatorname{Mod}_{2,\sigma} \Gamma^{\bullet}_{\mathrm{LR}} \cdot \operatorname{Mod}_{2,\sigma} \Gamma^{\circ}_{\mathrm{BT}} = 1, \tag{15.2}$$

where σ are the canonical weights defined in (15.1).

Proof. It will be enough to show that $\operatorname{Adm}\Gamma^{\circ}_{\mathrm{BT}}$ and $\operatorname{Adm}\Gamma_{\mathrm{cut}}(A,B)$ are isometrically isomorphic, as explained in Example 13.2.

Assume that, $\eta \in \operatorname{Adm}\Gamma_{\mathrm{cut}}(A,B)$. Given an edge $e^{\circ} \in G^{\circ}$, define

$$\hat{\eta}(e^{\circ}) := \eta(e^{\bullet}), \tag{15.3}$$

where $e^{\bullet}$ is the diagonal that is orthogonal to e°. We claim that $\hat{\eta} \in \operatorname{Adm}\Gamma^{\circ}_{\mathrm{BT}}$. If that's the case, then that establishes one direction. Moreover, note that $\sigma^{-1}(e^{\circ}) = \sigma(e^{\bullet})$.

Indeed, let

$$\gamma := w_0 \; e^{\circ}_1 \; w_1 \; e^{\circ}_2 \; \cdots \; w_{n-1} \; e^{\circ}_n \; w_n \tag{15.4}$$

be a path in $\Gamma^{\circ}_{\mathrm{BT}}$, i.e., so that $w_0 \in C \cap V^{\circ}$, $w_n \in D \cap V^{\circ}$, $w_j \in V^{\circ} \setminus (C \cup D)$, for $j = 1, \ldots, n-1$, and each $e^{\circ}_j = \{w_{j-1}, w_j\}$ is an edge in G°. Write $w_0 = x_0$ and $w_n = x_n + i$ where $x_0, x_n \in [0,1]$. The Jordan curve

$$J := \gamma \cup \{x_0 - it : t \le 0\} \cup \{x_n + is : s \ge 1\}$$

divides $\mathbb{C}$ into two simply connected domains Ω^+ and Ω^-, where Ω^- contains the half-space $\{\operatorname{Re} z < 0\}$ and Ω^+ contains the half-space $\{\operatorname{Re} z > 1\}$. Let

$$S := (A \cap V^{\bullet}) \cup (\Omega^- \cap V^{\bullet}).$$

We claim that S is an AB-cut. By definition, $(A \cap V^{\bullet}) \subset S$. So it is enough to show that

$$(B \cap V^{\bullet}) \cap S = \emptyset.$$

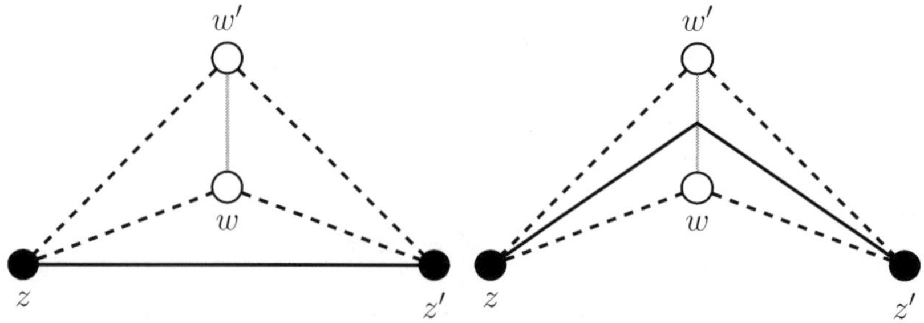

Figure 15.4 Faces can be non-convex.

Let $1+bi \in B \cap V^{\bullet}$, with $b \in [0,1]$. Note that the ray $R = \{x+ib : x > 1\}$ is contained in Ω^+. Therefore, whether $1+bi$ belongs to γ or not, we see that it cannot be contained in Ω^-, and hence $1+ib \notin S$.

For $j = 1, \ldots, n$, let Q_j be the face corresponding to e_j° and set $e_j^{\bullet} = \{z_j, z_j'\}$ be the other diagonal. Since J is a locally analytic Jordan curve, we can assume that $z_j \in \Omega^-$ and $z_j' \in \Omega^+$. Note that if the face is not convex, then the diagonal that would fall outside the face needs to be bent into two segments so that it goes through the midpoint of the other diagonal, see Figure 15.4.

From now on when we mention a diagonal and write $\{z, z'\}$ or $\{w, w'\}$, it may be broken appropriately as in Figure 15.4 if needed.

It follows that $\{e_j^{\bullet}\}_{j=1}^n \subset \delta S$. We claim conversely that $\{e_j^{\bullet}\}_{j=1}^n \supset \delta S$. Indeed, assume $\{z, z'\} \in \delta S$ with $z \in S$ and $z' \notin S$. Two cases arise:
1) If $z \in \Omega^- \cap V^{\bullet}$, then $z' \in \Omega^+ \cup \gamma$. Let $Q = [z, w, z', w']$ be the face for which $\{z, z'\}$ is a diagonal. If $z' \in \Omega^+$, then the edge $\{w, w'\}$ must be contained in γ, see the right panel in Figure 15.4. If $z' \in \gamma$, then either w or w' also belong to γ, see Figure 15.5.

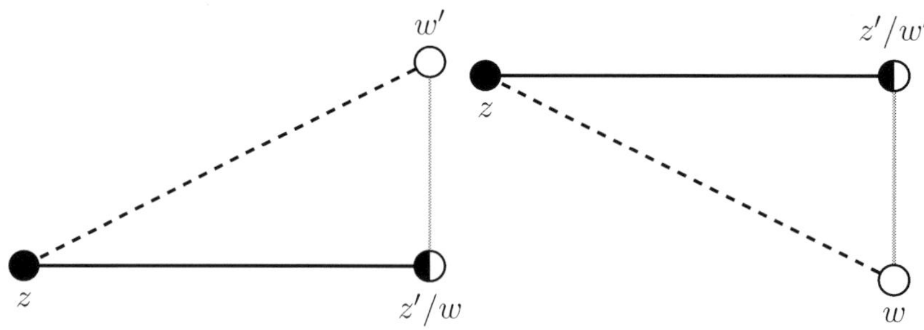

Figure 15.5 Degenerate case.

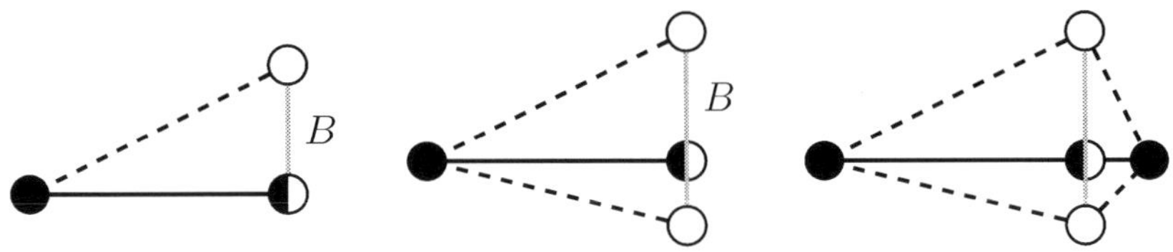

Figure 15.6 Degenerate faces may only abut other degenerate faces.

The reason is that only degenerate faces can share a diagonal edge with another degenerate face. So there are essentially three cases. Namely, either the degenerate vertex z'/w is the bottom right-hand corner of S, see the left panel in Figure 15.6; or z'/w lies on B, see the middle panel in Figure 15.6; or z'/w is an interior point in S, and then it can only be contained in a single path, see the right panel in Figure 15.6. Note that in this latter case, each one of the gray edges, as well as the black edges, should be thought as two parallel edges, one for each corresponding face.

2) If $z \in (A \cap V^\bullet) \setminus (\Omega^- \cap V^\bullet)$, then $z \in A \cap \gamma$. In this case both w and w' belong to $A \cap \gamma$ and z' doesn't. Indeed, there are three main cases: either $\{w, w'\}$ is contained in A and z belongs to its relative interior, see the first panel in Figure 15.7; alternatively, two degenerate faces meet at z on A as in the middle panel of Figure 15.7; or, finally, z may be one of the two corners of S that lie in A, see the last panel in Figure 15.7. Thus, we have seen that in every case the edge $\{w, w'\}$ belongs to γ, and this shows that $\{e_j^\bullet\}_{j=1}^n = \delta S$. Since S is an AB-cut, it must contain a minimal AB-cut and hence be admissible, meaning that

$$\sum_{j=1}^{n} \hat{\eta}(e_j^\circ) \geq 1. \tag{15.5}$$

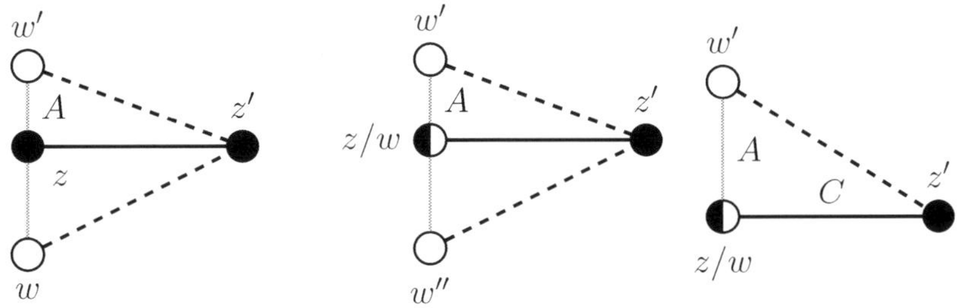

Figure 15.7 Last case.

Since γ was chosen arbitrarily, this means that $\hat{\eta} \in \operatorname{Adm} \Gamma^{\circ}_{\mathrm{BT}}$.

Conversely, assume that $\hat{\eta} \in \operatorname{Adm} \Gamma^{\circ}_{\mathrm{BT}}$ and define η using (15.3). We want to show that $\eta \in \operatorname{Adm} \Gamma_{\mathrm{cut}}(A, B)$. Let $S \subset V^{\bullet}$ be a minimal AB-cut. Let H and K be the subgraphs of G induced by S and $V^{\bullet} \setminus S$ respectively. Then $A \cup H$ and $B \cup K$ are disjoint compact sets in $\mathbb{C}$, in particular, they are a positive distance away. Thus, the set $U = S \setminus (A \cup H) \cup (B \cup K)$ is a separating set for A and B in S, meaning that every curve joining A and B in S must intersect U. By [38, Lemma 2.4], U contains a simple curve α starting on C and ending on D, but otherwise contained in $\operatorname{int}(S)$. Note that $\alpha \cap V^{\bullet} = \emptyset$. Let D be the union of all (closed) faces of G that intersect α. Let Q and Q' be any two faces in D. If $Q \cap Q' \neq \emptyset$, then

$$(Q \cap Q') \cap V^{\circ} \neq \emptyset.$$

This implies that there is a subgraph L, and hence a path $\gamma \subset L$ in G° that separates A and B in S. Write γ as in (15.4). Then, with the previous notations, $\{e_j^{\bullet}\}_{j=1}^{n} \subset \delta S$. Since $\hat{\eta}$ is admissible, (15.5) must hold. But this implies

$$\sum_{j=1}^{n} \eta(e_j^{\bullet}) \geq 1.$$

Therefore, since S was arbitrary, we have that η is admissible for $\Gamma_{\mathrm{cut}}(A, B)$.

Finally, we prove (15.2). Since $\Gamma^{\circ}_{\mathrm{BT}}$ is an admissible Fulkerson dual family for $\Gamma^{\bullet}_{\mathrm{LR}}$, we have that $\operatorname{Adm} \Gamma^{\circ}_{\mathrm{BT}}$ and $\operatorname{Adm} \Gamma_{\mathrm{cut}}(A, B)$ are isometrically isomorphic. Then,

$$\operatorname{Mod}_{2,\sigma} \Gamma^{\circ}_{\mathrm{BT}} = \operatorname{Mod}_{2,\sigma^{-1}} \Gamma_{\mathrm{cut}}(A, B)$$

The reason the weights are related that way is because objects in $\Gamma^{\circ}_{\mathrm{BT}}$ use dual edges $e^{\circ} = \{w, w'\}$, while objects in $\Gamma_{\mathrm{cut}}(A, B)$ use primal edges $e^{\bullet} = \{z, z'\}$. However, by (15.1), we have

$$\sigma(e^{\circ}) = \frac{1}{\sigma(e^{\bullet})} = \sigma^{-1}(e^{\bullet}).$$

□

15.4 MODULUS IS ALWAYS ONE

Here, we will give a proof of Theorem 15.1.

Let $S = [0, 1]^2 \subset \mathbb{C}$ be the unit square. Set $A = \{0\} \times [0, 1]$ to be the left vertical side and $B = \{1\} \times [0,1]$ the right vertical side (similarly, let C be the bottom side and D be the top side.) We want to solve the Dirichlet problem, both in the continuum and in the discrete case, when the boundary values are set to be identically zero on A and identically equal to one on B (respectively, identically zero on C and identically equal to one on D). The continuous solutions in this case are

$$h(z) := \operatorname{Re} z \qquad \text{and} \qquad \tilde{h}(z) := \operatorname{Im} z.$$

By abuse of notation, we continue to use h and $\tilde{h}$ for the restriction of these functions to $V^\bullet$ and V° respectively. Also, we let $h^\bullet$ be the discrete solution to the Dirichlet problem from left to right on $V^\bullet$, and let $f^\bullet = \sigma dh^\bullet$ be the resulting current flow. Likewise, h° is the discrete solution to the Dirichlet problem from left to right on V°. Finally, $\tilde{h}^\bullet$ and $\tilde{h}^\circ$ are the corresponding discrete solution of the Dirichlet problem from bottom to top on $V^\bullet$ and V° respectively.

We first state the following fact.

Lemma 15.1. *Let G be an orthodiagonal map in $\mathbb{C}$ as in Definition 15.1. The functions $h(z) = \operatorname{Re} z$ and $\tilde{h}(z) = \operatorname{Im} z$ are always discrete harmonic on the interior nodes of the primal or dual graph.*

Moreover, when G is tiling the square S as above, then h is further harmonic on the horizontal sides C and D, and $\tilde{h}$ is harmonic on the vertical sides A and B.

Proof. Suppose first that $z \in V^\bullet \setminus \partial G$. Then, z is contained within an interior face Q° of the dual graph G°. In particular, there is a cycle of dual edges surrounding z. Let $w_0, w_1, \cdots, w_k$ be the vertices incident to Q° in counterclockwise order, with $w_0 = w_k$. Consider the corresponding faces of the orthodiagonal map: $Q_j = (z, w_{j-1}, z_j, w_j)$ with $j = 1, \ldots, k$. By definition, we have

$$w_j - w_{j-1} = i(z_j - z)\sigma(z, z_j), \qquad \forall j = 1, \ldots, k. \tag{15.6}$$

Then,

$$0 = \sum_{j=1}^{k}(w_j - w_{j-1}) = i\sum_{j=1}^{k} \sigma(z, z_j)(z_j - z).$$

If we now take real and imaginary parts we find that

$$\sum_{j=1}^{k} \sigma(z, z_j)(h(z) - h(z_j)) = 0 \qquad \text{and} \qquad \sum_{j=1}^{k} \sigma(z, z_j)(\tilde{h}(z) - \tilde{h}(z_j)) = 0.$$

Namely, both h and $\tilde{h}$ satisfy discrete harmonicity as in (14.2).

Now, assume that the orthodiagonal map is tiling the square S, and that, for instance, $z \in C$. Then, the edges e° such that the corresponding orthogonal edge $e^\bullet$ contains z form a path with both endpoints in $C \cap V^\circ$. Let $w_0, w_1, \cdots, w_k$ be the vertices of this path, indexed so that connecting w_k to w_0 in the outer face of G° would give a counterclockwise cycle. Then

$$w_k - w_0 = \sum_{j=1}^{k}(w_j - w_{j-1}) = i\sum_{j=1}^{k}\sigma(z, z_j)(z_j - z).$$

In particular, if we take the imaginary part of both sides, we get

$$0 = \sum_{j=1}^{k}\sigma(z, z_j)(\operatorname{Re} z_j - \operatorname{Re} z) = \sum_{j=1}^{k}\sigma(z, z_j)(h(z_j) - h(z)).$$

A similar argument works if $z \in D$. □

Proof of Theorem 15.1. In view of Lemma 15.1 and by uniqueness of the solution to the Dirichlet problem (which follows from the maximum principle), we have $h = h^\bullet$ on $V^\bullet$ and $h = h^\circ$ on V°. Likewise, $\tilde{h} = \tilde{h}^\bullet$ on $V^\bullet$ and $\tilde{h} = \tilde{h}^\circ$ on V°.

Now, write $Q = (z, w, z', w')$ for the generic face of the orthodiagonal map that tiles the square S. We have $w' - w = i(z' - z)\sigma(z, z')$. We abbreviate this identity as

$$W = i\sigma(Z)Z. \tag{15.7}$$

Next we sum the identity (15.7) over all the faces in the orthodiagonal map:

$$\begin{aligned}
0 &= \sum_Q |W - i\sigma(Z)Z|^2\,\sigma(W)\\
&= \sum_Q |W|^2\,\sigma(W) + |Z|^2\,\sigma(Z) - 2\operatorname{Im}\left(W\overline{Z}\right)\\
&= \mathcal{E}^\circ(h^\circ) + \mathcal{E}^\circ(\tilde{h}^\circ) + \mathcal{E}^\bullet(h^\bullet) + \mathcal{E}^\bullet(\tilde{h}^\bullet) - 2\sum_Q \sigma(Z)|Z|^2\\
&= \mathcal{E}^\circ(h^\circ) + \mathcal{E}^\circ(\tilde{h}^\circ) + \mathcal{E}^\bullet(h^\bullet) + \mathcal{E}^\bullet(\tilde{h}^\bullet) - 4\sum_Q \operatorname{Area}(Q)
\end{aligned}$$

where we used (15.7) and the fact that $\sigma(W)\sigma(Z) = 1$.

In other words,

$$\mathcal{E}^\circ(h^\circ) + \mathcal{E}^\circ(\tilde{h}^\circ) + \mathcal{E}^\bullet(h^\bullet) + \mathcal{E}^\bullet(\tilde{h}^\bullet) = 4.$$

On the other hand, by Fulkerson duality (Proposition 15.1),

$$\mathcal{E}^\bullet(h^\bullet)\mathcal{E}^\circ(\tilde{h}^\circ) = 1 \qquad \text{and} \qquad \mathcal{E}^\circ(h^\circ)\mathcal{E}^\bullet(\tilde{h}^\bullet) = 1.$$

Therefore,

$$\left(\mathcal{E}^\bullet(h^\bullet) + \frac{1}{\mathcal{E}^\bullet(h^\bullet)}\right) + \left(\mathcal{E}^\circ(h^\circ) + \frac{1}{\mathcal{E}^\circ(h^\circ)}\right) = 4.$$

But this implies that

$$\mathcal{E}^\bullet(h^\bullet) = \mathcal{E}^\circ(h^\circ) = 1,$$

because $x + \frac{1}{x} \geq 2$ for all $x \geq 0$, with equality if and only if $x = 1$. Likewise, we get

$$\mathcal{E}^\bullet(\tilde{h}^\bullet) = \mathcal{E}^\circ(\tilde{h}^\circ) = 1.$$

This ends the proof of Theorem 15.1. □

15.5 ORTHODIAGONAL APPROXIMATIONS OF PLANAR DOMAINS

Consider the polygonal domain in Figure 15.8. We first produce a Delaunay triangulation of the domain, using the software package Triangle [73]. This is a standard technique in computational geometry. The circumcenters of the triangles form a dual grid known as the Voronoi diagram. In the middle panel of Figure 15.8, we show the same triangulation along with its Voronoi dual. Note that the edges of the triangulation and the edges connecting two circumcenters are orthogonal. Hence, they can be thought as the orthogonal diagonals of a quadrilateral. This is a special case of a more general family of graphs that are planar and whose faces are quadrilaterals with orthogonal diagonals. Such graphs are known as *orthodiagonal maps*, see also [74, Section 5.1]. We have drawn the corresponding orthodiagonal map in the right panel of Figure 15.8. We also note that the edge weights σ must be defined in a very specific way, on the edges of the orthodiagonal map. Consider now the arcs *ab* and *cd* on the boundary of the domain in the left panel of Figure 15.8. We are interested in computing the modulus of the path family Γ

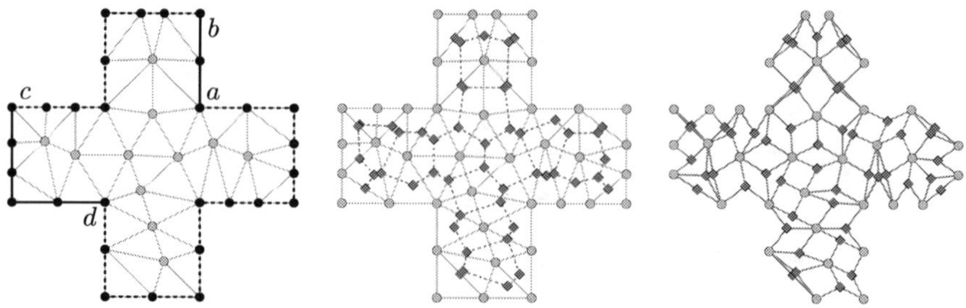

Figure 15.8 Left: a Delaunay triangulation of a cross domain; Center: the triangulation shown together with its dual Voronoi graph; Right: the corresponding orthodiagonal map.

connecting ab and cd. In particular, we want to study the optimal pmfs that arise and their convergence properties if we let the mesh size of this approximating map go to zero. This is a difficult problem in general because of the non-uniqueness of the optimal pmf's. In this case, however, it's possible to identify a special (unique) optimal pmf μ^* which is carried by a family of non-crossing paths connecting ab and cd. One must first find the discrete harmonic function (*the potential*) that is 0 on cd and 1 on ab. Then, using this potential, we devised an algorithm that produces a family of non-crossing paths and also their corresponding probabilities. We have drawn a selection of these non-crossing paths in Figure 15.9. Using the non-crossing path decomposition, we are able

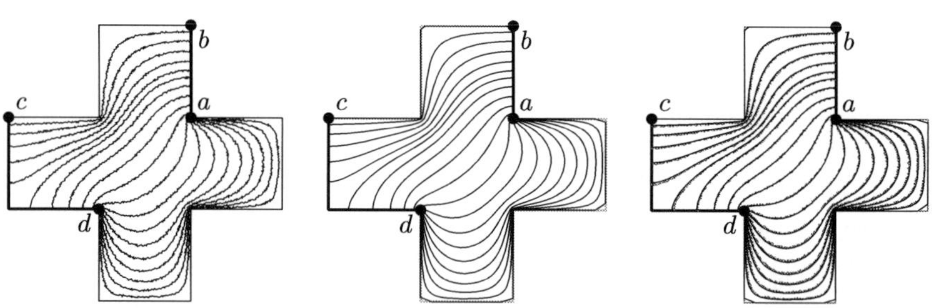

Figure 15.9 Left: non-crossing curves for a triangulation of the cross domain; Center: the corresponding curves as computed by the Schwarz-Christoffel Toolbox in Matlab [35]; Right: the superposition of both sets of curves.

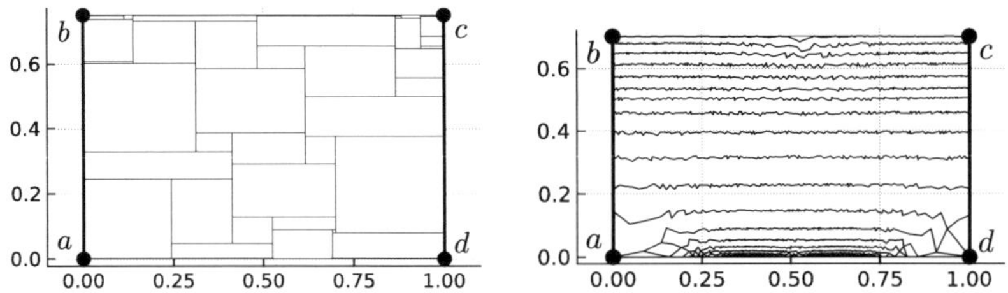

Figure 15.10 Left: the rectangle packing associated with a triangulation of the cross domain shown in Figure 15.8; Right: the images of the non-crossing curves from Figure 15.9 in coordinates given by the rectangle packing. The rectangle on the left is taller because the coarser triangulation has a higher modulus.

to assign a two-dimensional coordinate to each vertex in the graph. The x-coordinate of each vertex is simply the value of the discrete potential h on that vertex. To obtain the y-coordinate of a vertex v, we locate the first path in the sequence passing through v and sum the flows of all previous paths. This creates a rectangle packing as shown in the left panel of Figure 15.10. This choice of coordinates provides a mapping on the vertices of the triangulation that approximates the conformal map to the rectangle. The right panel of Figure 15.10 shows the non-crossing paths in these coordinates. Except near a and d, where there is large distortion, these curves are nearly horizontal, as we would expect.

One of the main results in [10] is the following.

Theorem 15.2 ([10]). *Let (G_n, A_n, B_n) be a sequence of orthodiagonal maps with identified boundary arcs, which is approximating a Jordan domain G in the plane with boundary arcs A and B. Consider the family of all paths in G_n from A_n to B_n, and let Γ_n be the unique minimal subfamily with non-crossing paths obtained from our algorithm, with corresponding pmf μ_n. Then, as $n \to \infty$, μ_n converges to the transverse measure μ on the extremal curves (horizontal trajectories) from A to B in Ω (under weak convergence using the Hausdorff metric as the underlying metric).*

Our proof depends crucially on the work of Gurel-Gurevich, Jerison, and Nachmias [47], which we modify for our purposes.

In view of Theorem 15.2, we may ask if there are other situations where such a convergence results can be established, for instance, considering domains in higher-dimensional Euclidean spaces or more general metric spaces.

Another possibility is to change the exponent $p = 2$ to other values of $p \in [1, \infty]$. That is what we will address next.

15.6 p-MODULUS ON ORTHODIAGONAL MAPS

As we have seen in the previous section, orthodiagonal maps are especially well-suited for modulus. Here, we explore what happens when $p \neq 2$.

Fix $1 < p, q < \infty$, with $(p-1)(q-1) = 1$. Then, the *primal network* $(G^\bullet, \sigma)$ and the *dual network* (G°, σ) must be given the following canonical edge weights:

$$\sigma(e_Q^\bullet) = \frac{|w - w'|}{|z - z'|^{p-1}} \quad \text{and} \quad \sigma(e_Q^\circ) = \frac{|z - z'|}{|w - w'|^{q-1}}. \tag{15.8}$$

Note that these weights have the property that

$$\sigma(z, z')^{\frac{1}{p}} \sigma(w, w')^{\frac{1}{q}} = 1. \tag{15.9}$$

Recall that the (p, σ)-modulus $\mathrm{Mod}_{p,\sigma} \Gamma^\bullet(L, R)$ equals the p-effective conductance $\mathrm{eff}\mathcal{C}^\bullet_{p,\sigma}(L, R)$, which is the value of the p-Dirichlet problem (14.4) when using the primal $(G^\bullet, \sigma)$, and the family $\mathcal{H}$ consists of all potentials $h \in \mathbb{R}^{V^\bullet}$ such that $h_{|L} = 0$ and $h_{|R} = 1$. On the other hand, as noted in Remark 14.5, the p-effective resistance $\mathrm{eff}\mathcal{R}^\circ_{p,\hat{r}}(B, T)$ is the value of the p-Thomson problem (14.15) when using the dual network $(G^\circ, \hat{r})$, where $\hat{r} = \sigma^{1-p}$, and the family $\mathcal{F}$ consists of all unit BT-flows.

Proposition 15.2. *The following identity holds:*

$$\mathrm{eff}\mathcal{C}^\bullet_{p,\sigma}(L, R) = \mathrm{eff}\mathcal{R}^\circ_{p,\hat{r}}(B, T),$$

where $\hat{r} := \sigma^{1-p}$.

Proposition 15.2 generalizes from the case $p = 2$ an unpublished result of Asaf Nachmias. The first proof we give here is different, in that it uses Fulkerson duality for modulus.

First proof of Proposition 15.2. We have the following string of identities:

$$\begin{aligned}\mathrm{eff}\mathcal{C}^{\bullet}_{p,\sigma}(L,R) &= \mathrm{Mod}_{p,\sigma}\,\Gamma^{\bullet}(L,R) && (15.10)\\ &= \mathrm{Mod}_{q,\hat{\sigma}}\,\Gamma_{\mathrm{cut}}(L,R)^{-p/q} && (15.11)\\ &= \mathrm{Mod}_{q,\sigma}\,\Gamma^{\circ}(B,T)^{-p/q} && (15.12)\\ &= \mathrm{eff}\mathcal{C}^{\circ}_{q,\sigma}(B,T)^{-p/q} && (15.13)\\ &= \mathrm{eff}\mathcal{R}^{\circ}_{p,\hat{r}}(B,T). && (15.14)\end{aligned}$$

Equation (15.10) and (15.13) follow from Theorem 14.5. Equation (15.11) is Fulkerson duality where $\hat{\sigma} := \sigma^{-q/p}$. Equation (15.12) follows from the fact that $\mathrm{Adm}\,\Gamma_{\mathrm{cut}}(L,R) = \mathrm{Adm}\,\Gamma^{\circ}(B,T)$ as subsets of $\mathbb{R}^n_{\geq 0}$, with n equal to the number of bounded faces of G, see Proposition 15.1, and the fact that

$$\hat{\sigma}(z,z') = \sigma(z,z')^{-q/p} = \sigma(w,w'),$$

which follows from (15.9). Equation (15.14) follows from (14.27). □

The first proof of Proposition 15.2 is based on the connection with the modulus developed in the previous sections and the Fulkerson duality for the modulus. However, there is also a more direct argument which we sketch below.

Second proof of Proposition 15.2. In one direction, let $h^{\bullet}$ be the optimal potential for $\mathrm{eff}\mathcal{C}^{\bullet}_{p,\sigma}(L,R)$. Then, for every face $Q = [z,w,z',w']$, define the flow

$$f(w,w') := h^{\bullet}(z') - h^{\bullet}(z).$$

One checks that for $w \in V^{\circ}$ we have $\mathrm{Div}_f(w) = 0$, if $z \notin B \cup T$. So f is a BT-flow.

Moreover, let $\gamma^{\bullet}$ be a simple path connecting L and R in $G^{\bullet}$, and assume that $\gamma^{\bullet}$ is a path of steepest ascent for $h^{\bullet}$. This yields a BT-cut by turning every edge $e^{\bullet} \in \gamma^{\bullet}$ into an edge $e^{\circ} \in C \subset E^{\circ}$. Moreover, the flow across C gives

$$\mathrm{Val}\,f = \sum_{e^{\circ}\in C} |f(e^{\circ})| = \sum_{e^{\bullet}\in\gamma^{\bullet}} |\nabla h^{\bullet}(e^{\bullet})| = 1.$$

Therefore, by definition of $\mathrm{eff}\mathcal{R}^{\circ}_{p,\hat{r}}(B,T)$, we have

$$\begin{aligned}
\mathrm{eff}\mathcal{R}^{\circ}_{p,\hat{r}}(B,T) \leq \mathcal{E}^{\circ}_{p,\hat{r}}(f) &= \sum_{Q} \sigma(w,w')^{1-p}|f(w,w')|^{p} \\
&= \sum_{Q} \sigma(z,z')\,|h^{\bullet}(z') - h^{\bullet}(z)|^{p} && \text{(by (15.9))} \\
&= \mathcal{E}^{\bullet}_{p,\sigma}(h^{\bullet}) = \mathrm{eff}\mathcal{C}^{\bullet}_{p,\sigma}(L,R).
\end{aligned}$$

Conversely, let f° be optimal for $\mathrm{eff}\mathcal{R}^{\circ}_{p,\hat{r}}(B,T)$. For every face $Q = [z,w,z',w']$, define the function

$$\rho(z,z') := |f^{\circ}(w,w')|.$$

Next define a potential for $z \in V^{\bullet}$, by setting

$$h(z) := \inf_{\gamma^{\bullet}:L\rightsquigarrow z} \sum_{e^{\bullet}\in\gamma^{\bullet}} \rho(e^{\bullet}).$$

Here the infimum is taken over all the paths connecting L to z in $G^{\bullet}$. Then, we check that $h = 0$ on L and $h \geq 1$ on R (see below). Moreover,

$$|h(z') - h(z)| = \rho(z,z').$$

So,

$$\begin{aligned}
\mathrm{eff}\mathcal{C}^{\bullet}_{p,\sigma}(L,R) \leq \mathcal{E}^{\bullet}_{p,\sigma}(h) &= \sum_{Q} \sigma(z,z')|h(z') - h(z)|^{p} \\
&\sum_{Q} \sigma(z,z')\rho(z,z')^{p} = \sum_{Q} \sigma(w,w')^{1-p}|f^{\circ}(w,w')|^{p} \\
&= \mathrm{eff}\mathcal{R}^{\circ}_{p,\hat{r}}(B,T).
\end{aligned}$$

To show that $h \geq 1$ on R we argue by contradiction. Suppose, that there is a simple path $\gamma^{\bullet}$ connecting L to R in $G^{\bullet}$ such that

$$\sum_{e^{\bullet}\in\gamma^{\bullet}} \rho(e^{\bullet}) < 1.$$

By turning the edges in $\gamma^{\bullet}$ we get a BT-cut such that

$$1 > \sum_{e^{\circ}=\{w,w'\}} |f^{\circ}(w,w')|.$$

But this contradicts the fact that f° must be a unit BT-flow.

The fact that $h = 0$ on B follows by definition. □

Finally, we establish a type of discrete Cauchy-Riemann equations.

Proposition 15.3. *Let $h^\bullet \in \mathbb{R}^{V^\bullet}$ be the (p,σ)-capacitary function such that $h^\bullet_{|L} = 0$, $h^\bullet_{|R} = 1$, and $\mathcal{E}_{p,\sigma}(h^\bullet) = \mathrm{eff}\mathcal{C}^\bullet_{p,\sigma}(L,R)$. Also, let $\tilde{g}^\circ \in \mathbb{R}^{V^\circ}$ be the (q,σ)-capacitary function such that $\tilde{g}^\circ_{|B} = 0$, $\tilde{g}^\circ_{|T} = 1$, and $\mathcal{E}_{q,\sigma}(\tilde{g}^\circ) = \mathrm{eff}\mathcal{C}^\circ_{q,\sigma}(B,T)$. Define the function*

$$\tilde{h}^\circ := (\mathrm{eff}\mathcal{C}^\bullet_{p,\sigma}(L,R))\tilde{g}^\circ$$

Then, on each face $Q = [z,w,z',w']$, the following identity holds:

$$\frac{\tilde{h}^\circ(w') - \tilde{h}^\circ(w)}{|w'-w|} = \left(\frac{|h^\bullet(z') - h^\bullet(z)|}{|z'-z|}\right)^{p-2} \cdot \frac{h^\bullet(z') - h^\bullet(z)}{|z'-z|}. \tag{15.15}$$

Proof. As in the second proof of Proposition 15.2, for every face $Q = [z,w,z',w']$, define the flow

$$f(w,w') := h^\bullet(z') - h^\bullet(z).$$

One checks that for $w \in V^\circ$ we have $\mathrm{Div}_f(w) = 0$, if $z \notin B \cup T$. So f is an BT-flow, and, as before, one shows that $\mathrm{Val}\, f = 1$. Then,

$$\begin{aligned}
\mathcal{E}^\circ_{p,\hat{r}}(f) &= \sum_Q \sigma(w,w')^{1-q} |f(w,w')|^p \\
&= \sum_Q \sigma(z,z')\, |h^\bullet(z') - h^\bullet(z)|^p && \text{(by (15.9))} \\
&= \mathcal{E}^\bullet_{p,\sigma}(h^\bullet) = \mathrm{eff}\mathcal{C}^\bullet_{p,\sigma}(L,R) \\
&= \mathrm{eff}\mathcal{R}^\circ_{p,\hat{r}}(B,T). && \text{(by Proposition 15.2)}
\end{aligned}$$

In other words, f is optimal for the p-Thomson problem. Hence, we can apply Theorem 14.6 and define $\phi^* := |f|^{p-2} f$, so that ϕ^* admits a potential $\tilde{g} \in \mathbb{R}^{V^\circ}$, i.e., with

$$\hat{r}(w,w')\phi^*(w,w') = \tilde{g}(w') - \tilde{g}(w)$$

on every face $Q = [z,w,z',w']$, such that $\tilde{g}_{|B} = 0$, $\tilde{g}_{|T} = \mathrm{eff}\mathcal{R}^\circ_{p,\hat{r}}(B,T)$, and such that $\tilde{g}$ is $(q,\tilde{\sigma})$-harmonic. Here, $\tilde{\sigma} = \hat{r}^{1-q} = \sigma^{(1-p)(1-q)} = \sigma$. In particular, $\tilde{g}^\circ = \tilde{g}/\mathrm{eff}\mathcal{R}^\circ_{p,\hat{r}}(B,T)$. Thus, by Proposition 15.2 again, $\tilde{h}^\circ = \tilde{g}$ and, on every face $Q = [z,w,z',w']$, we have

$$\begin{aligned}
\tilde{g}(w') - \tilde{g}(w) &= \sigma(w,w')^{1-p}\, |h^\bullet(z') - h^\bullet(z)|^{p-2}\, (h^\bullet(z') - h^\bullet(z)) \\
&= \sigma(z,z')\, |h^\bullet(z') - h^\bullet(z)|^{p-2}\, (h^\bullet(z') - h^\bullet(z))
\end{aligned}$$

which can be rewritten as in (15.15). □

In order to obtain a counterpart to (15.15), we need to introduce the complementary network $G^\dagger$, where $V(G^\dagger)^\bullet = V(G)^\circ$ and $V(G^\dagger)^\circ = V(G)^\bullet$. In particular, the canonical weights for $G^\dagger$ are equal to the ones for G where p and q are switched. Indeed,

$$\sigma^\dagger(w', w) = \frac{|z'-z|}{|w'-w|^{p-1}} \qquad \text{while} \qquad \sigma(w', w) = \frac{|z'-z|}{|w'-w|^{q-1}}.$$

We will continue to use the notation $\mathcal{E}^\bullet$ and $\mathcal{E}^\circ$ to mean the energy calculated on $V(G)^\bullet$ and $V(G)^\circ$ respectively. With these notations, Proposition 15.3 has the following corollary.

Corollary 15.1. *Let $h^\circ \in \mathbb{R}^{V(G^\dagger)^\bullet}$ be the $(p,\sigma^\dagger)$-capacitary function such that $h^\circ_{|L} = 0$, $h^\circ_{|R} = 1$, and $\mathcal{E}^\circ_{p,\sigma^\dagger}(h^\circ) = \mathrm{eff}\mathcal{C}^\circ_{p,\sigma^\dagger}(L,R)$. Also, let $\tilde{g}^\bullet \in \mathbb{R}^{V(G^\dagger)^\circ}$ be the $(q,\sigma^\dagger)$-capacitary function such that $\tilde{g}^\bullet_{|B} = 0$, $\tilde{g}^\bullet_{|T} = 1$, and $\mathcal{E}^\bullet_{q,\sigma^\dagger}(\tilde{g}^\bullet) = \mathrm{eff}\mathcal{C}^\bullet_{q,\sigma^\dagger}(B,T)$. Define the function*

$$\tilde{h}^\bullet := (\mathrm{eff}\mathcal{C}^\circ_{p,\sigma^\dagger}(L,R))\tilde{g}^\bullet.$$

Then, on each face $Q = [w, z', w', z]$, the following identity holds:

$$\frac{\tilde{h}^\bullet(z') - \tilde{h}^\bullet(z)}{|z'-z|} = -\left(\frac{|h^\circ(w') - h^\circ(w)|}{|w'-w|}\right)^{p-2} \cdot \frac{h^\circ(w') - h^\circ(w)}{|w'-w|}. \tag{15.16}$$

We now apply (15.15) and (15.16) to rewrite the following quantity:

$$\begin{aligned}
&\mathcal{E}^\bullet_{p,\sigma}(h^\bullet) + \mathcal{E}^\circ_{p,\sigma^\dagger}(h^\circ) \\
&= \sum_{Q=[z,w,z',w']} \frac{|w'-w|}{|z'-z|^{p-1}} |h^\bullet(z') - h^\bullet(z)|^p \\
&+ \sum_{Q=[w,z',w',z]} \frac{|z'-z|}{|w'-w|^{p-1}} |h^\circ(w') - h^\circ(w)|^p \\
&= 2 \sum_{Q=[z,w,z',w']} \frac{h^\bullet(z') - h^\bullet(z)}{|z'-z|} \cdot \frac{\tilde{h}^\circ(w') - \tilde{h}^\circ(w)}{|w'-w|} \mathrm{Area}(Q) \\
&- 2 \sum_{Q=[z,w,z',w']} \frac{h^\circ(w') - h^\circ(w)}{|w'-w|} \cdot \frac{\tilde{h}^\bullet(z') - \tilde{h}^\bullet(z)}{|z'-z|} \mathrm{Area}(Q)
\end{aligned}$$

This can be rewritten as:

$$\frac{\mathcal{E}^{\bullet}_{p,\sigma}(h^{\bullet}) + \mathcal{E}^{\circ}_{p,\sigma^{\dagger}}(h^{\circ})}{2}$$
$$= \sum_{Q=[z,w,z',w']} \frac{h^{\bullet}(z') - h^{\bullet}(z)}{|z'-z|} \cdot \frac{\tilde{h}^{\circ}(w') - \tilde{h}^{\circ}(w)}{|w'-w|} \text{Area}(Q)$$
$$- \sum_{Q=[z,w,z',w']} \frac{h^{\circ}(w') - h^{\circ}(w)}{|w'-w|} \cdot \frac{\tilde{h}^{\bullet}(z') - \tilde{h}^{\bullet}(z)}{|z'-z|} \text{Area}(Q) \qquad (15.17)$$

EXERCISES

15.1 *Orthodiagonal maps are bipartite.* Let $G = (V, E)$ be an orthodiagonal map. Then, G is a finite connected plane graph so that each inner face is a quadrilateral. This is enough to show that G is a bipartite graph and hence V admits a *two-coloring*, i.e., a color assignment, either black or white, so that no two neighbors are the same color.

(i) Show that a connected graph is bipartite if and only if it has no cycles of odd length.

(ii) Use part (i) and the Euler characteristic formula to conclude that G is bipartite.

15.2 *p-Modulus is not always one.* As in the setup of Section 15.4, let $S = [0,1]^2 \subset \mathbb{C}$ be the unit square. Set $A = \{0\} \times [0,1]$ to be the left vertical side and $B = \{1\} \times [0,1]$ the right vertical side (similarly, let C be the bottom side and D be the top side.) We also fix $1 < p, q < 1$ so that $pq = p + q$.

(i) Recall that, with the notation in (15.7), the equations defining the canonical weights in (15.8) can be written as $\sigma(Z) = \frac{|W|}{|Z|^{p-1}}$ and $\sigma(W) = \frac{|Z|}{|W|^{q-1}}$. Show that

$$W = i\sigma(Z)|Z|^{p-2}Z. \qquad \text{(P15.1)}$$

(ii) Use (P15.1) to show that the function $F(z) := z$ is always discrete p-harmonic on the interior nodes of the primal or dual graph.

By the non-linearity of the p-harmonic equation, this does not mean that the real and imaginary part of F are also p-harmonic.

(iii) Next, use the fact that on each face Q the identity (P15.1) gives $|W| = \sigma(Z)|Z|^{p-1}$, and the fact that $\sigma(Z) = \sigma(W)^{1-p}$, to compute the discrete energy of F. Namely, show that

$$\mathcal{E}_p^\bullet(F) = \mathcal{E}_q^\circ(F) = 2\,\mathrm{Area}(S).$$

(iv) Show that the functions $h(z) := \operatorname{Re} z$ and $\tilde{h}(z) := \operatorname{Im} z$ satisfy the discrete q-harmonic Cauchy-Riemann equations, meaning that on every face $Q = [z, w, z', w']$

$$\begin{cases} \frac{h(Z)}{|Z|} = \left(|\nabla\tilde{h}|^{q-2}\right)\frac{\tilde{h}(W)}{|W|} \\ \frac{h(W)}{|W|} = -\left(|\nabla\tilde{h}|^{q-2}\right)\frac{\tilde{h}(Z)}{|Z|} \end{cases}$$

where $Z = z' - z$, $W = w' - w$, and

$$\nabla\tilde{h} = \begin{bmatrix} \frac{\tilde{h}(Z)}{|Z|} \\ \frac{\tilde{h}(W)}{|W|} \end{bmatrix}$$

In general, h and $\tilde{h}$ are note as the optimal discrete capacitary functions and the discrete p-modulus is not equal to one in general, as the following example shows.

(v) Fix $1 < p, q < \infty$ with $(p-1)(q-1) = 1$. Consider the tiling of the square given in Figure 15.11, with canonical weights on the diagonals of a face $Q = [z, w, z', w']$ given by

$$\sigma(z, z') := \frac{|w' - w|}{|z' - z|^{p-1}} \qquad \text{and} \qquad \sigma(w, w') := \frac{|z' - z|}{|w' - w|^{q-1}}.$$

Recall that in this general p-modulus case the parallel rule for two conductances c_1 and c_2 in parallel is

$$\sigma_1 + \sigma_2,$$

while the serial rule is

$$\left[\sigma_1^{1-q} + \sigma_2^{1-q}\right]^{1-p}.$$

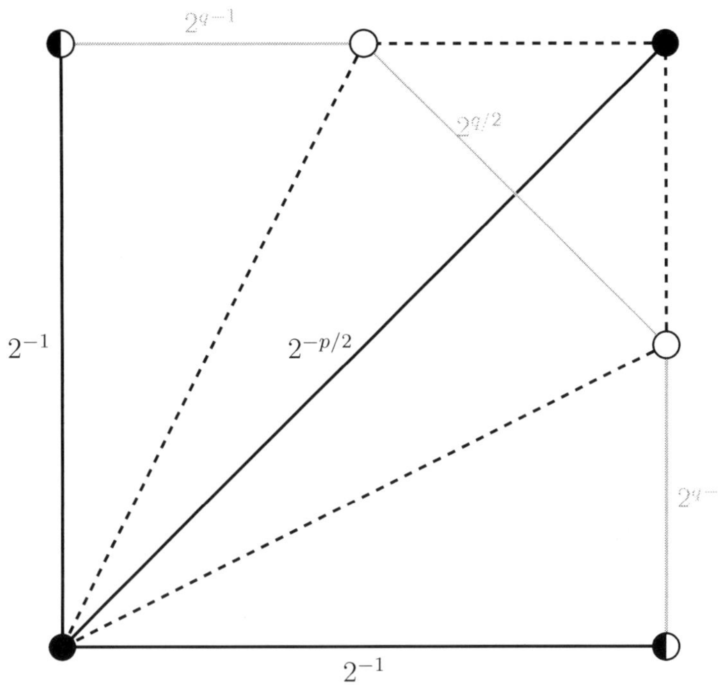

Figure 15.11 Counter example in the general p case.

Show that

$$\operatorname{Mod}_p \Gamma^{\bullet}(L,R) = 2^{-1} + 2^{-p/2}$$

and

$$\operatorname{Mod}_q \Gamma^{\circ}(B,T) = \left[2^{-1} + 2^{-p/2}\right]^{1-q}.$$

Also, if h° is the capacitary function, then show that

$$h^{\circ}(1+i/2) = \left[1 + 2^{1-p/2}\right]^{-1}.$$

CHAPTER 16

Modulus Metrics

As we have seen the modulus of connecting families of paths is related to classical metrics such as the effective resistance metric and the shortest path metric. Here, we show that p-modulus can be used to embed these classical metrics in a one-parameter family of metrics we call modulus metrics.

16.1 MODULUS METRICS

Let $G = (V, E)$ be a connected simple unweighted graph. The fact that effective resistance $d_{\mathrm{ER}}(a, b) := \mathrm{eff}\mathcal{R}(a, b)$ is a metric on V has several known proofs, see Remark 3.5 and Section 6.5. Here, we embed the effective resistance metric d_{ER} in a one-parameter family of modulus metrics δ_p depending on $p \in [1, \infty]$. In particular, Theorem 16.1 (i) and (iii) below will provide a new proof, a modulus proof, that effective resistance is a metric on graphs.

In fact, we may even assume that $\sigma \in \mathbb{R}^E_{>0}$ is a set of edge weights. Given $a, b \in V$, let $\Gamma(a, b)$ be the connecting family of all paths between a and b. Fix $1 < p < \infty$ and define the (p, σ)-*modulus metric* on G as

$$\delta_{p,\sigma}(a, b) := \begin{cases} \mathrm{Mod}_{p,\sigma}(\Gamma(a, b))^{-\frac{1}{p-1}} & \text{if } a \neq b, \\ 0 & \text{if } a = b. \end{cases} \tag{16.1}$$

We saw in Theorem 14.1 that in the case of connecting families the (p, σ)-modulus of $\Gamma(a, b)$, $\mathrm{Mod}_{p,\sigma}(\Gamma(a, b))$, satisfies:

- $\mathrm{Mod}_{\infty,1}(\Gamma(a, b))^{-1} = \ell(\Gamma(a, b))$ is the (unweighted) shortest path length;
- $\mathrm{Mod}_{2,\sigma}(\Gamma(a, b))^{-1} = \mathrm{eff}\mathcal{R}(a, b)$ is the effective resistance metric;

DOI: 10.1201/9781003024866-20

- $\mathrm{Mod}_{1,\sigma}(\Gamma(a,b))^{-1} = \mathrm{MC}(a,b)^{-1}$ is the reciprocal of min-cut.

In all three cases, if G is a connected graph, these are distances (also known as metrics, see Definition 1.4). The fact that the shortest path $d_{\mathrm{SP}}(a,b) := \ell(\Gamma(a,b))$ is a metric on V is well known and follows easily from the definition. The fact that $d_{\mathrm{MC}}(a,b) := \mathrm{MC}(a,b)^{-1}$ is an ultra-metric (i.e., that the sum can be replaced by the maximum in the triangle inequality) is left as an exercise, or see [6], where a proof is given.

Theorem 16.1. *Given a weighted, connected, simple graph $G = (V, E, \sigma)$, let $\delta_{p,\sigma}$ be the modulus metrics defined in (16.1). Then, the following hold,*

(i) for all $p \in [1,\infty]$ and $\sigma \in \mathbb{R}^E_{>0}$, $\delta_{p,\sigma}$ is a metric on V;

(ii) if we let p tend to infinity, we recover the shortest path metric: $\lim_{p\uparrow\infty} \delta_{p,\sigma} = d_{\mathrm{SP}}$;

(iii) for $p = 2$, we recover the effective resistance metric: $\delta_2 = d_{\mathrm{ER}}$;

(iv) for $1 < p < 2$, $\mathrm{Mod}_{p,\sigma}(\Gamma(a,b))^{-1}$ is a metric and it tends to $d_{\mathrm{MC}}(a,b)$ as $p \to 1$;

(v) if we let p tend to 1, δ_p converges to the limit

$$\lim_{t\to\infty} (d_{\mathrm{MC}}(a,b))^t = \begin{cases} 0 & \text{if } d_{\mathrm{MC}}(a,b) < 1, \\ 1 & \text{if } d_{\mathrm{MC}}(a,b) = 1, \\ \infty & \text{if } d_{\mathrm{MC}}(a,b) > 1. \end{cases}$$

(vi) Finally, for every $\epsilon > 0$ and every $p \in (1,\infty)$ there is a connected graph for which $\delta_p^{1+\epsilon}$ is not a metric.

Proof. See Exercise 16.1, and Exercise 16.2 as well. □

For unweighted graphs, the limit in Theorem 16.1 (v) essentially decomposes the graph into its 2-edge-connected components. All nodes in the same component are distance zero from one another while nodes in different components are at distance one.

In the rest of this chapter, we mention a few possible directions of inquiry involving modulus metrics.

16.2 ANTISNOWFLAKING

Given an arbitrary metric, the operation of *snowflaking* yields a new metric on the same set. Specifically, suppose d is a metric on X and let $0 < \epsilon < 1$. Then, d^ϵ is also a metric on X. Therefore, given a metric d on X, we may ask whether d is the snowflaked version of some other metric d'. In other words, we may ask whether there exists an exponent $t > 1$ such that d^t is also a metric. When such a $t > 1$ exists, we shall call the resulting metric $d' = d^t$ an *antisnowflaking* of d.

For a finite graph $G = (V, E)$, suppose a, b and c are distinct nodes in V. If $d(a, b) < d(a, c) + d(c, b)$, then the inequality also holds with d replaced by d^t for sufficiently small $t > 1$. We call such a triple of points (a, b, c) a *proper triangle*. On the other hand, if $d(a, b) = d(a, c) + d(c, b)$, then, for arbitrarily small $t > 1$, the function d^t violates the triangle inequality. We refer to such a triple as a *flat triangle*. Thus, a metric d on a finite graph G can be antisnowflaked if and only if G contains no flat triangles.

We define the *antisnowflaking exponent* of a metric d as

$$\mathrm{ASFE}(d) := \sup\{t \geq 1 : \ d^t \text{ is a metric}\}. \tag{16.2}$$

For instance, if d is an ultrametric, then $\mathrm{ASFE}(d) = \infty$.

Let $\mathcal{G}$ be a given family of finite, connected unweighted graphs $G = (V, E)$, and let $\delta_{p,G}$ be the corresponding p-modulus metrics defined in (16.1). We set the smallest antisnowflaking exponent for all the p-modulus metrics, over all graphs ranging in $\mathcal{G}$, as

$$s_{\mathcal{G}}(p) := \inf\{\mathrm{ASFE}(\delta_{p,G}) : G \in \mathcal{G}\}. \tag{16.3}$$

For instance, if $\mathcal{G}_c$ is the family of all connected graphs, then Theorem 16.1 (vi) can be interpreted as saying that

$$s_{\mathcal{G}_c}(p) \equiv 1, \qquad \text{for all } p \in (1, \infty).$$

At the other extreme, Exercise 16.3 shows that the modulus metrics on complete graphs are ultrametrics. Therefore, if we let $\mathcal{K}$ be the family of all complete graphs, then

$$s_{\mathcal{K}}(p) \equiv \infty, \qquad \text{for all } p \in (1, \infty).$$

What about other families of graphs? For instance, recall that for $n \geq 2$, the *hypercube* H_n is the graph whose nodes are strings of 0 and 1 of length n, and two such strings are connected by an edge if they differ in exactly one position. If $\mathcal{H}$ is the family of all hypercubes, can one estimate $s_{\mathcal{H}}(p)$ as a function of p?

16.3 METRIC EMBEDDINGS

Here we follow the exposition in [6] with some modifications. A mapping $f : (X, d_X) \to (Y, d_Y)$ of one metric space into another is called an *isometric embedding* or *isometry* if $d_Y(f(x), f(y)) = d_X(x, y)$ for all $x, y \in X$. Two metric spaces are isometric if there exists a bijective isometry between them.

Theorem 16.2 (Schoenberg, 1935). *Given a finite metric space $X = \{x_0, ..., x_m\}$ and an integer $n \in \mathbb{N}$, X embeds isometrically into $\mathbb{R}^n$ if and only if the matrix $M \in \mathbb{R}^{m\times m}$ whose entries are*

$$d(x_i, x_0)^2 + d(x_0, x_j)^2 - d(x_i, x_j)^2$$

is positive semi-definite and of rank less than or equal to n.

We will call the matrix M the "Schoenberg matrix". As an example, consider the square in Figure 16.1. We fix the node a and derive the Schoenberg matrix so we can analyze the embeddings of this graph when the nodes are endowed with modulus metrics δ_p. By symmetry, these metrics have the property that the distance between two neighboring nodes of the square is a constant $\alpha > 0$ and the distance between diagonally opposite nodes is some other constant $\beta > 0$.

Let the columns and rows of the matrix M represent nodes b, c, d respectively, we get a Schoenberg matrix

$$M = \begin{bmatrix} 2\alpha^2 & \beta^2 & 2\alpha^2 - \beta^2 \\ \beta^2 & 2\beta^2 & \beta^2 \\ 2\alpha^2 - \beta^2 & \beta^2 & 2\alpha^2 \end{bmatrix}$$

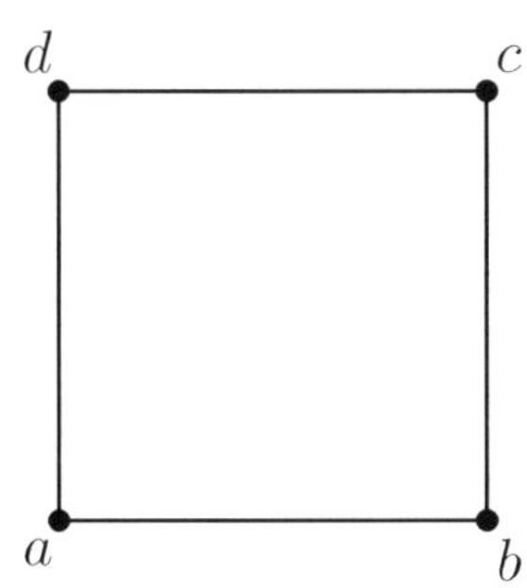

Figure 16.1 The square graph

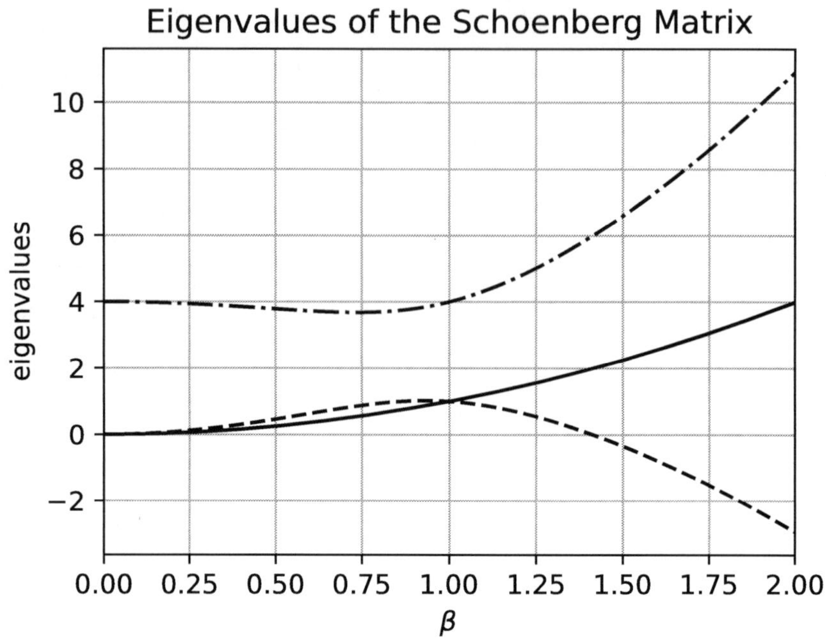

Figure 16.2 Schoenberg eigenvalues for the square

Note that for the triangle inequality to hold, we also want the condition,

$$\beta \leq 2\alpha.$$

Without loss of generality, we can normalize the edge distance to be 1, setting $\alpha = 1$, and then plot how the eigenvalues of M change with β. In Figure 16.2, we have plotted the eigenvalues of M as the normalized parameter which we still call β varies from 0 to 2. We see that when $\beta > \sqrt{2} \approx 1.4$, the matrix M starts having negative eigenvalues and thus fails to be positive semi-definite. For β in $(0, \sqrt{2})$ the square is embeddable in $\mathbb{R}^3$ and for $\beta = \sqrt{2}$ it is embeddable in $\mathbb{R}^2$. We describe these embeddings by fixing one edge of the square at $(1/2, 0, 0)$ and $(-1/2, 0, 0)$, while the opposite edge is horizontal at some height $h > 0$ and by symmetry is also centered on the vertical axis. Since all four edges have unit length, as the top horizontal edge twists about the vertical axis by an angle $\theta \geq 0$, the height of this edge must decrease. More precisely, let $\theta \in [0, \pi]$, then the opposite edge is the segment connecting the points

$$\left(\frac{1}{2}\cos\theta, \frac{1}{2}\sin\theta, \cos\frac{\theta}{2}\right) \qquad \text{and} \qquad \left(-\frac{1}{2}\cos\theta, -\frac{1}{2}\sin\theta, \cos\frac{\theta}{2}\right).$$

With this β is computed as the distance between diagonally opposite nodes:

$$\beta = \sqrt{1+\cos\theta}.$$

When $\theta = 0$, $\beta = \sqrt{2}$, and the square is embedded in the xz-plane. For $\theta \in (0,\pi)$, the embedding is three-dimensional and h tends to 0 as θ tends to π. When $\theta = \pi$, $\beta = 0$ and d stops being a metric.

To connect this to modulus metrics, note that

$$\delta_p(a,b) = (1+3^{1-p})^{-1/(p-1)} \qquad \text{and} \qquad \delta_p(a,c) = 2^{(p-2)/(p-1)}.$$

Therefore, the ratio β/α for the case of the δ_p metrics is

$$f(p) := \left(2^{(p-2)}(1+3^{(1-p)})\right)^{1/(p-1)}$$

A computation shows that f is increasing, thus as p decreases to 1 the ratio β/α decreases to $f(1) = 1$. Hence, the metric δ_p on the square graph is embeddable in $\mathbb{R}^3$ for $1 \le p \le p_0$, for some value $p_0 \approx 2.48$.

One may ask if this is a general phenomenon for arbitrary graphs. Moreover, the topic of metric embeddings is vast, for instance, see [63, Chapter 15]. We believe modulus metrics could be useful to study in this context.

16.4 EPIDEMIC HITTING TIMES

Finally, we mention a connection between modulus metrics and epidemic modeling. Perhaps the simplest epidemic model is the SI model, where nodes are either infected or susceptible. In this model, once a node is infected it stays infected forever, and as soon as infection occurs all the edges incident to that node start an exponentially distributed clock, so that the infection will be transmitted along that edge when the clock goes off. Given a simulation started by infecting a node a, one may measure the time it takes for the infection to spread to another given node b. Averaging over many repeated simulations gives rise to what is known as the *Epidemic Hitting Time* $\mathrm{EHT}(a,b)$.

It was observed in [43] that $\mathrm{EHT}(a,b)$ is also a metric on the graph G, in the sense of Definition 1.4, so that in particular, it is symmetric and satisfies the triangle inequality. Moreover, using deep results of Lyons et al. [61], it is shown in [43] that

$$\mathrm{eff}\mathcal{R}(a,b) \le \mathrm{EHT}(a,b) \le d_{\mathrm{SP}}(a,b). \qquad (16.4)$$

In view of Theorem 16.1, this seems to indicate that finding the right metric to capture the underlying structure of the graph can be useful in the study of disease spreading on networks. Moreover, the work of Lovasz [59] can be used, see [4, Section 8], to show that the lower bound for epidemic hitting time in terms of effective resistance is robust in the sense that it continues to hold even for non-Markovian versions of the SI process, namely, in situations where the infection times are not distributed exponentially.

Note that, since $\text{eff}\mathcal{R}$ is the modulus metric δ_2 and d_{SP} is δ_∞, it would be interesting to know whether the bounds in (16.4) can be made tighter. In other words, is it possible to find $2 < p_1 < p_2 < \infty$, so that

$$\delta_{p_1} \leq \text{EHT} \leq \delta_{p_2}?$$

EXERCISES

16.1 *Modulus metrics.* The goal of this exercise is to prove Theorem 16.1. Given the definitions, in Section 16.1, do the following.

(i) For all $p \in [1, \infty]$ and $\sigma \in \mathbb{R}^E_{>0}$, use Fulkerson duality (13.15), monotonicity and subadditivity from Proposition 10.4, and subordination from Proposition 10.5, to show that $\delta_{p,\sigma}$ is a metric on V.

(ii) Show that $\lim_{p \uparrow \infty} \delta_{p,\sigma} = d_{\text{SP}}$.

(iii) Show that $\delta_2 = d_{\text{ER}}$.

(iv) Show that for $1 < p < 2$, $\text{Mod}_{p,\sigma}(\Gamma(a,b))^{-1}$ is a metric and it tends to $d_{\text{MC}}(a,b)$ as $p \to 1$.

(v) First, show that an arbitrary positive power of an ultrametric is also an ultrametric, so $(d_{\text{MC}})^t$ is a metric for any $t > 0$. Then, using (11.3) and (11.4), show that as $p \downarrow 1$, δ_p converges to the limit

$$\lim_{t \to \infty} (d_{\text{MC}}(a,b))^t = \begin{cases} 0 & \text{if } d_{\text{MC}}(a,b) < 1, \\ 1 & \text{if } d_{\text{MC}}(a,b) = 1, \\ \infty & \text{if } d_{\text{MC}}(a,b) > 1. \end{cases}$$

Also, show that for unweighted graphs, this limit essentially decomposes the graph into its 2-edge-connected components. All nodes in the same component are distance zero from one another while nodes in different components are at a distance one.

(vi) Finally, show that for every $\epsilon > 0$ and every $p \in (1, \infty)$ there is a connected graph for which $\delta_p^{1+\epsilon}$ is not a metric.

16.2 *Proving the triangle inequality via modulus of potentials* Let $G = (V, E)$ be a finite simple graph. Fix $a \neq b \in V$. Let $\Gamma := \Gamma_{\text{pot}}(a, b)$ be the family of all *ab-potentials*, namely, $h \in \Gamma$ if and only if h is a function $h : V \to \mathbb{R}$ with $h(a) \leq 0$ and $h(b) \geq 1$. We define the edge usage of the potential $h \in \Gamma$ for an edge $e = \{x, y\}$ as

$$\mathcal{N}(h, e) := |h(x) - h(y)|.$$

(i) Let $\Gamma_{\text{cut}}(a, b)$ be the family of all minimal *ab*-cuts defined in Example 13.1. Show that $\Gamma_{\text{cut}}(a, b) \subset \Gamma_{\text{pot}}(a, b)$. Hence deduce that,

$$\text{Adm}\,\Gamma_{\text{pot}}(a, b) \subset \text{Adm}\,\Gamma_{\text{cut}}(a, b).$$

(ii) Let $\gamma = x_0 x_1 \cdots x_n \in \Gamma_s(a, b)$ be an *ab*-path. Identify γ with the vector $\mathcal{N}(\gamma, \cdot)^T$ in $\mathbb{R}^E$, and show that γ is admissible for $\Gamma_{\text{pot}}(a, b)$. Hence, deduce that $\Gamma_{\text{path}}(a, b) \subset \text{Adm}\,\Gamma$.

Next, taking the dominant of both sides we get that,

$$\text{Adm}\,\Gamma_{\text{cut}}(a, b) = \text{Dom}\,\Gamma_{\text{path}}(a, b) \subset \text{Adm}\,\Gamma_{\text{pot}}(a, b).$$

Thus, we have shown that

$$\text{Adm}\,\Gamma_{\text{cut}}(a, b) = \text{Adm}\,\Gamma_{\text{pot}}(a, b),$$

and this implies that

$$\text{Mod}_q(\Gamma) = \text{Mod}_q(\Gamma_{\text{cut}}(a, b)).$$

for all $q \in (1, \infty)$.

(iii) Suppose $G = (V, E, \sigma)$ is a weighted, connected, simple graph. Recall the quantity $\delta_{p,\sigma}$ defined in Exercise 16.1. Use $\Gamma_{\text{pot}}(a, b)$ to show that $\delta_{p,\sigma}$ is a metric on V.

16.3 *Modulus metrics on the complete graph* Let $G = K_n$ be the complete graph on the vertex set $V = [n] = \{1, 2, \ldots, n\}$, see Figure 16.3 for the example of K_6. Fix a parameter $p \in (1, \infty)$.

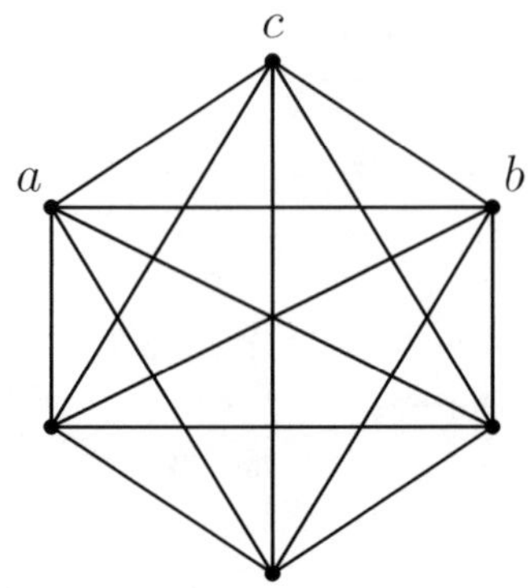

Figure 16.3 K_6-Complete graph on 6 nodes.

(i) Show that, by symmetry, $\text{Mod}_p(\Gamma(a,b))$, and hence $d_p(a,b)$, does not depend on the choice of $a \neq b$ in V. Therefore, deduce that d_p is an ultrametric on complete graphs.

(ii) Fix $a \neq b$ in V. We want to compute $d_p(a,b)$. To that end, argue that the density ρ^*, defined as $\rho^*(a,x) = 1/2 = \rho^*(b,x)$ for every $x \neq a, b$, $\rho^*(a,b) = 1$, and $\rho^* = 0$ otherwise, is extremal for $\Gamma(a,b)$ in K_N.
Hint: consider the subfamily $\tilde{\Gamma}$ of simple paths consisting of $[a,\ b]$ and $[a,\ x,\ b]$ for any $x \neq a, b$.

(iii) Show that $\text{Mod}_p(\Gamma(a,b)) = 1 + 2(N-2)\frac{1}{2^p}$ and hence, deduce that

$$d_p(a,b) = \left(1 + \frac{N-2}{2^{p-1}}\right)^{-q/p}.$$

V

The Family of Spanning Trees

CHAPTER 17

Loops and Trees

IN THIS CHAPTER, we revisit the theory of random walks on graphs developed in Chapter 6. In particular, we reexamine the stopped Green's function from Section 5.5 and analyze it in more depth following the paper of G. Lawler [54]. This will demonstrate that loops and trees on a graph are intimately connected. We invite the reader to consult [54] for the historical development of these ideas and the wider context in which they were developed. See also the more recent monograph [55].

17.1 GREEN'S FUNCTIONS REVISITED

We begin with a concrete example. Let $G = (V, E)$ be a connected finite graph, as in Figure 17.1. Consider a subset $A \subset V$ so that $\partial A = V \setminus A$. Here, ∂A consists of all nodes not in A that have a neighbor in A. A typical example for this situation is when $A = V \setminus \{b\}$ for some node $b \in V$. For instance, in Figure 17.1, we have $A = \{1, 2, 3, 4\}$ and $\partial A = \{0\}$.

Let P be the transition matrix for Simple Random Walk on the graph G. In our example,

$$P = \begin{bmatrix} 0 & \frac{1}{2} & 0 & \frac{1}{2} & 0 \\ \frac{1}{3} & 0 & \frac{1}{3} & \frac{1}{3} & 0 \\ 0 & \frac{1}{3} & 0 & \frac{1}{3} & \frac{1}{3} \\ \frac{1}{4} & \frac{1}{4} & \frac{1}{4} & 0 & \frac{1}{4} \\ 0 & 0 & \frac{1}{2} & \frac{1}{2} & 0 \end{bmatrix} \tag{17.1}$$

DOI: 10.1201/9781003024866-22

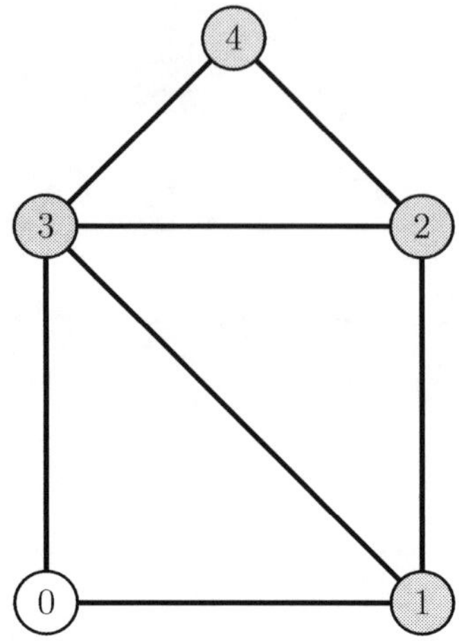

Figure 17.1 A connected finite graph with a subset A of the form $A = V \setminus \{b\}$.

Let $Q = P[A, A]$ be the submatrix of P obtained by restricting the rows and columns to the nodes in A. In our case, Q is obtained from P by erasing the first row and the first column:

$$Q = \begin{bmatrix} 0 & \frac{1}{3} & \frac{1}{3} & 0 \\ \frac{1}{3} & 0 & \frac{1}{3} & \frac{1}{3} \\ \frac{1}{4} & \frac{1}{4} & 0 & \frac{1}{4} \\ 0 & \frac{1}{2} & \frac{1}{2} & 0 \end{bmatrix} \tag{17.2}$$

Note that the rows of Q add up to at most one, in the sense that

$$\sum_{y \in A} Q(x, y) \leq 1, \qquad \forall x \in A,$$

and the inequality is strict whenever $x \sim b$. In other words, we are exactly in the situation of Exercise 5.7. In particular, we deduce that $I - Q$ is invertible and

$$(I - Q)^{-1} = \sum_{n=0}^{\infty} Q^n.$$

Note that for $x, y \in A$, the entries $Q^n(x, y)$ count the number of walks that stay in A and go from x to y in n steps.

Let $\mathcal{K}_A(x,y)$ be the set of all walks starting at x and ending at y that stay in A. Given a walk $\omega = [x_0 = x, x_1, \ldots, x_{n-1}, x_n = y] \in \mathcal{K}_A(x,y)$, its probability is

$$p(\omega) := P(x_0, x_1) \cdots P(x_{n-1}, x_n) = Q(x_0, x_1) \cdots Q(x_{n-1}, x_n). \quad (17.3)$$

Consider the exit time $\tau_{\partial A} := \inf\{j \geq 0 : X_j \in \partial A\}$. Recall from (5.13) that the Green function stopped the first time the walk exits A is defined as:

$$\begin{aligned} G_A(x,y) &= \mathbb{E}_x\,[\text{n. of visits at } y \text{ before exiting } A] \\ &= \sum_{n=0}^{\infty} \mathbb{P}_x(X_n = y, \tau_{\partial A} > n). \end{aligned}$$

By adding up all the walk probabilities, we see that

$$G_A(x,y) = \sum_{n=0}^{\infty} Q^n(x,y) = (I - Q)^{-1}(x,y) = \sum_{\omega \in \mathcal{K}_A(x,y)} p(\omega). \quad (17.4)$$

Note that

$$G_A(x,x) = 1 + \sum_{\omega \in \mathcal{K}_A(x,x)} p(\omega), \quad (17.5)$$

because the constant walk at x does count as a visit at x.

This approach can be used, for instance, to give another proof of the fact that

$$G_A(x,x) = \mathbb{E}_x\,[\text{n. of visits at } x \text{ before exiting } A] = \frac{1}{1 - f_x} \quad (17.6)$$

where

$$f_x := \mathbb{P}_x(\tau_x^+ < \tau_{\partial A}) \quad (17.7)$$

is the *first return probability* at x. As we have seen in Chapter 7, in the case when $x = a \neq b$ and $A = V \setminus \{b\}$, this can be deduced by combining Proposition 6.1 and Corollary 6.1. However, here we will describe an argument using walk probabilities.

First, we need some notation. Given a walk $\omega = [x_0, \ldots, x_n]$, write $|\omega| = n$ for the number of steps that ω takes. Also, if $\omega_1 = [x_0, \ldots, x_n]$ and $\omega_2 = [y_0, \ldots, y_m]$ are two walks with $x_n = y_0$, write

$$\omega_1 \cdot \omega_2 = [x_0, \ldots, x_n = y_0, \ldots, y_m]$$

for the *concatenation* of the two walks. Define an *elementary loop* at x to be a walk ℓ such that

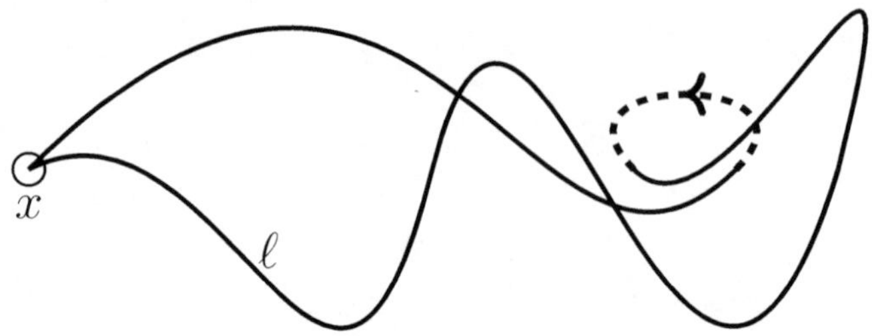

Figure 17.2 An elementary loop.

(i) $|\ell| > 0$;

(ii) ℓ visits x only two times, at the beginning and at the end.

As in [54], write $\tilde{\mathcal{L}}_x^1(A)$ for the set of all elementary loops at x that stay in A. Note that the first return probability f_x defined in (17.7) can be written as

$$f_x = \sum_{\omega \in \tilde{\mathcal{L}}_x^1(A)} p(\omega). \tag{17.8}$$

Also, write $\tilde{\mathcal{L}}_x^k(A)$ for the set of loops of the form

$$\ell = \ell^1 \cdot \cdots \cdot \ell^k, \qquad \text{with } \ell^j \in \tilde{\mathcal{L}}_x^1(A),\ j = 1, \ldots, k,$$

So that

$$\mathcal{K}_A(x, x) = \bigcup_{k=0}^{\infty} \tilde{\mathcal{L}}_x^k(A) \tag{17.9}$$

where $\tilde{\mathcal{L}}_x^0(A)$ consists exclusively of the trivial loop $[x]$.

To see why (17.6) holds, note that a loop $\ell \in \mathcal{K}_A(x, x)$ has length $|\ell| \geq 1$, if and only if,

$$\ell = \omega_1 \cdot \omega_2 \tag{17.10}$$

where $\omega_1 \in \tilde{\mathcal{L}}_x^1$ and $\omega_2 \in \mathcal{K}_A(x, x)$. This can be seen by letting ω_1 follow ℓ until the first return time at x and then switching to ω_2. Therefore,

$$\begin{aligned}
G_A(x, x) &= 1 + \sum_{\ell \in \mathcal{K}_A(x,x)} p(\ell) && \text{(by (17.5))} \\
&= 1 + \sum_{\omega_1 \in \tilde{\mathcal{L}}_x^1(A)} \sum_{\omega_2 \in \mathcal{K}_A(x,x)} p(\omega_1) p(\omega_2) && \text{(by (17.10))} \\
&= 1 + \left(\sum_{\omega_1 \in \tilde{\mathcal{L}}_x^1(A)} p(\omega_1) \right) \left(\sum_{\omega_2 \in \mathcal{K}_A(x,x)} p(\omega_2) \right) && \text{(product of sums)} \\
&= 1 + f_x G_A(x, x) && \text{(by (17.8)).}
\end{aligned}$$

17.2 LOOP ERASED RANDOM WALK

We first describe an algorithm called *Loop Erasure* (LE) that takes as input a walk $\omega = [x_0, \ldots, x_n]$, erases its loops and outputs a self-avoiding walk $\eta = [y_0, \ldots, y_m]$ with $y_0 = x_0$ and $y_m = x_n$. This is accomplished by the following algorithm:

(i) Let $j_0 := \max\{k : \omega(k) = x_0\}$, namely, j_0 is the time of the last visit at x_0, and set

$$\eta(0) = \omega(0) = \omega(j_0);$$

(ii) For $i \geq 1$, if $j_{i-1} < n$, let $j_i := \max\{k : \omega(k) = \omega(j_{i-1} + 1)\}$, the time of the last visit at $\omega(j_{i-1} + 1)$, which is the node the random walk visits after $\eta(i-1) = \omega(j_{i-1})$. Then, set

$$\eta(i) = \omega(j_{i-1} + 1) = \omega(j_i);$$

(iii) as soon as $j_m = n$, stop.

Note that by construction, $\eta(0) = \omega(0)$ and $\eta(m) = \omega(n)$.

Let $\mathcal{K}_A$ be the set of walks in A and $\overline{\mathcal{K}}_A$ be the set of walks in A that are allowed to start or end on ∂A, but do not visit ∂A otherwise. Similarly, $\mathcal{W}_A$ and $\overline{\mathcal{W}}_A$ are the corresponding sets of self-avoiding walks. A measure on $\overline{\mathcal{W}}_A$ can be defined as a push-forward using the map LE:

$$\hat{p}(\eta) := \sum_{\omega \in \overline{\mathcal{K}}_A :\ \mathrm{LE}(\omega) = \eta} p(w), \qquad \text{for } \eta \in \overline{\mathcal{W}}_A. \tag{17.11}$$

The next result gives a formula for the push-forward measure $\hat{p}$ in terms of the stopped Green function, and it is Proposition 3.1 in [54].

Proposition 17.1. *Assume $\eta = [y_0, \ldots, y_m] \in \mathcal{W}_A$. Then,*

$$\hat{p}(\eta) = p(\eta) \prod_{j=0}^{m} G_{A_j}(y_j, y_j),$$

where $A_j = A \setminus \{y_0, \ldots, y_{j-1}\}$.

Proof. See Exercise 17.1. □

The following is a fundamental result due to G. Lawler. It is Proposition 3.3 in [54]. Later in Section 17.6, we will rephrase this result in terms of electrical networks and modulus and give an alternate proof.

Theorem 17.1. *Let* $\{x_1, \ldots, x_n\} \subset A$ *and set* $A_1 = A$ *and* $A_j = A \setminus \{x_1, \ldots, x_{j-1}\}$, *for* $j = 2, \ldots, n$. *Define*

$$F_{x_1 \ldots x_n}(A) := \prod_{j=1}^{n} G_{A_j}(x_j, x_j). \tag{17.12}$$

Then, for any permutation $\sigma : [n] \to [n]$,

$$F_{x_1 \ldots x_n}(A) = F_{\sigma(x_1) \ldots \sigma(x_n)}(A).$$

Proof. First assume that $n = 2$. We want to show that $F_{x_1 x_2}(A) = F_{x_2 x_1}(A)$, namely:

$$G_A(x_1, x_1) G_{A \setminus \{x_1\}}(x_2, x_2) = G_A(x_2, x_2) G_{A \setminus \{x_2\}}(x_1, x_1) \tag{17.13}$$

For simplicity, write $\mathcal{K}_A(x_1^*, x_2)$, $\mathcal{K}_A(x_1, x_2^*)$, and $\mathcal{K}_A(x_1^*, x_2^*)$, to mean the subfamilies of $\mathcal{K}_A(x_1, x_2)$ consisting of walks that visit the starred nodes only once. Note that the walk ω in $\mathcal{K}_A(x_1, x_2)$ can be decomposed in a unique way as:

$$\omega = \ell_1 \cdot \omega^* \cdot \ell_2, \tag{17.14}$$

by looping at x_1 in $A \setminus \{x_2\}$, until the last time the walk visits x_1 before visiting x_2, then traveling from x_1 to x_2 so as to visit x_1 and x_2 only once, and finally looping at x_2 in all of A, see Figure 17.3. Therefore, we can write this decomposition as

$$\mathcal{K}_A(x_1, x_2) = \mathcal{K}_{A \setminus \{x_2\}}(x_1, x_1) \cdot \mathcal{K}_A(x_1^*, x_2^*) \cdot \mathcal{K}_A(x_2, x_2). \tag{17.15}$$

On the other hand, $\mathcal{K}_A(x_1, x_2)$ can also be decomposed in a unique way as:

$$\mathcal{K}_A(x_1, x_2) = \mathcal{K}_A(x_1, x_1) \cdot \mathcal{K}_A(x_1^*, x_2^*) \cdot \mathcal{K}_{A \setminus \{x_1\}}(x_2, x_2). \tag{17.16}$$

by looping at x_1 inside all of A until the last time the walk visits x_1, then traveling from x_1 to x_2 so as to visit x_1 and x_2 only once, and finally looping at x_2 in $A \setminus \{x_1\}$.

If we let $k := p(\mathcal{K}_A(x_1^*, x_2^*))$, then (17.15) implies that

$$G_A(x_1, x_2) = k G_{A \setminus \{x_2\}}(x_1, x_1) G_A(x_2, x_2),$$

while (17.16) implies that

$$G_A(x_1, x_2) = k G_A(x_1, x_1) G_{A \setminus \{x_1\}}(x_2, x_2).$$

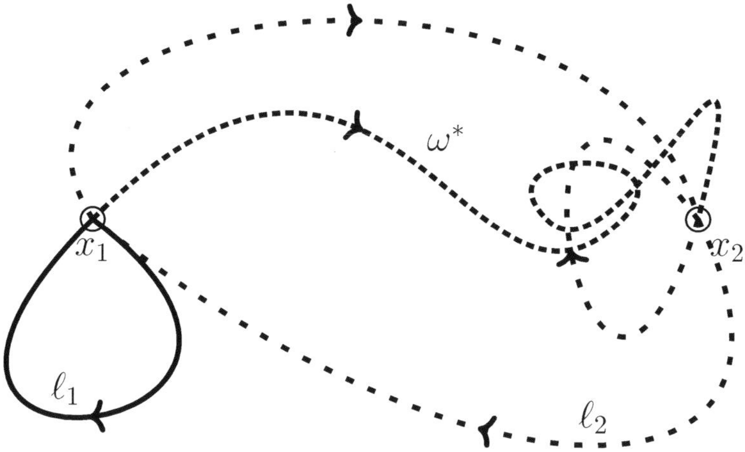

Figure 17.3 A walk from x_1 to x_2 can be decomposed by first tracing the solid black loop ℓ_1 at x_1 that stays in $A \setminus \{x_2\}$, until the last time the walk visits x_1 before visiting x_2, then following the densely dashed walk ω^* from x_1 to x_2 that visits x_1 and x_2 only once, and finally using the loosely dashed loop ℓ_2 at x_2 that can visit all of A.

This shows that the statement holds when $n = 2$.

In the general $n > 2$ case, it will be enough to show the case when σ is a transposition, and without loss of generality, it is enough to transpose x_1 and x_n. To see how it works, assume first that $n = 3$:

$$\begin{aligned}F_{x_1x_2x_3}(A) =& G_A(x_1, x_1)G_{A\setminus\{x_1\}}(x_2, x_2)G_{A\setminus\{x_1,x_2\}}(x_3, x_3)\\ =& G_A(x_2, x_2)G_{A\setminus\{x_2\}}(x_1, x_1)G_{A\setminus\{x_2,x_1\}}(x_3, x_3)\\ =& G_A(x_2, x_2)G_{A\setminus\{x_2\}}(x_3, x_3)G_{A\setminus\{x_2,x_3\}}(x_1, x_1)\\ =& F_{x_2x_3x_1}\end{aligned}$$

and by reversing this procedure, we can move the last node to the first position as well. The general case works similarly. □

As a corollary of the proof of Theorem 17.1, we obtain the following generalization of Green function symmetry.

Corollary 17.1. *We have*

$$\frac{G_A(x_1, x_2)}{p\left(\mathcal{K}_A(x_1^*, x_2^*)\right)} = \frac{G_A(x_2, x_1)}{p\left(\mathcal{K}_A(x_2^*, x_1^*)\right)}$$

In view of the results in Exercise 6.2, this implies that when $A = V \setminus \{b\}$:

$$\frac{p\left(\mathcal{K}_A(x_1^*, x_2^*)\right)}{\deg_c(x_2)} = \frac{p\left(\mathcal{K}_A(x_2^*, x_1^*)\right)}{\deg_c(x_1)}$$

Moreover, Theorem 17.1 implies that the quantity $F_\eta(A)$ defined in (17.12), with $\eta = \{x_1, \ldots, x_n\} \subset A$, only depends on η as a subset of A and not on its order. In particular, when choosing all of A, the quantity $F_A(A)$ is well-defined and has a surprising connection with the determinant of $I - Q$, which can be thought as a reduced walk Laplacian. This result is Proposition 3.5 in [54].

Proposition 17.2. *We have:*

$$F_A(A) = \det G_A = \frac{1}{\det(I - Q)}.$$

Proof. See Exercise 17.2. □

17.3 WILSON'S ALGORITHM

The idea of Wilson's algorithm is to build a random spanning tree of a graph using loop-erased random walks. Here, it will be useful to extend the loop erasure algorithm and the results of the previous sections to more general Markov chains on a graph. In other words, we may allow different transition matrices P, that are not the one for the simple random walk as in (17.1). But, we will still be able to restrict it to a subset $A \subset V$ and form a matrix Q as in (17.2). In this case, we will still call the resulting process a loop-erased random walk.

Given a finite simple graph $G = (V, E)$, enumerate the vertices as $V = \{x_0, \ldots, x_n\}$. Then, given a Markov chain with transition matrix P, do the following:

(i) Let $\eta^0 := \{x_0\}$ and set $A_1 := A = V \setminus \eta^0$. Then, pick a loop erased random walk η^1 starting at x_1 and stop it the first time it visits x_0. Finally, set $A_2 := A \setminus \eta^1$;

(ii) on the k-th iteration, with $k \geq 2$, if $A_k = \emptyset$, stop;

(iii) else, let j be the smallest index so that $x_j \in A_k$, and pick a loop erased random walk η^k starting at x_j and stop it the first time it reaches $\eta^0 \cup \cdots \cup \eta^{k-1}$. Namely, $\eta^k \in \mathcal{W}_{A_k}(x_j, \eta^0 \cup \cdots \cup \eta^{k-1})$. Then, set $A_{k+1} := A_k \setminus \eta^k$.

Moreover, at each iteration with $k \geq 1$, we may orient the path η^k from x_j to $\eta^0 \cup \cdots \cup \eta^{k-1}$. This gives rise to an oriented rooted tree $\vec{T} = \{\vec{e}_j\}_{j=1}^{|V|-1}$, and it allows us to give it the weight:

$$p(\vec{T}; x_0) = \prod_{j=1}^{n} p(\vec{e_j}). \tag{17.17}$$

On the other hand, forgetting about the root and the orientation gives rise to an underlying spanning tree T of G. The next result shows that the resulting probability distribution on the family of spanning trees Γ_G can be computed in terms of (17.17) and the determinant of the reduced walk Laplacian. This is Proposition 4.7 in [54].

Theorem 17.2. *Let Γ_G be the family of spanning trees for the finite simple graph G. Then Wilson's algorithm picks a tree $T \in \Gamma_G$ with probability*

$$p(\vec{T}; x_0) F_A(A)$$

where $\vec{T}$ is the oriented rooted tree T with root x_0 and with the induced orientation towards the root. Also, $A = V \setminus \{x_0\}$. Moreover,

$$\sum_{T \in \Gamma_G} p(\vec{T}; x_0) = \frac{1}{F_A(A)} = \det(I - Q). \tag{17.18}$$

Proof. Wilson's algorithm picks a tree with probability

$$\hat{p}(\eta^1) \cdots \hat{p}(\eta^k)$$

for some k. By Proposition 17.1, we can write it as

$$p(\eta^1) F_{\eta^1}(A) p(\eta^2) F_{\eta^2}(A \setminus \eta^1) \cdots p(\eta^k) F_{\eta^k}(A \setminus (\eta^1 \cup \cdots \cup \eta^{k-1})), \tag{17.19}$$

which, after some reorganization, using Theorem 17.1, is seen to be equal to $p(\vec{T}; x_0) F_A(A)$.

Finally, adding all these probabilities and using Proposition 17.2, yields (17.18). □

We will mainly be interested in the case when $G = (V, E)$ is a weighted graph with given edge weights $\sigma \in \mathbb{R}^E_{>0}$, and the corresponding Markov chain is the weighted random walk as described in Section 6.1.

Wilson's algorithm can then be used to sample *weighted uniform spanning trees* (UST_σ). To describe this distribution, note that every spanning tree $\gamma \in \Gamma_G$ acquires a weight

$$\sigma[\gamma] := \prod_{e\in E} \sigma(e)^{\mathcal{N}(\gamma,e)}. \tag{17.20}$$

This defines a pmf $\mu_\sigma \in \mathcal{P}(\Gamma_G)$ so that $\mu_\sigma(\gamma)$ is proportional to $\sigma[\gamma]$. We say that μ_σ is the weighted uniform distribution UST_σ.

In the unweighted case, this is the probability distribution that for $T \in \Gamma_G$ assigns

$$\mathbb{P}_{\mathrm{UST}}\left(\underline{\gamma} = T\right) \equiv \frac{1}{|\Gamma_G|}.$$

Corollary 17.2. *Let $G = (V, E)$ be a simple finite connected graph, with edge weights $\sigma \in \mathbb{R}^E_{>0}$, and let Γ_G be the family of spanning trees of G. Fix a node $x_0 \in V$, let $A = V \setminus \{x_0\}$ and do the Wilson algorithm as described above using the transition matrix of the weighted random walk as described in Section 6.1. Then, the resulting random trees are* UST_σ, *meaning that they are sampled with the weighted uniform distribution.*

Proof. If Wilson's algorithm is performed using the transition matrix P for the weighted random walk on G, as given in (6.1), then, by Theorem 17.2, a tree $T \in \Gamma_G$ is sampled with probability proportional to $P(\vec{T}, x_0)$, as given in (17.17). Note that, in the oriented tree $\vec{T}$, each node $x \neq x_0$ has exactly one outgoing edge $\vec{e}_x$, otherwise we could follow two distinct direct paths from x to x_0. Therefore, using (5.6), we see that

$$P(\vec{T}, x_0) = \prod_{x\in V\setminus\{x_0\}} \frac{\sigma(\vec{e}_x)}{\deg_\sigma(x)} = \sigma[T] \cdot \prod_{x\in V\setminus\{x_0\}} \frac{1}{\deg_\sigma(x)}.$$

In particular, for every $T \in \Gamma_G$, this is proportional to the tree weight $\sigma[T]$. □

17.4 EDGE PROBABILITIES OF UNIFORM SPANNING TREES

There is a surprising connection between random uniform spanning trees and effective resistance. This result is sometimes attributed to Kirchhoff, as is the Matrix-Tree Theorem below, see Section 17.5.

Theorem 17.3. *Let $G = (V, E)$ be a simple finite connected graph, with edge weights $\sigma \in \mathbb{R}^E_{>0}$, and let Γ_G be the family of spanning trees of G endowed with the weighted uniform probability. Fix an edge $e = \{x, y\} \in$*

E and write $\mathbb{P}_{\mathrm{UST}_\sigma}\left(e \in \underline{\gamma}\right)$ *for the probability that a randomly chosen tree contains the edge e. Also, recall that hooking up nodes x and y to a battery so as to flow one unit of current* $f_{x,y}$ *through the graph from x to y gives rise to the effective resistance* $\mathrm{eff}\mathcal{R}_\sigma(x, y)$*, see Section 2.3. Then,*

$$\mathbb{P}_{\mathrm{UST}_\sigma}\left(e \in \underline{\gamma}\right) = \sigma(x, y)\mathrm{eff}\mathcal{R}(x, y) = \frac{\mathrm{eff}\mathcal{R}(x, y)}{r(x, y)},$$

where $r(x, y) = \sigma(x, y)^{-1}$ *is the edge resistance.*

Proof. We perform Wilson's algorithm on the weighted random walk (WRW), with y equal to the root and x the starting point. Let η^1 be the first loop erased walk from x to y. Since η^1 is a path, there are two cases: either $\eta^1 = [x, y]$ and hence $e \in \eta^1$, or $e \notin \eta^1$. Therefore, $\mathbb{P}_{\mathrm{UST}_\sigma}\left(e \in \underline{\gamma}\right)$ is equal to

$$\mathbb{E}_x\,(\text{number of times the WRW jumps from } x \text{ to } y, \text{ before hitting } y)\,,$$

because the random variable inside the expectation can only take the value 0 or 1.

Now recall from (6.10) and (6.11), that the unit current flow $f_{x,y}$ from x to y is equal to the expected *net* amount of times that the random walk started at x crosses from x to y before visiting y. However, the number of times this stopped random walk crosses from y to x is zero. Thus, we have that

$$\mathbb{P}_{\mathrm{UST}_\sigma}\left(e \in \underline{\gamma}\right) = f_{x,y}(x, y).$$

Recall that the voltage potential $h = h_{x,y}$ to create a unit current flow from x to y, satisfies $h(x) = 0$ and $h(y) = \mathrm{eff}\mathcal{R}_\sigma(x, y)$. Therefore, by Ohm's Law:

$$f_{x,y}(x, y) = \sigma(x, y)(h(y) - h(x)) = \sigma(x, y)\mathrm{eff}\mathcal{R}_\sigma(x, y).$$

□

17.5 THE MATRIX-TREE THEOREM

Wilson's algorithm can also be used to deduce another Kirchhoff theorem that computes the number of spanning trees of a graph in terms of the determinant of the Laplacian.

Corollary 17.3 (Matrix-Tree Theorem). *Let $G = (V, E)$ be a simple finite connected graph, with edge weights $\sigma \in \mathbb{R}^E_{>0}$, and let Γ_G be the family of spanning trees of G. Fix a node $x_0 \in V$, let $A = V \setminus \{x_0\}$ and let L_σ be the weighted combinatorial Laplacian for G. Write $L_\sigma[A, A]$ for the submatrix obtained by erasing the row and column in L_σ corresponding to the node x_0. Finally, for each tree $\gamma \in \Gamma_G$ let $\sigma[\gamma]$ be the weight defined in (17.20). Then,*

$$\sum_{\gamma \in \Gamma_G} \sigma[\gamma] = \det L_\sigma[A, A].$$

In particular, this shows that $L_\sigma[A, A]$ is always invertible, a fact that we already saw in Exercise 3.10, and moreover, its determinant doesn't depend on which node x_0 we choose.

Proof. First, let $D := \max \deg_\sigma(G)$ be the maximum (generalized) degree in G and pick a number $\epsilon < 1/D$. Then, define a transition matrix

$$P_\epsilon(x, y) = \begin{cases} \epsilon\sigma(x, y) & \text{if } x \sim y; \\ 1 - \epsilon \deg_\sigma(x) & \text{if } x = y. \end{cases}$$

In matrix form, if W is the weighted adjacency matrix of G, D is the diagonal matrix of generalized degrees, and $L_\sigma = D - W$ is the (weighted) combinatorial Laplacian, then we have that

$$P_\epsilon = (I - \epsilon D) + \epsilon W = I - \epsilon(D - W) = I - \epsilon L_\sigma.$$

Or in other words,

$$\epsilon L_\sigma = I - P_\epsilon. \tag{17.21}$$

Note that, if we use P_ϵ in Wilson's algorithm, then

$$P(\vec{T}; x_0) = \epsilon^{N-1} \sigma[T], \tag{17.22}$$

where $N = |V|$, and moreover,

$$\begin{aligned} \epsilon^{N-1} \sum_{T \in \Gamma_G} \sigma[T] &= \sum_{T \in \Gamma_G} P(\vec{T}; x_0) && \text{(by (17.22))} \\ &= \det(I - P_\epsilon[A, A]) && \text{(by Theorem 17.2)} \\ &= \det(\epsilon L_\sigma[A, A]) && \text{(by (17.21))} \\ &= \epsilon^{N-1} \det(L_\sigma[A, A]) \end{aligned}$$

□

Another version of the matrix-tree theorem says that the number of spanning trees is related to the product of the non-zero eigenvalues of the Laplacian.

Proposition 17.3. *With the notations of Corollary 17.3, assume that the eigenvalues of the Laplacian L_σ are ordered as follows, $\lambda_1 = 0 < \lambda_2 \leq \cdots \leq \lambda_N$, where $N = |V|$. Then, we have*

$$\sum_{\gamma \in \Gamma_G} \sigma[\gamma] = \frac{\lambda_2 \cdots \lambda_N}{N}.$$

Proof. See Exercise 17.3. □

17.6 CONNECTIONS WITH MODULUS

Proposition 17.4. *Let $G = (V, E)$ be a simple connected graph, and suppose $a \neq b$ are two distinct nodes $a, b \in V$. Let $\Gamma = \Gamma(a, b)$ be the family of paths from a to b. Performing loop-erased random walk from a to b, gives rise to a pmf μ on Γ. Let $\gamma := [x_0, x_1, \ldots, x_n]$ with $x_0 = a$ and $x_n = b$ be one such walk. Set $A := A_0 := V \setminus \{b\}$ and $A_j := A \setminus \{x_0, \ldots, x_{j-1}\}$ for $j = 1, \ldots, n-1$. Then,*

$$\mu(\gamma) = \prod_{j=0}^{n-1} \frac{1}{\mathrm{Mod}_2\left(\Gamma(x_j, \{x_0, \ldots, x_{j-1}, b\})\right)}, \tag{17.23}$$

with the understanding that for $j = 0$, the notation $\{x_0, \ldots, x_{j-1}, b\}$ denotes the set $\{b\}$.

Note that the connecting family $\Gamma(x_j, \{x_0, \ldots, x_{j-1}, b\})$ will in general connect a node to a set of nodes, so we proceed as in Remark 14.2 to compute its modulus.

Proof of Proposition 17.4. The probability of γ is

$$\begin{aligned}
\mu(\gamma) &= \hat{p}(\gamma) = p(\gamma) F_\gamma(A) && \text{(by Proposition 17.1)} \\
&= \prod_{j=0}^{n-1} p(x_j, x_{j+1}) G_{A_j}(x_j, x_j) && \text{(by (17.12))} \\
&= \prod_{j=0}^{n-1} \frac{1}{\deg(x_j)} G_{A_j}(x_j, x_j) && \text{(by (5.6))}
\end{aligned}$$

To compute the stopped Green's function, first we identify all the nodes $\{x_0, \dots, x_{j-1}\}$ with the node b. This results in a new graph H with possible multiple edges, where we have removed the self-loops. Use Corollary 6.1 in H and get,

$$G_{\mathcal{T}_b}(x_j, x_j) = \deg(x_j)\mathrm{eff}\mathcal{R}(x_j, b) = \frac{\deg(x_j)}{\mathrm{Mod}_2\left(\Gamma(x_j, b); H\right)}$$

where the last equality follows from Theorem 14.1 (iii). Note that the degree of x_j is unchanged when we pass from G to H. Also, the modulus $\mathrm{Mod}_2\left(\Gamma(x_j, b); H\right)$ for paths from x_j to b in H, is equal to the modulus $\mathrm{Mod}_2\left(\Gamma(x_j, \partial A_j); G\right)$ from x_j to $\partial A_j = \{x_0, \dots, x_{j-1}, b\}$ in G. This can be seen from the fact that the edges of H are in one-to-one correspondence with the set of edges e in G such that $e \cap A_j \neq \emptyset$.

Therefore, we see that the degrees cancel and we get formula (17.23).

□

Relatedly, we can also rewrite the fundamental "switch identity" (17.13) in terms of modulus, as follows. Given a graph G and three nodes $a, b, c \in V$, we think of $A = V \setminus \{b\}$, $x_1 = a$ and $x_2 = c$. Then, by Corollary 6.1,

$$G_A(x_1, x_1)G_{A\setminus\{x_1\}}(x_2, x_2) = G_A(x_2, x_2)G_{A\setminus\{x_2\}}(x_1, x_1)$$

becomes

$$\begin{aligned}
&\frac{\deg(a)}{\mathrm{Mod}_2\left(\Gamma(a, b)\right)}\frac{\deg(c)}{\mathrm{Mod}_2\left(\Gamma(c, \{a, b\})\right)} \\
&= \frac{\deg(c)}{\mathrm{Mod}_2\left(\Gamma(c, b)\right)}\frac{\deg(a)}{\mathrm{Mod}_2\left(\Gamma(a, \{c, b\})\right)}.
\end{aligned}$$

Hence, simplifying, we get

$$\begin{aligned}
&\mathrm{Mod}_2\left(\Gamma(a, b)\right)\mathrm{Mod}_2\left(\Gamma(c, \{a, b\})\right) \\
&= \mathrm{Mod}_2\left(\Gamma(c, b)\right)\mathrm{Mod}_2\left(\Gamma(a, \{c, b\})\right). \qquad (17.24)
\end{aligned}$$

The *switch identity* (17.24) can also be proved by induction on the size of the graph G using the star-mesh transformation (2.11) for effective conductance and Theorem 14.1 (iii). For more details, see [26].

EXERCISES

17.1 *A formula for the measure on loop-erased walks.* The goal in this exercise is to prove Proposition 17.1. Given a connected graph $G = (V, E)$, let $A \subset V$ be such that $\partial A = V \setminus A$. Fix a path $\eta = [y_0, \ldots, y_m] \in \mathcal{W}_A$ and define the decreasing sets $A_j := A \setminus \{y_0, \ldots, y_{j-1}\}$, for $j = 1, \ldots, m$. Also, set $A_0 = A$.

(i) Let $\omega = [x_0, \ldots, x_n]$ be a walk with loop erasure $\mathrm{LE}(\omega) = \eta$. Show that ω can be decomposed as a concatenation

$$\omega = \ell_0 \cdot [y_0, y_1] \cdot \ell_1 \cdots [y_{m-1}, y_m] \cdot \ell_m, \tag{P17.1}$$

where $\ell_j \in \mathcal{K}_{A_j}(y_j, y_j)$, for $j = 0, \ldots, m$.
Hint: Consider the sequence of "last times" produced by the loop-erasure algorithm.

(ii) Conversely, show that every choice of loops ℓ_j as in (i) gives rise to a walk ω such that $\mathrm{LE}(\omega) = \eta$.

(iii) With the notation in (17.3), use the decomposition in (i) and show that

$$p(\omega) = p(\eta)p(\ell_0) \cdots p(\ell_m).$$

(iv) Finally, with the notation in (17.6) and (17.11), show that

$$\hat{p}(\eta) = p(\eta) \prod_{j=0}^{m} G_{A_j}(y_j, y_j).$$

17.2 *A formula for the determinant of the reduced walk Laplacian.* The goal in this exercise is to prove Proposition 17.2. Choose an enumeration of A, for instance, $A = \{x_1, \ldots, x_n\}$. Let $F_A(A) = \prod_{j=1}^{n} G_{A_j}(x_j, x_j)$, where $A_j := A \setminus \{x_1, \ldots, x_{j-1}\}$, with the understanding that $A_1 = A$. We want to show that

$$F_A(A) = \det G_A = \frac{1}{\det(I - Q)}. \tag{P17.2}$$

The proof will proceed by induction.

(i) The base case is when $A = \{x\}$ and $Q = [q]$ with $q \in [0, 1]$ being the probability of staying at x. Verify (P17.2) in this case.

Now, assume the statement is true for $n - 1$. Set $A' = \{x_2, \ldots, x_n\}$ and write $F_A(A) = G_A(x_1, x_1)F_{A'}(A')$.

(ii) Consider the function on A defined by $v(\cdot) := G_A(\cdot, x_1)$. Show that, for $y \in A$,

$$v(y) = \sum_{z \in A} Q(y,z)v(z) + \delta_{x_1}(y).$$

In other words, deduce that $(I - Q)v = \delta_{x_1}$.
Hint: Use linearity of expectation as in (5.13) and the Markov property.

Next recall Cramer's rule for solving a system of linear equations of the form $Mx = b$, where $x = [x_1 \cdots x_n]^T$, the matrix M is $n \times n$ and invertible, and b is an $n \times 1$ vector. Then, for $i = 1, \ldots, n$

$$x_i = \frac{\det M_i}{\det M}$$

where M_i is the matrix M where the i-th column is replaced by the non-homogeneous term b.

(iii) Use Cramer's rule to finish the induction.

17.3 *Kirchhoff's Matrix-Tree Theorem.* The goal in this exercise is to prove Proposition 17.3. Let $G = (V, E)$ be a graph with edge weights $\sigma \in \mathbb{R}^E_{>0}$. Write $N = |V|$, and $[N] = \{1, 2, \ldots, N\}$. Consider the combinatorial Laplacian L_σ of G, with eigenvalues $\lambda_1 = 0 < \lambda_2 \leq \cdots \leq \lambda_N$. Consider the elementary symmetric function

$$e_k(L_\sigma) := \sum_{\{i_1,\ldots,i_k\} \in \binom{[N]}{k}} \lambda_{i_1} \cdots \lambda_{i_k}$$

(i) Use $M := xI + L_\sigma$ and the determinant formula

$$\det(M) = \sum_{\sigma \in \Sigma_N} \operatorname{sign}(\omega) M(1, \omega(1)) \cdots M(N, \omega(N))$$

where Σ_N is the set of permutations on $[N]$, to show that

$$e_k(L) = \sum_{S \in \binom{[N]}{k}} \det\left(L_\sigma[S, S]\right).$$

(ii) Prove that for any $k \in [N]$, if $A = [N] \setminus \{k\}$, then

$$\det(L_\sigma[A, A]) = \lambda_2 \cdots \lambda_N.$$

and hence, by Corollary 17.3, show that

$$\sum_{\gamma \in \Gamma_G} \sigma[\gamma] = \frac{1}{N}\lambda_2 \cdots \lambda_N.$$

17.4 *Loop erased walk on the path graph.* Let G_n be the path graph with vertex set $V_n = [n] = \{1, 2, \ldots, n\}$. Let γ be the loop-erased random walk starting at 1 and ending at n. We want to compute $\mu(\gamma)$ using Proposition 17.4.

(i) Use the serial rule to compute $\text{eff}\mathcal{R}(1, n)$.

(ii) For $j = 2, 3, \ldots, n-1$, let $\gamma_{j-1} = \{1, \ldots, j-1\}$. Show that, for $j = 2, \ldots, n-1$,

$$\begin{aligned}\text{Mod}_2(j, \gamma_{j-1} \cup \{n\}; G_n) &= \text{Mod}_2(2, \{1, n-j+2\}; G_{n-j+2}) \\ &= \frac{n-j+1}{n-j}.\end{aligned}$$

where the notation $\text{Mod}_2(a, b; G)$ means 2-modulus from a to b within the graph G, and where a and b are subsets of V.

(ii) Thus, verify that $\mu(\gamma) = 1$.

17.5 *Loop erased walk on the complete graph.* Let $G_n = K_n$ be the complete graph on the vertex set $V = [n] = \{1, \ldots, n\}$. Suppose $\gamma = [1, 2, \ldots, n]$ is a loop-erased random walk starting at 1 and ending at n.

(i) Compute $\text{eff}\mathcal{R}(1, n)$.
Hint: See Exercise 16.3.

(ii) For $j = 2, 3, \ldots, n-1$, let $\gamma_{j-1} = \{1, 2, \ldots, j-1\}$. Show that, for $j = 2, \ldots, n-1$,

$$\begin{aligned}\text{Mod}_2\left(j, \gamma_{j-1} \cup \{n\}; K_n\right) &= \text{Mod}_2\left(j, n; \left(\hat{K}_{n-j+1}, \sigma_j\right)\right) \\ &= \frac{1}{4}\left(-j^2 + (n+2)j + n - 1\right),\end{aligned}$$

where $\left(\hat{K}_{n-j+1}, \sigma_j\right)$ is a weighted complete graph on the nodes $\{j, j+1, \ldots, n\}$ such that $\sigma_j(x, n) = j$, for $x \neq n$, and $\sigma_j = 1$ otherwise.

(iii) Finally, compute $\mu(\gamma)$ as defined in Proposition 17.4, when $n = 3, 4$, and 5.

CHAPTER 18

Spanning Tree Modulus

WE EXPLORE the theory of modulus in the case when the family of objects is the family of all spanning trees of a graph. As we will see, spanning tree modulus incorporates several combinatorial and probabilistic aspects of graphs. We follow [5,7,11,29,64,80,81].

18.1 THE FAMILY OF SPANNING TREES

In this chapter, a graph $G = (V, E)$ will be allowed to have multi-edges. However, if self-loops arise, they will be removed. Write Γ_G for the family of all spanning trees of G. See Figure 18.1 for the example when G is the diamond graph. In this case, there are 8 spanning trees, where $\gamma_1, \gamma_2, \gamma_3, \gamma_4$ don't use the diagonal, and $\gamma_5, \gamma_6, \gamma_7, \gamma_8$ do.

If we enumerate the edges of G, then we can write down the usage matrix $\mathcal{N} = [\mathcal{N}(\gamma_i, e_j)]_{\substack{i=1,\ldots,8\\ j=1,\ldots,5}}$. For a fixed tree γ_j, the usage vector $\mathcal{N}(\gamma_j, \cdot)$ is its indicator function in $\{0,1\}^E \subset \mathbb{R}^E$. It will sometimes be useful to identify this vector with γ_j itself. So we may think of the family Γ_G as a subset of vectors in $\mathbb{R}^E$ and write $\Gamma_G \subset \mathbb{R}^E$.

Recall from Theorem 13.5 that the Fulkerson blocker for the family Γ_G is known. Namely, let Φ be the family of feasible partitions for G. Recall that $P \in \Phi$, if $P = (V_1, \ldots, V_{k_P})$ is a partition of the vertex set V, so that each part V_j induces a connected subgraph $G_j = G[V_j]$. In particular, k_P is the number of parts, and each feasible partition gives rise to a cut set E_P consisting of all the edges that are incident to two distinct parts. With these notations, the Fulkerson blocker $\hat{\Gamma} \subset \mathbb{R}^E$ of

 DOI: 10.1201/9781003024866-23

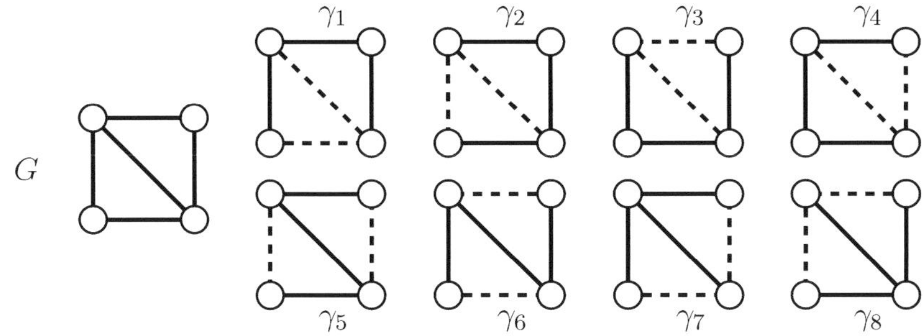

Figure 18.1 Spanning trees for the diamond graph.

Γ_G is the set of usage vectors of the form

$$\phi = \frac{1}{k_P - 1} \mathbb{1}_{E_P}, \tag{18.1}$$

for $P \in \Phi$.

Continuing with our example above, in Figure 18.2, the trivial partition P_1 where every node is a part will correspond to a fractional usage vector, since $k_{P_1} - 1 = 3$. The rest are all partitions with $k_P = 2$, and hence have integer usage.

18.2 p-MODULUS OF SPANNING TREES

We consider the unweighted p-modulus of the family of spanning trees. It turns out that for the spanning tree modulus the only cases that matter are the $p = 1$ and $p = 2$ cases.

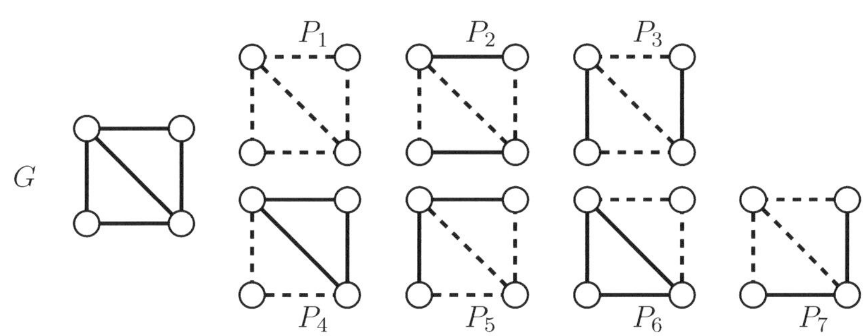

Figure 18.2 Feasible partitions for the diamond graph.

When $p = 1$, the 1-modulus of Γ_G is related to the classical packing problem for spanning trees and to the notion of strength of a graph. Indeed, note that the family Γ_G is a clutter in the sense of Section 13.9. In particular, the dual problem (13.30) shows that $\mathrm{Mod}_1\,\Gamma_G$ is equal to the *Packing Tree* (PT) problem:

$$\text{maximize the number of non-overlapping spanning trees in } G \tag{18.2}$$

On the other hand, if $G = (V, E)$ is a graph and $\sigma \in \mathbb{R}^E_{>0}$ are edge weights, then the *strength* of G is defined as

$$S_\sigma(G) := \min_{P\in\Psi} \frac{\sigma(E_P)}{k_P - 1}, \tag{18.3}$$

where Ψ is the family of all partitions of V. If G is unweighted then we simply write

$$S(G) := \min_{P\in\Psi} \frac{|E_P|}{k_P - 1}, \tag{18.4}$$

The intuition is that each partition determines a cutset E_P and the number of parts k_P measures the damage of removing the edges in E_P. In other words, the strength tells us that in order to create a certain number of parts, we must cut a definite amount of edges.

Recall, from Proposition 13.5, that

$$\mathrm{Mod}_1\,\Gamma_G = \min_{\hat\gamma\in\hat\Gamma} \mathbf{1}^T\hat\gamma = \min_{P\in\Phi} \frac{|E_P|}{k_P - 1}.$$

One may observe that the family of all partitions Ψ can be replaced by the family of feasible partitions Φ in definitions (18.3) and (18.4). See Exercise 18.2. So, from the point of view of Fulkerson duality, the 1-modulus is also equal to the strength of G.

When $p = \infty$, since all spanning trees have the same number of edges, by Proposition 11.1, we get that

$$\mathrm{Mod}_\infty\,\Gamma_G = \frac{1}{|V| - 1}.$$

Hence, the $p = \infty$ case is not interesting.

Furthermore, as we will see in Section 18.10, the solution for the case when $p \in (1, \infty) \setminus \{2\}$ can be deduced from the $p = 2$ case. Therefore, the case $p = 2$ will constitute the bulk of this chapter.

18.3 THE FAIREST EDGE USAGE PROBLEM

We want to analyze the probabilistic interpretation of the spanning tree modulus. From now on, we will assume that $p = 2$. As we have seen in Theorem 13.1, the probabilistic interpretation of modulus entails the minimization of a norm-like quantity involving the expected edge usage $\mathbb{E}_\mu\left[\mathcal{N}(\underline{\gamma}, e)\right]$ as μ varies over all possible pmfs on Γ_G. Note that since the usage vectors for spanning trees are in $\{0,1\}^E$, the expected edge usages become *edge probabilities* $\mathbb{P}_\mu\left(e \in \underline{\gamma}\right)$. We write

$$\eta_\mu(e) := \mathbb{P}_\mu\left(e \in \underline{\gamma}\right) = \sum_{\gamma \in \Gamma_G} \mu(\gamma)\mathcal{N}(\gamma, e) = (\mu^T \mathcal{N})(e). \tag{18.5}$$

If we use the fact that every spanning tree has exactly $|V| - 1$ edges, we find that

$$\sum_{e \in E} \eta_\mu(e) = \sum_{e \in E} \mathbb{P}_\mu\left(e \in \underline{\gamma}\right) = |V| - 1. \tag{18.6}$$

Indeed, by (18.5),

$$\sum_{e \in E} \eta_\mu(e) = \sum_{e \in E} \sum_{\gamma \in \Gamma_G} \mu(\gamma)\mathcal{N}(\gamma, e) = \sum_{\gamma \in \Gamma_G} \mu(\gamma) \sum_{e \in E} \mathcal{N}(\gamma, e) = |V| - 1.$$

Furthermore, by Corollary 13.1, the 2-modulus of the family of spanning trees involves minimizing the quantity

$$\mu^T \mathcal{N}\mathcal{N}^T \mu = \sum_{e \in E} \mathbb{P}_\mu\left(e \in \underline{\gamma}\right)^2 = \sum_{e \in E} \eta_\mu(e)^2 \tag{18.7}$$

over all possible pmfs $\mu \in \mathcal{P}(\Gamma_G)$.

Thus, we can equivalently minimize the variance

$$\begin{aligned} \operatorname{Var}(\eta_\mu) &:= \frac{1}{|E|} \sum_{e \in E} \left(\eta_\mu(e) - \frac{|V| - 1}{|E|}\right)^2 \\ &= \frac{1}{|E|} \sum_{e \in E} \eta_\mu(e)^2 - \left(\frac{|V| - 1}{|E|}\right)^2 \end{aligned} \tag{18.8}$$

In other words, the spanning tree modulus finds optimal pmfs on Γ_G so that all edges are used the same amount, if possible.

Thus, the *Fairest Edge Usage* problem (FEU) is the following problem:

$$\begin{aligned} \underset{\mu \in \mathbb{R}^{\Gamma_G}}{\text{minimize}} \quad & \operatorname{Var}(\eta_\mu) \\ \text{subject to} \quad & \eta_\mu = \mathcal{N}^T \mu \\ & \mu \geq 0, \text{ and } \mathbf{1}^T \mu = 0. \end{aligned} \tag{18.9}$$

A pmf $\mu \in \mathcal{P}(\Gamma)$ is optimal for FEU if and only if it is optimal for the probabilistic interpretation of modulus. In particular, an optimal pmf always exists, but is not necessarily unique. We will say that an optimal pmf μ for FEU is a *fair pmf*, and every tree in its support, i.e., $\gamma \in \Gamma_G$ such that $\mu(\gamma) > 0$, will be called a *fair tree*. Also, we will write FEU(G) to mean the value of the FEU problem, and we will let Γ_G^f be the family of all fair trees for G.

Observe that the objective function in (18.7) can be rewritten as

$$\begin{aligned} &\mu^T \mathcal{N}\mathcal{N}^T \mu \\ &= \sum_{e \in E} \mathcal{N}^T \mu(e)^2 = \sum_{e \in E} \left(\sum_{\gamma \in \Gamma_G} \mu(\gamma)\mathcal{N}(\gamma, e) \right) \left(\sum_{\gamma' \in \Gamma_G} \mu(\gamma')\mathcal{N}(\gamma', e) \right) \\ &= \sum_{e \in E} \sum_{\gamma, \gamma' \in \Gamma_G} \mu(\gamma)\mu(\gamma')\mathcal{N}(\gamma, e)\mathcal{N}(\gamma', e) \\ &= \sum_{\gamma, \gamma' \in \Gamma_G} \mu(\gamma)\mu(\gamma') \sum_{e \in E} \mathcal{N}(\gamma, e)\mathcal{N}(\gamma', e) = \sum_{\gamma, \gamma' \in \Gamma_G} \mu(\gamma)\mu(\gamma') \left|\gamma \cap \gamma'\right| \\ &= \mathbb{E}_\mu \left|\underline{\gamma} \cap \underline{\gamma}'\right| \end{aligned}$$

Here, $|\gamma \cap \gamma'|$ is the number of edges that are in common between γ and γ'. We interpret $\mathbb{E}_\mu \left|\underline{\gamma} \cap \underline{\gamma}'\right|$ as the *expected overlap* when sampling two independent trees using the pmf μ. In other words, the probabilistic interpretation of modulus is equivalent to the *Minimum Expected Overlap* problem (MEO) of finding an optimal pmf $\mu \in \mathcal{P}(\Gamma_G)$, which minimizes the expected overlap of two independent trees. We will write MEO(G) to mean the value of the MEO problem.

The following theorem connects all four problems: the modulus of spanning trees, the modulus of feasible partitions, the fairest edge usage problem and the minimum expected overlap problem. See also, Corollary 13.2.

Theorem 18.1. *Let $G = (V, E)$ be a graph, let Γ_G be the family of spanning trees of G, and let $\tilde{\Gamma}$ be a Fulkerson dual family for Γ_G, in the sense of Definition 13.4. Furthermore, let $\rho \in \mathbb{R}^E_{\geq 0}$, $\eta \in \mathbb{R}^E_{\geq 0}$ and $\mu \in \mathcal{P}(\Gamma_G)$.*

Then, ρ, η, and μ are optimal, respectively, for $\mathrm{Mod}_2(\Gamma_G)$, $\mathrm{Mod}_2(\tilde{\Gamma})$, and the FEU or MEO problems, if and only if, the following conditions are satisfied:

(i) $\rho \in \mathrm{Adm}(\Gamma_G)$ and $\eta = \mathcal{N}^T\mu$;

(ii) ρ and η are parallel, meaning that $\rho(e) = M\eta(e)$, for all $e \in E$, for some constant $M > 0$;

(iii) $\mu(\gamma)(1 - \ell_\rho(\gamma)) = 0$, for all $\gamma \in \Gamma_G$.

In particular, the constant M in (ii) equals $\mathrm{Mod}_2(\Gamma_G)$ and its reciprocal $1/M$ equal

$$\mathrm{Mod}_2(\tilde{\Gamma}) = \mathrm{MEO}(G) = \mathrm{FEU}(G) \cdot |E| + \left(\frac{|V|-1}{|E|}\right)^2 \tag{18.10}$$

Proof. Assume first that ρ, η, and μ are optimal for $\mathrm{Mod}_2(\Gamma_G)$, $\mathrm{Mod}_2(\tilde{\Gamma})$, and MEO or FEU respectively. Then, Theorem 13.4 and Fulkerson duality imply that ρ and η are parallel with constant $M = \mathrm{Mod}_2(\Gamma_G)$. This establishes (ii). Next, define $\eta_\mu := \mathcal{N}^T\mu$ as in (18.5). Then, $\mu^T\mathcal{N}\mathcal{N}^T\mu = \eta_\mu^T\eta_\mu$ is the minimum expected overlap, and therefore $\rho = \mathrm{Mod}_2(\Gamma_G)\eta_\mu$ by (13.4). In particular, by (ii) with M as above, we have $\eta = \eta_\mu$. Also, since ρ is optimal, we have $\rho \in \mathrm{Adm}\,\Gamma_G$. Thus, (i) holds as well. Finally, (iii) is the complementary slackness condition from Proposition 12.1 for the Lagrange multipliers, translated in terms of μ using the relationship established in the proof of Theorem 13.1.

Conversely, assume ρ, $\eta \in \mathbb{R}^E_{\geq 0}$ and $\mu \in \mathcal{P}(\Gamma_G)$ satisfy conditions (i), (ii), and (iii). Define $\Gamma' := \{\gamma \in \Gamma_G : \mu(\gamma) > 0\}$ and let $\mathcal{N}'$ be the usage matrix for Γ'. To show that ρ is optimal for $\mathrm{Mod}_2\,\Gamma_G$, we use Beurling's Criterion, Theorem 10.2, for the case $p = 2$, applied to the subfamily Γ'. Note that, by (iii), $\ell_\rho(\gamma) = 1$ for all $\gamma \in \Gamma'$ and by (i), $\rho \in \mathrm{Adm}\,\Gamma_G$. Therefore, the only remaining hypothesis we have to check in Theorem 10.2 is 2(b). So, assume $h \in \mathbb{R}^E$ and $\sum_{e \in E} h(e)\mathcal{N}(\gamma, e) \geq 0$ for every

$\gamma \in \Gamma'$. Then, we want to show that $h^T\rho \geq 0$. Indeed,

$$\begin{aligned} h^t\rho &= Mh^T\eta && \text{(by (ii))} \\ &= Mh^T\mathcal{N}^T\mu && \text{(by(i))} \\ &= Mh^T\mathcal{N}'^T\mu && \text{(by definition of } \Gamma'\text{)} \\ &= M\mu^T\mathcal{N}'h \geq 0, \end{aligned}$$

since $\mathcal{N}'h \geq 0$. Thus, we have shown that ρ is the extremal density for $\operatorname{Mod}_2 \Gamma_G$.

Next, let η^* be the extremal density for $\operatorname{Mod}_2 \tilde{\Gamma}$, and $\mu^* \in \mathcal{P}(\Gamma_G)$ an optimal pmf from MEO or FEU. Then, by (13.16), we have $\rho = \operatorname{Mod}_2(\Gamma_G)\eta^* = M\eta$. Hence,

$$\mathcal{N}^T\mu = \frac{\operatorname{Mod}_2 \Gamma_G}{M}\mathcal{N}^T\mu^*. \tag{18.11}$$

In particular, this will imply that $M = \operatorname{Mod}_2 \Gamma_G$. Assume not, then either $M > \operatorname{Mod}_2 \Gamma_G$, or $\operatorname{Mod}_2 \Gamma_G > M$. For simplicity, we introduce the following notation. Given two vectors $u, v \in \mathbb{R}^E$, write $u > v$ to mean that, for every $e \in E$, we have $u(e) \geq v(e)$, but there are edges e' such that $u(e') > v(e')$. Then, assuming that $M \neq \operatorname{Mod}_2 \Gamma_G$, equation (18.11) implies that either $\mathcal{N}^T\mu > \mathcal{N}^T\mu^*$, or $\mathcal{N}^T\mu^* > \mathcal{N}^T\mu$. However, the next result shows that these two alternatives are impossible. Hence, we must have $M = \operatorname{Mod}_2 \Gamma_G$.

Claim 18.1. *Let $G = (V, E)$ be a graph, let Γ_G be its family of spanning trees and let $\mathcal{N}$ be the corresponding usage matrix. Assume $\mu_1, \mu_2 \in \mathcal{P}(\Gamma_G)$. Then, $\mathcal{N}^T\mu_1 > \mathcal{N}^T\mu_2$ is an impossibility.*

Proof of Claim 18.1. Now, assume that $\mathcal{N}^T\mu_1 > \mathcal{N}^T\mu_2$, then

$$\mathbf{1}^T\mathcal{N}^T\mu_1 > \mathbf{1}^T\mathcal{N}^T\mu_2.$$

Equivalently,

$$\mu_1^T(\mathcal{N}\mathbf{1}) > \mu_1^T(\mathcal{N}\mathbf{1}).$$

But, for spanning trees, $\mathcal{N}\mathbf{1} = (|V| - 1)\mathbf{1}$. Thus, simplifying, we have $1 = \mu_1^T\mathbf{1} > \mu_2^T\mathbf{1} = 1$, which is a contradiction. □

Furthermore, since $M = \operatorname{Mod}_2 \Gamma_G$, equation (18.11) implies that $\eta = \eta^*$, and hence is the extremal density for $\operatorname{Mod}_2 \tilde{\Gamma}$. Moreover, by (13.4), μ is optimal for MEO.

Finally, (18.10) follows from (18.8) □

Example 18.1 (The diamond graph). Let us compute the spanning tree modulus in the case of the diamond graph, as in Figure 18.1.

First notice that the uniform distribution (UST) on Γ_G uses the diagonal half of the time, while each edge of the square gets used with probability $\frac{5}{8}$. So the diagonal is, in a sense, under-utilized by UST.

We claim that in this case, it is possible to define a pmf μ^* on Γ_G that uses every edge fairly. In other words, we claim that $\mathrm{FEU}(G) = 0$. To describe μ^*, imagine you duplicate each edge of the square, and instead triplicate the diagonal. Next, consider the uniform distribution on the resulting multigraph $\hat{G}$ and then project it down onto trees for the original graph G. We then get the probabilities listed in Figure 18.3. Note that the original graph G had 8 spanning trees, while the new multigraph has 80 trees, however, each tree for the multigraph corresponds to a unique tree on the original graph. With these new probabilities, the trees that use the diagonal have a bit of extra mass and we get that

$$\mathbb{P}_{\mu^*}\left(\text{diagonal} \in \underline{\gamma}\right) = \frac{12}{20} = \frac{3}{5},$$

while

$$\mathbb{P}_{\mu^*}\left(\text{side} \in \underline{\gamma}\right) = \frac{3}{10} + \frac{6}{20} = \frac{3}{5}.$$

So μ^* is optimal for the FEU problem. Thus, in this case $\eta(e) \equiv \frac{3}{5}$ and $\mathrm{Mod}_2\,\Gamma_G = \left(5\frac{3^2}{5^2}\right)^{-1} = \frac{5}{9}$. So $\rho(e) \equiv \frac{1}{3}$

Example 18.2 (An effective resistance upper bound). If we use the connection between UST and effective resistance described in Theorem 17.3, we obtain a lower bound for spanning tree modulus in terms of the path

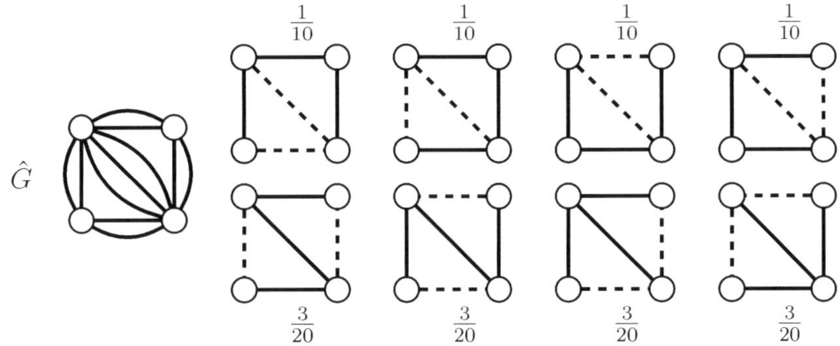

Figure 18.3 An optimal pmf for the diamond graph.

modulus of each edge. Moreover, if we use Proposition 3.2, we get a lower bound in terms of the algebraic connectivity of the graph.

Corollary 18.1. *Let Γ_G be the family of all spanning trees of a given graph G. Then,*

$$\mathrm{Mod}_2(\Gamma_G) \geq \left(\sum_{e=\{x,y\}\in E} \mathrm{Mod}_2(\Gamma(x,y))^{-2} \right)^{-1}$$

In particular, if λ_2 is the algebraic connectivity of G, then

$$\mathrm{Mod}_2(\Gamma_G) \geq \frac{(\lambda_2)^2}{4|E|}.$$

Proof. See Exercise 18.4. □

18.4 HOMOGENEOUS GRAPHS

We say that a graph $G = (V, E)$ is *homogeneous (for spanning tree modulus)* if and only if, $\mathrm{FEU}(G) = 0$, or equivalently, if and only if, there is a pmf μ on the family of spanning trees, for which the edge probabilities are constant, i.e.,

$$\mathbb{P}_\mu\left(e \in \underline{\gamma}\right) \equiv \frac{|V|-1}{|E|}.$$

Although the term homogeneous is overused in mathematics, we will often drop the qualifier 'for spanning tree modulus'.

Homogeneous graphs can be characterized combinatorially. We define the *denseness* of G as

$$\theta(G) := \frac{|E|}{|V|-1}. \tag{18.12}$$

This notion is known in the literature as *fractional arboricity*, or 1-*density*, see for instance [15] and [22].

If $A \subset E$, we interpret A as an edge-induced subgraph of G with $V(A)$ consisting of the nodes incident to edges in A. Then, the *maximum denseness* of G is

$$D(G) := \max_{A\subset E} \theta(A). \tag{18.13}$$

Similarly, if $\sigma \in \mathbb{R}^E_{>0}$ are positive edge weights, then we set

$$\theta_\sigma(G) := \frac{\sigma(E)}{|V|-1}. \quad \text{and} \quad D_\sigma(G) := \max_{A\subset E} \theta_\sigma(A). \tag{18.14}$$

The following results are due to Huy Truong [80].

Theorem 18.2. *Let $G = (V, E)$ be a graph and let Γ_G be its family of spanning trees. Let η^* be the extremal density for the 2-modulus of a Fulkerson dual family of Γ_G, as in Theorem 18.1. Set $\eta_{\max} := \max_{e\in E} \eta^*(e)$ and $\eta_{\min} := \min_{e\in E} \eta(e)$. Then,*

$$\eta_{\max}^{-1} = S(G) \le \theta(G) \le D(G) = \eta_{\min}^{-1},$$

where $S(G)$ is the strength of G as defined in (18.3), $\theta(G)$ is the denseness of G as in (18.12), and $D(G)$ is the maximum denseness of G as in (18.13).

Theorem 18.3. *Let $G = (V, E)$ be a graph. Let $S(G)$ be the strength of G as defined in (18.3) and $D(G)$ is the maximum denseness of G as in (18.13). Then, the following are equivalent:*

(i) G is homogeneous;

(ii) $S(G) = D(G)$;

(iii) $S(G) = \theta(G)$;

(iv) $D(G) = \theta(G)$.

If we compute the spanning tree modulus, we get the optimal density ρ^* for $\mathrm{Mod}_2\,\Gamma_G$. Then, using Theorem 18.1 (ii), we may get the optimal density η^*. As we saw in Theorem 18.3, the values $\eta_{\min}$ and $\eta_{\max}$ recover the strength and the maximum denseness of G. In fact, Huy Truong has shown [80] that every value of η^* has a combinatorial interpretation.

Example 18.3 (The house graph). As an example of a homogeneous graph, we will examine the house graph, see Figure 18.4. In order to get a spanning tree we must break the two cycles S and T. One way to do this is to choose an edge from $T = \{c, r_R, r_L\}$ and one from $E \setminus T = \{p, w_R, w_L\}$. This results in $3 \cdot 3 = 9$ spanning trees, which we have organized into a 'magic square' in Figure 18.5. We will explain the significance of the magic square below. However, there are also two extra trees, which arise when we remove the ceiling c and one of the roof edges r_L or r_R. Namely, in this case, we remove two edges from T, see Figure 18.6. That gives a total of 11 spanning trees for the house graph.

To see that the house graph is homogeneous, consider the pmf μ_1 which is uniform on the 3 trees in the top row of the magic square. A quick count shows that each edge is used 2 out of 3 times. Therefore, the edge probability vector for μ_1 is

$$\eta_{\mu_1}(e) = \mathbb{P}_{\mu_1}\left(e \in \underline{\gamma}\right) \equiv \frac{2}{3}.$$

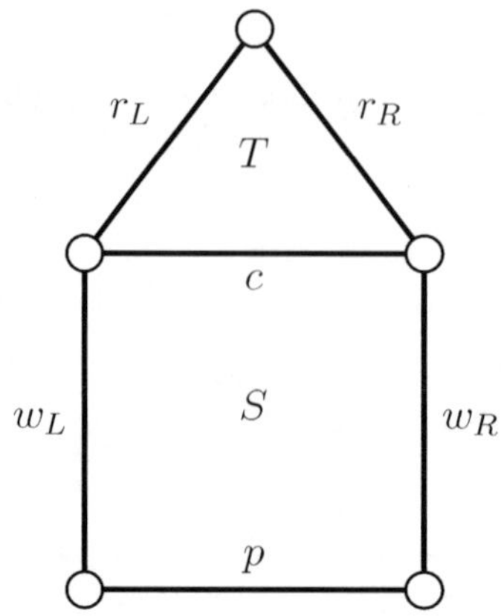

Figure 18.4 The house graph: p is the pavement, c is the ceiling, w_L and w_R are the walls and r_L and r_R form the roof. Also, T is the triangle $\{c, r_R, r_L\}$ and S is the square $\{p, w_R, c, w_L\}$.

This means that the house graph is homogeneous.

Note that we could have taken the uniform distribution restricted to any one of the other rows or columns in the magic square, giving six optimal pmfs $\mu_1, \ldots, \mu_6$ for the FEU problem.

Lemma 18.1. *Let $\mathcal{S} \subset \mathcal{P}(\Gamma_G)$ be the set of optimal solutions for the* FEU *problem in (18.9). Then, $\mathcal{S}$ is a non-empty convex set in $\mathbb{R}^{\Gamma_G}$.*

Proof. The fact that $\mathcal{S}$ is non-empty follows from the probabilistic interpretation of modulus, see, for instance, Theorem 12.2 (i) and Proposition 13.4 (a).

Convexity follows from the fact that if $\mu_t = t\mu_1 + (1-t)\mu_2$, with $t \in (0,1)$, then

$$\eta_{\mu_t}(e) = \mathcal{N}^T \mu_t = t\eta_{\mu_1} + (1-t)\eta_{\mu_2}.$$

Moreover, if μ_1 and μ_2 are optimal, then $\eta_{\mu_1} = \eta_{\mu_2}$, because the edge probabilities solve the dual modulus problem, where we know there is uniqueness of the extremal density. Therefore, $\eta_{\mu_t} = \eta_{\mu_1}$ as well, and so $\mu_t \in \mathcal{S}$. □

Next, we argue that any optimal pmf μ must be supported on the nine trees in the magic square. Indeed, let's add up the edge probabilities

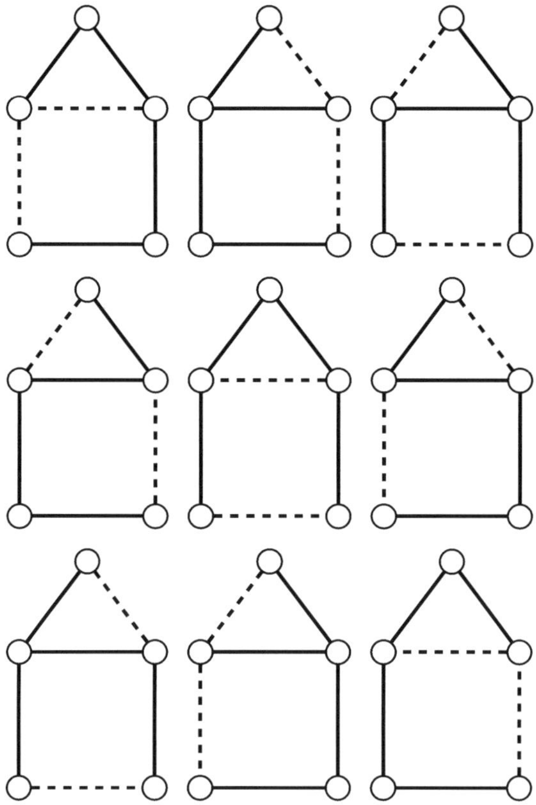

Figure 18.5 Josh Donnoe's magic square.

for the subgraph T in Figure 18.4:

$$\begin{aligned}\sum_{e\in T}\eta_\mu(e) &= \sum_{e\in T}\mathcal{N}^T\mu(e) = \sum_{e\in T}\sum_{\gamma\in\Gamma_G}\mu(\gamma)\mathcal{N}(\gamma,e)\\ &= \sum_{\gamma\in\Gamma_G}\mu(\gamma)\sum_{e\in T}\mathcal{N}(\gamma,e) = \sum_{\gamma\in\Gamma_G}\mu(\gamma)|T\cap\gamma|\\ &= \mathbb{E}_\mu\left|T\cap\underline{\gamma}\right|.\end{aligned}$$

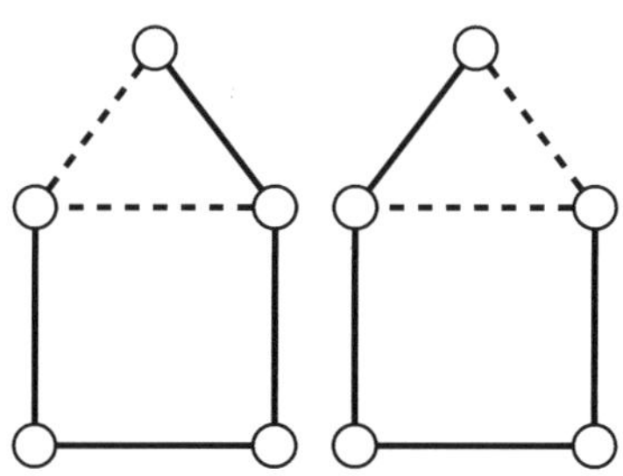

Figure 18.6 Two extra trees.

Since we know that the graph is homogeneous, if μ is optimal for the FEU problem, then, by Theorem 18.2, we must have $\eta_\mu(e) = 2/3 = \theta(G)^{-1}$. On the other hand, since T is a cycle, for any tree $\gamma \in \Gamma_G$, we have $|T \cap \gamma| \le 2$. Therefore, we get that, whenever $\mu(\gamma) > 0$, we must also have

$$|T \cap \gamma| = 2.$$

In other words, the two extra trees in Figure 18.6 are in a sense forbidden: they cannot be in the support of any optimal pmf μ, because they use only one edge from T. The next result is due to Huy Truong, see [80], and is an application of Hall's marriage theorem.

Theorem 18.4 ([80]). *A pmf $\mu \in \mathcal{P}(\Gamma_G)$ is optimal for the* FEU *problem on the house graph if and only if it is in the convex hull of the six pmfs $\mu_1, \ldots, \mu_6$, that are uniform on the rows and columns of the magic square.*

See Example 18.6 below, for the proof in a special case.

18.5 STRICTLY HOMOGENEOUS GRAPHS

We have seen that both the diamond graph in Figure 18.1 and the house graph in Figure 18.4 are homogeneous graphs. However, in the case of the diamond graph, we were able to define an optimal pmf by replacing the edges by an appropriate amount of multiedges and then considering the uniform distribution on the resulting multigraph, and finally restricting it back to the original graph. In order to formalize this property, we consider weighted uniform pmf's. Given positive edgeweights $\sigma \in \mathbb{R}^E_{>0}$, every spanning tree $\gamma \in \Gamma_G$ acquires a weight

$$\sigma[\gamma] := \prod_{e \in E} \sigma(e)^{\mathcal{N}(\gamma,e)}. \tag{18.15}$$

This defines a pmf $\mu_\sigma \in \mathcal{P}(\Gamma_G)$ so that $\mu_\sigma(\gamma)$ is proportional to $\sigma[\gamma]$. We say that μ_σ is the *weighted uniform distribution* UST_σ.

The operation of replacing the edges of a simple graph with multiedges and then restricting the uniform distribution on the multigraph back onto the original graph is equivalent to considering a weighted UST_σ where $\sigma(e)$ is an integer representing the number of multiedges in the resulting multigraph.

Therefore, we say that a graph $G = (V, E)$ is *strictly homogeneous* iff there are positive edge weights $\sigma \in \mathbb{R}^E_{>0}$ so that the corresponding pmf μ_σ has constant edge probabilities.

Next, we look at denseness again. Given a graph $G = (V, E)$, let $\mathcal{H}$ be the set of all vertex-induced subgraphs of G with at least one edge. We say that G is *strictly dense* iff for every subgraph $H \in \mathcal{H}$ with $H \neq G$, we have

$$\theta(H) < \theta(G).$$

Note that the house graph in Figure 18.4 is not strictly homogeneous, because the triangle T is a vertex-induced subgraph with at least one edge and

$$\theta(T) = \frac{2}{3} = \theta(G).$$

The following results are from [11, 64].

Theorem 18.5. *A graph G is strictly dense if and only if it is strictly homogeneous and biconnected.*

Example 18.4. Let G consist of two triangles joined along one vertex to give rise to a bow-tie. Then, G is not biconnected because it has an articulation point. But G is homogeneous because picking UST as the pmf on Γ_G gives rise to a constant edge usage.

In Exercise 18.7, you will show that a strictly dense graph is necessarily biconnected.

The harder implication in Theorem 18.5 is to show that strictly dense implies strictly homogeneous, namely, to come up with a weighted UST_σ which is optimal for FEU. The key to finding these optimal weights σ is to consider a different problem involving the determinant of the Laplacian and entropy maximization, which we describe next.

18.6 MINIMIZING THE DETERMINANT OF THE LAPLACIAN

This section is based on [11]. Let $G = (V, E)$ be a connected graph, possibly with multi-edges, but no self-loops. Suppose $\sigma \in \mathbb{R}^E_{>0}$ are positive edge weights, then the weighted Laplacian L_σ, as defined in (3.14), has eigenvalues

$$\lambda_1 = 0 < \lambda_2 \leq \cdots \leq \lambda_{|V|}.$$

We define the *determinant of the Laplacian* as the product of the non-zero eigenvalues:

$$\det{}' L_\sigma := \prod_{j=2}^{|V|} \lambda_j.$$

By Proposition 17.3, this determinant is proportional to the total mass

$$Z_\sigma := \sum_{\gamma\in\Gamma_G} \sigma[\gamma].$$

The main object of study in [11] is to minimize the determinant of the weighted Laplacian, as in the following *Minimum Laplacian Determinant* problem.

$$\begin{aligned} \underset{\sigma>0}{\text{minimize}} \quad & \sum_{\gamma\in\Gamma} \sigma[\gamma] \\ \text{subject to} \quad & \prod_{e\in E} \sigma(e) = 1, \end{aligned} \tag{18.16}$$

Depending on the graph G this problem splits into three main cases:

(i) The minimum is zero.

(ii) The minimum is bounded away from zero, but is not attained.

(iii) The minimum is bounded away from zero, and there is an optimal set of weights $\sigma^* \in \mathbb{R}^E_{>0}$ that attains the minimum.

The main result in [11] shows that these three cases are determined by the property of homogeneity.

Theorem 18.6. *Let G be a biconnected graph. Then,*

(a) The minimum in (18.16) is attained by an optimal $\sigma^* \in \mathbb{R}^E_{>0}$ and is thus positive if and only if the graph G is strictly homogeneous.

(b) The minimum in (18.16) is positive but not attained if and only if the graph G is homogeneous but not strictly homogeneous.

(c) The minimum in (18.16) is zero if and only if the graph G is not homogeneous.

A key to understanding the Minimum Laplacian Determinant problem (18.16) is its dual formulation. First, we observe that (18.16) becomes a convex optimization problem, if we change variables, by setting $r(e) := \log \sigma(e)$, for all $e \in E$. Using Lagrangian duality, it is possible to show, see [11], that the dual problem for the minimum determinant of the Laplacian in (18.16) is related to the following *Entropy Maximization* problem. For a pmf $\mu \in \mathcal{P}(\Gamma_G)$, define the *Shannon entropy* (or information theoretic entropy) of μ as

$$H(\mu) := -\sum_{\gamma\in\Gamma_G} \mu(\gamma)\log\mu(\gamma).$$

Also, consider the convex cone

$$\mathcal{U} := \left\{ u \in \mathbb{R}^{\Gamma_G}_{\geq 0} : \sum_{\gamma\in\Gamma_G} \mathcal{N}(\gamma, e)u(\gamma) = (|V|-1)u(\Gamma_G), \quad \forall e \in E \right\}.$$

Then, [11, Lemma 3.8] shows that, if $\mathcal{U}$ is trivial, then the infimum in (18.16) is zero. Otherwise, the infimum is equal to $\exp(H(\mu^*))$, where μ^* is the unique maximum-entropy pmf in $\mathcal{U}$.

18.7 MAXIMUM DENSENESS CORES

We are interested in a more precise version of the maximum denseness problem, which we call the *strict denseness* problem. Let $\mathcal{H}(G)$ be the family of all subgraphs $H \subset G$ such that H is:

(i) non-trivial, meaning that it has at least one edge;

(ii) connected;

(iii) vertex-induced.

We say $H \in \mathcal{H}(G)$ solves the strict denseness problem if $\theta(H) = D(G)$ and any proper subgraph H' of H satisfies $\theta(H') < \theta(H)$. In other words, H solves the maximum denseness problem and is a non-trivial vertex-induced connected subgraph that is itself strictly dense. When this happens, we say that H is a *minimal core* for G. Also, we define the *shrunk graph* G/H obtained by identifying every vertex in H to a single vertex v_H, and removing all resulting self-loops, but keeping all resulting multi-edges. The following deflation result is similar to [5, Theorem 5.7] but uses minimal cores instead.

Theorem 18.7. *Let $G = (V, E)$ be a biconnected graph, possibly with multi-edges, but no selfloops. Then, either*

(i) G is strictly dense, or

(ii) G admits a minimal core $H_0 \in \mathcal{H}(G)$, which is itself strictly homogeneous. Moreover, in this case,

(a) H_0 has the fair restriction property *with respect to G, meaning that for every fair tree $\gamma \in \Gamma_G^f$, the restriction $\gamma \cap H_0$ is a fair spanning tree in H_0;*

(b) for any fair tree γ of G, the projection of $\gamma \setminus H_0$ onto the shrunk graph G/H_0 is a fair spanning tree of G/H_0;

(c) fair trees of the core H_0 can be coupled with fair trees of the shrunk graph G/H_0 to produce fair trees of G.

Proof. See Exercise 18.8. □

Note that if G satisfies Theorem 18.7 (ii), then G may be homogeneous or not. However, if G is homogeneous and not strictly homogeneous, then it must admit a proper subgraph with the same denseness. This prompted the following observation by Jason Clemens, see [29].

Corollary 18.2. *If G, as in Theorem 18.7, is homogeneous and the numbers E and $|V|-1$ are relatively prime, then G must be strictly homogeneous.*

Proof. If not, by Theorem 18.7 (ii), G admits a proper subgraph H_0 with the same 1-density. Hence, the fraction $|E(G)|/(|V(G)|-1)$ can be non-trivially simplified to $|E(H_0)|/(|V(H_0)|-1)$. □

Example 18.5. Note that for the house graph G in Example 18.3 we have:

$$\theta(G) = D(G) = \theta(T) = \frac{3}{2},$$

where T is the triangle corresponding to the roof in Figure 18.4. Therefore, the house graph is homogeneous, but not strictly homogeneous, and the roof T is a minimal core for G.

18.8 BEURLING PARTITIONS

Here we follow [81]. Let $G = (V, E)$ be a connected graph. Let Φ be the dual family of all feasible partitions P of G, with usage matrix $\widetilde{\mathcal{N}}$ defined as in (18.1). Let η^* be the optimal density for $\mathrm{Mod}_2(\Phi)$. A feasible partition P of G is said to be a *Beurling partition* if it has minimal usage with respect to η^*, equivalently,

$$\sum_{e\in E} \widetilde{\mathcal{N}}(P, e)\eta^*(e) = 1.$$

In other words, P is a Beurling partition if and only if $\eta^*(E_P) = k_P - 1$.

If $P = \{V_1, V_2, ..., V_{k_P}\}$ is a feasible partition, then the *shrunk graph* G^P of the partition P is obtained by identifying all the nodes in V_i as a single node v_i and removing all resulting self-loops. The graph G^P has k_P nodes and the edge set E_P. Since G is connected, the shrunk graph G^P is also connected. Furthermore, every spanning tree γ of G must connect all the subsets V_i. Therefore, the restriction of γ onto E_P forms a connected spanning subgraph of G^P.

Finally, define the *weight* of a partition P as follows,

$$w(P) = \frac{|E_P|}{k_P - 1}. \tag{18.17}$$

Then, we have the following *serial rule for spanning tree modulus*, see [81, Theorem 1.15].

Theorem 18.8. *Let $G = (V, E)$ be a connected graph. Let Γ_G be the spanning tree family of G. Let Φ be the family of all feasible partitions P of G, with usage vectors defined as in (18.1). Let η^* be the optimal density for* $\mathrm{Mod}_2(\Phi)$. *Let $P = \{V_1, V_2, ..., V_{k_P}\}$ be a Beurling partition of G. Let $G_i = (V_i, E_i)$ be the subgraph induced by V_i for $i = 1, 2, \ldots, k_P$ and let G^P be the shrunk graph of the partition P. Then,*

(i) The minimum expected overlap problem MEO *splits as follows:*

$$\mathrm{MEO}(G) = \mathrm{MEO}(G_1) + \cdots + \mathrm{MEO}(G_{k_P}) + \mathrm{MEO}\left(G^P\right);$$

(ii) A pmf $\mu \in \mathcal{P}(\Gamma_G)$ is optimal for $\mathrm{MEO}(G)$ *if and only if its marginal pmfs $\mu_i \in \mathcal{P}(\Gamma_{G_i})$, $i = 1, 2, \ldots, k_P$ are optimal for* $\mathrm{MEO}(G_i)$ *respectively and its marginal pmf $\mu_P \in \mathcal{P}(\Gamma_{G^P})$ is optimal for* $\mathrm{MEO}(G^P)$ *(where we have identified E_P with $E(G^P)$);*

(iii) Conversely, given pmfs $\nu_i \in \mathcal{P}(\Gamma_{G_i})$ that are optimal for $\mathrm{MEO}(G_i)$ *for $i = 1, 2, \ldots, k_P$ and given a pmf ν_P that is optimal for* $\mathrm{MEO}(G^P)$, *define a product measure in $\mathcal{P}(\Gamma)$ as follows,*

$$(\nu_1 \oplus \nu_2 \oplus \cdots \oplus \nu_k)(\gamma) := \nu_1(\zeta_1)\nu_2(\zeta_2)\ldots\nu_k(\zeta_k),$$

for all $\gamma = \bigcup_{i=1}^{k} \zeta_i$ where $\zeta_i \in \Gamma_i$, $i = 1, 2, \ldots, k$. Then, $\nu_1 \oplus \nu_2 \oplus + \cdots \oplus \nu_{k_P} \oplus \nu_P$ is an optimal pmf in $\mathcal{P}(\Gamma_G)$ for $\mathrm{MEO}(G)$;

(iv) The restriction of η^* onto E_i is optimal for $\mathrm{Mod}_2(\widehat{\Gamma_{G_i}})$ *for $i = 1, 2, \ldots, k_P$ and the restriction of η^* onto E_P is optimal for* $\mathrm{Mod}_2(\widehat{\Gamma_{G_P}})$.

18.9 DEFLATION OF INHOMOGENEOUS GRAPHS

There are two main deflation processes that can be performed on an inhomogeneous graph. Both depend on the solution of the spanning tree modulus problem. For more information see [5,7,81]. The optimal density η^* for $\text{Mod}_2\,\Phi$ can be used to compute two special Beurling partitions.

First, let

$$E_{max} := \left\{ e \in E : \eta^*(e) = \max_{e' \in E} \eta^*(e') =: \eta^*_{max} \right\}.$$

Then, there exists a Beurling partition P_{max} such that $E_{P_{max}} = E_{max}$. Moreover, P_{max} is a critical partition for the strength problem and

$$\eta^*_{max} = \frac{1}{w(P_{max})} = \frac{1}{S(G)}. \tag{18.18}$$

Since η^* is constant in $E_{P_{max}}$, by Theorem 18.8, the restriction of η^* on $E_{P_{max}}$ is optimal for $\text{Mod}_2(\widehat{\Gamma}_{G_{P_{max}}})$. Therefore, the shrunk graph $G^{P_{max}}$ is homogeneous.

Assume that

$$P_{max} = \{V_1, V_2, ..., V_{k_{P_{max}}}\}$$

Let $G_i = (V_i, E_i)$ be the subgraph induced by V_i for $i = 1, 2, \ldots, k_{P_{max}}$. Then, we have

$$\text{MEO}\left(G^{P_{max}}\right) = \frac{(k_{P_{max}} - 1)^2}{|E_{P_{max}}|},$$

and,

$$\text{MEO}(G) = \text{MEO}(G_1) + \cdots + \text{MEO}(G_{k_{P_{max}}}) + \frac{(k_{P_{max}} - 1)^2}{|E_{P_{max}}|}.$$

Iterating this procedure on each G_i gives a decomposition of the graph G into a sequence of shrunk graphs and subgraphs, eventually terminating when all final subgraphs are homogeneous and can not be decomposed further. By tracing back through the deflation process, we can construct an optimal pmf for $\text{MEO}(G)$ using optimal pmfs for the intermediate shrunk graphs.

Secondly, let

$$E_{min} := \left\{ e \in E : \eta^*(e) = \min_{e' \in E} \eta^*(e') =: \eta^*_{min} \right\},$$

and let H_{min} be the subgraph induced by E_{min}. Next, pick a connected component $H = (V_H, E_H)$ of H_{min}. Let G/H be the shrunk graph of the partition P_H which is formed by V_H and all singletons v where $v \notin V_H$. Then, it can be shown that P_H is a Beurling partition and we have

$$\text{MEO}(H) = \frac{(|V_H| - 1)^2}{|E_H|},$$

and

$$\text{MEO}(G) = \text{MEO}(G/H) + \frac{(|V_H| - 1)^2}{|E_H|}. \tag{18.19}$$

Iterating this procedure gives a decomposition of G into a sequence of shrunk graphs, eventually terminating when a final homogeneous graph is shrunk to a single vertex. By tracing back through the deflation process, one can build an optimal pmf for MEO(G) using optimal pmfs for the intermediate steps.

Example 18.6. Recall the six pmfs $\mu_1, \ldots, \mu_6$ that are uniform on the rows and columns of the magic square in Figure 18.5. As we have seen, if $\mathcal{S}$ is the set of optimal pmf's, then $\mu_1, \ldots, \mu_6 \in \mathcal{S}$. Therefore, by Lemma 18.1, we have $\text{co}(\mu_1, \ldots, \mu_6) \subset \mathcal{S}$. Theorem 18.4 claims that conversely, $\mathcal{S} \subset \text{co}(\mu_1, \ldots, \mu_6)$. Here, we examine the case when an optimal $\mu \in \mathcal{S}$ is supported on 3 trees. Then, μ must be one of $\mu_1, \ldots, \mu_6$. Indeed, the rows and columns of the magic square are the only triples of trees with the property that when we restrict them to the roof T or the quotient graph G/T, we get all the trees of the triangle. This can be seen by the following counting argument. To build a triple of trees with this property, we first chose a tree from the roof and then a tree from the quotient, that's $3 \cdot 3 = 9$ choices. Next, we pick a tree from the roof that hasn't been chosen yet, and likewise from the quotient graph. That's $2 \cdot 2 = 4$ choices for a total of 36 choices. However, the triple of trees we get should be unordered, so we may divide by 3! and get 6 different such triples of trees.

Now, we use the fact that the uniform distribution is the unique optimal pmf for the triangle, see Exercise 18.3. By Theorem 18.8 (ii), the marginals of μ after restricting to the roof T and the quotient graph G/T must both be uniform. However, if μ is supported on a triple of trees that does not restrict to all the trees from the roof, for instance, then, it cannot be uniform.

18.10 OTHER VALUES OF p

Here, we use the deflation process described in the previous section in the context of 2-modulus to compute $\mathrm{Mod}_p \Gamma_G$, for $p \in (1,\infty) \setminus \{2\}$. We thank Huy Truong for valuable input in writing up this result.

Theorem 18.9. *Let $G = (V,E)$ be a graph and let $H_1, H_2, \ldots, H_k$ be the sequence of densest subgraphs that appear in the deflation process for 2-modulus, when using the minimum of η^*, as in (18.19). Let $p \in (1,\infty) \setminus \{2\}$ and let q be the Hölder conjugate exponent with $(p-1)(q-1) = 1$. Set*

$$\rho := \frac{(\eta^*)^{q-1}}{\mathcal{E}_q(\eta^*)}. \tag{18.20}$$

Then, $\rho \in \mathrm{Adm}\,\Gamma_G$ and is extremal for p-modulus. In particular,

$$\mathrm{Mod}_p \Gamma_G = \left(\sum_{j=1}^{k} \frac{(|V_{H_j}| - 1)^q}{|E_{H_j}|^{q-1}} \right)^{1-p}. \tag{18.21}$$

Proof. Let $H_1, H_2, \ldots, H_k$ be the sequence of subgraphs arising from the deflation process. Note that η^* is constant on each H_j and equals $\frac{|V_{H_j}|-1}{|E_{H_j}|}$. However, η^* may equal $\eta^*_{\min}$ on the first several components $H_1, \ldots, H_{j_1}$. In particular, by (18.19),

$$(\mathrm{Mod}_2 \Gamma_G)^{-1} = \sum_{j=1}^{k} \frac{(|V_{H_j}| - 1)^2}{|E_{H_j}|}.$$

Let p and q be conjugate exponents. Then, the q-energy of η^* is:

$$\mathcal{E}_q(\eta^*) = \sum_{j=1}^{k} |E_{H_j}| \left(\frac{|V_{H_j}| - 1}{|E_{H_j}|} \right)^q \tag{18.22}$$

Define a density ρ by the formula (18.20). We want to check that $\rho \in \mathrm{Adm}\,\Gamma_G$. To that end, let $\gamma \in \Gamma_G$ satisfy

$$\ell_\rho(\gamma) = \min_{\gamma' \in \Gamma_G} \ell_\rho(\gamma'). \tag{18.23}$$

Assume that $|\gamma \cap H_1| < |V_{H_1}| - 1$. Then, since H_1 is connected, $\gamma \cap H_1$ is a forest with at least two components. Let $e_+ = \{x, y\}$ be an edge in $H_1 \setminus \gamma$, so that x and y belong to distinct components of $\gamma \cap H_1$. Since γ is a tree, it contains a path p connecting x and y. Also, since x and

y belong to distinct components of $\gamma \cap H_1$, it must be the case that the path p visits an edge e_- in $E \setminus E_{H_{\min}}$. Next, we swap e_+ and e_- to get a new spanning tree $\gamma' = (\gamma \cup \{e_+\}) \setminus \{e_-\}$. However, since $q - 1 > 0$,

$$\rho(e_-) = \frac{(\eta^*(e_-))^{q-1}}{\mathcal{E}_q(\eta^*)} > \frac{(\eta^*(e_+))^{q-1}}{\mathcal{E}_q(\eta^*)} = \rho(e_+).$$

Therefore,

$$\ell_\rho(\gamma') = \ell_\rho(\gamma) + \rho(e_+) - \rho(e_-) < \ell_\rho(\gamma),$$

but this contradicts (18.23). This means that $\gamma \cap H_1$ is a spanning tree of H_1 and, in particular, is a subtree of γ. So, if we shrink H_1, then $\gamma \setminus H_1$ becomes a spanning tree for the shrunk graph G/H_1. Moreover, since η^* does not change when we shrink the graph, we can repeat the argument. By induction, we get that every tree $\gamma \in \Gamma_G$ uses at least $|V_{H_j}| - 1$ edges in H_j. Therefore,

$$\begin{aligned}
\ell_\rho(\gamma) &\geq \frac{1}{\mathcal{E}_q(\eta^*)} \sum_{j=1}^{k} (|V_{H_j}| - 1) \left(\frac{|V_{H_j}| - 1}{|E_{H_j}|} \right)^{q-1} \\
&= \frac{1}{\mathcal{E}_q(\eta^*)} \sum_{j=1}^{k} |E_{H_j}| \left(\frac{|V_{H_j}| - 1}{|E_{H_j}|} \right)^{q} \\
&= 1,
\end{aligned}$$

where the last equality follows from (18.22).

In particular, since $\rho \in \operatorname{Adm} \Gamma_G$ and $\eta^* \in \operatorname{Adm} \widehat{\Gamma_G}$, where $\widehat{\Gamma_G}$ is the Fulkerson blocker family for Γ_G, we have

$$\begin{aligned}
\operatorname{Mod}_p \Gamma_G &\leq \mathcal{E}_p(\rho) \\
&= \sum_{e \in E} \rho(e)^p = \frac{1}{\mathcal{E}_q(\eta^*)^p} \sum_{e \in E} (\eta^*(e))^{qp-p} \\
&= \frac{1}{\mathcal{E}_q(\eta^*)^p} \sum_{e \in E} (\eta^*(e))^q = \mathcal{E}_q(\eta^*)^{1-p}. \\
&\leq \left(\operatorname{Mod}_q \widehat{\Gamma_G} \right)^{1-p}
\end{aligned} \tag{18.24}$$

But, Fulkerson duality for p-modulus, Theorem 13.4, says that

$$\left(\operatorname{Mod}_q \widehat{\Gamma_G} \right)^{1-p} = (\operatorname{Mod}_p \Gamma_G)^{-(q-pq)/p} = \operatorname{Mod}_p \Gamma_G.$$

Therefore, equality holds in (18.24) and we have shown that ρ is extremal for $\operatorname{Mod}_p \Gamma_G$. In particular, we have the following formula (18.21). □

EXERCISES

18.1 *Spanning tree modulus on the Paw graph.* Let G be the graph of Figure 18.7 and let Γ_G be the set of spanning trees, with usage function $\mathcal{N}(\gamma, e)$ equaling 1 if e is included in the tree γ and 0 otherwise.

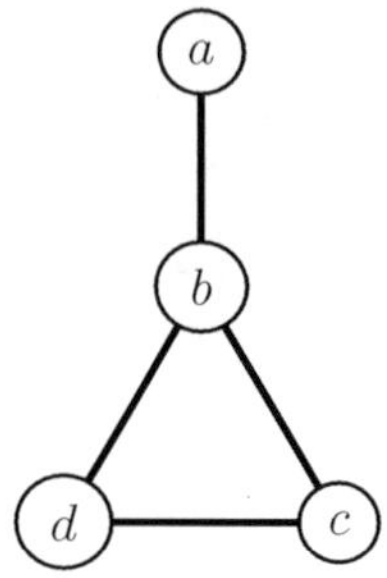

Figure 18.7 The Paw graph.

(i) Identify the spanning trees for G, enumerate the edges and Γ_G and write down the usage matrix $\mathcal{N}$ for Γ_G. Then, compute the overlap matrix $C = \mathcal{N}\mathcal{N}^T$. In particular, note that C does not depend on the enumeration of Γ.

(ii) Let ρ be the function taking the value $3/7$ on edge ab and $2/7$ elsewhere. Show that ρ is admissible for Γ_G.

(iii) Let μ be the uniform pmf on Γ_G. Compute the expected overlap $\mu^T C \mu$ and the edge probabilities $\eta = \mathcal{N}^T \mu$.

(iv) Show that ρ and μ are optimal and compute $\text{Mod}_2\, \Gamma_G$, first using Corollary 13.2, and then Theorem 18.1.

18.2 *Partitions vs. feasible partitions* Show that the family of all partitions Ψ can be replaced by the family of feasible partitions Φ in definition (18.3).
Hint: Given an arbitrary partition P, create a finer partition P' that is feasible.

18.3 UST *is the unique optimal pmf for the triangle*

Let T be the complete graph on 3 nodes.

(i) Use the UST to show that T is homogeneous.

(ii) Show that a pmf μ on Γ_T that is optimal for the FEU problem must be supported on all the trees of T, and finally show that we must have $\mu = \text{UST}$.

18.4 *A lower bounds for spanning tree modulus*

Let Γ_G be the family of spanning trees on a graph G. Let μ be the uniform distribution on Γ_G. Use Corollary 13.1, Theorem 17.3, and Proposition 3.2 to give a proof of Corollary 18.1.

18.5 *Second-order edge probabilities.*

Let Γ_G be the family of spanning trees of a graph G and let $\mu \in \mathcal{P}(\Gamma_G)$ be a pmf on Γ_G. Define the *second-order edge probabilities* for edges $e, e' \in E$,

$$\eta_\mu(e, e') := \mathbb{P}_\mu(e, e' \in \underline{\gamma}) = \sum_{\gamma \in \Gamma} \mu(\gamma)\mathcal{N}(\gamma, e)\mathcal{N}(\gamma, e').$$

(i) Show that the diagonal elements $\eta_\mu(e, e)$ recover the edge probability $\eta_\mu(e)$ defined in (18.5).

(ii) Show that the row sums are

$$\sum_{e' \in E} \eta_\mu(e, e') = (N - 1)\eta_\mu(e).$$

(iii) Consider the symmetric *association matrix*

$$S_\mu(e, e') := \eta_\mu(e, e') - \eta_\mu(e)\eta_\mu(e').$$

Show that the diagonal elements of S are non-negative.

(iv) Show that the row sums of the matrix S are always zero.

(v) Now assume that $\mu = \mu_\sigma$ is a σ-uniform measure. The goal will be to show that in this case the off-diagonal terms of the matrix S in (iii) are non-positive. This will follow from a known fact about uniform spanning trees known as *negative association*, namely, that if $e \neq e'$, then

$$\mathbb{P}_{\mu_\sigma}(e \in \underline{\gamma} | e' \in \underline{\gamma}) \leq \mathbb{P}_{\mu_\sigma}(e \in \underline{\gamma}). \qquad \text{(P18.1)}$$

The goal in this part is to prove (P18.1).
Recall that $\mathbb{P}_{\mu_\sigma}(e \in \underline{\gamma}) = \sigma(e)\mathcal{R}_\sigma(e)$. Now let H be the graph obtained from G by contracting the edge e'. Note that the spanning trees of H are in one-to-one correspondence with the spanning trees of G that contain e'. Deduce a formula for $\mathbb{P}_{\mu_\sigma}(e \in \underline{\gamma} | e' \in \underline{\gamma})$ in terms of effective resistance on H. Then, use Rayleigh monotonicity (see Exercise 2.8) to conclude.
Hint: H can be thought as G where $\sigma(e')$ is set to infinity.

In particular, in this case, the matrix S_{μ_σ} can be thought of as a Laplacian matrix on the graph $K_{|E|}$ with edge weights $\tilde{\sigma}(e, e') := -S(e, e')$.

18.6 *Negative association fails for fair trees.*

Do fair trees also satisfy the negative association property? The answer is "in general no", as the following example shows. Consider the diamond graph G in Figure 18.1. Note that G has 8 spanning trees, 4 of which contain the diagonal, and 6 of which are path graphs. The graph G is strictly dense and hence strictly homogeneous, so it admits a σ-uniform fair pmf. By Exercise 18.5 (v), this fair pmf does satisfy the negative association property.

(i) Consider a pmf μ, so that $\mu = \frac{1}{10}$ on each of the four trees that do not use the diagonal, and let $\mu = \frac{3}{10}$ on the two trees that are paths and do contain the diagonal. Show that μ is fair.

(ii) Enumerate the edges $e_1, \ldots, e_5$ so that e_5 is the diagonal, and e_1 and e_3 are opposite edges. Show that e_1 and e_3 are positively correlated.

18.7 *Strictly dense graphs are biconnected.*

First note that, if G is strictly dense, then Theorem 18.5 implies that G is homogeneous.

Next, suppose that G is not biconnected. Then, there exists a decomposition of G into two subgraphs $H_i = (V_i, E_i) \in \S_e(G)$ for $i = 1, 2$ with the property that $|V_1 \cap V_2| = 1$. That is, the graphs H_1 and H_2 are edge-disjoint and meet at a single articulation vertex.

(i) Show that each subgraph H_i, $i = 1, 2$, is homogeneous.

(ii) Show that $\theta(G)$ is a convex combination of the $\theta(H_i)$, $i = 1, 2$.

(iii) Use the fact that G is homogeneous to reach a contradiction.

18.8 *The deflation process using minimal cores.*

The goal in this exercise is to prove Theorem 18.7.

(i) Assume that G is not strictly dense. Show that G contains a proper vertex-induced subgraph H_0, such that $\theta(H_0) = D(G)$, and so that H_0 is strictly dense. In particular, by Theorem 18.5, H_0 is necessarily homogeneous and biconnected, and, $\theta(H_0) \geq \theta(G)$.

(ii) To prove (a) in Theorem 18.7 (ii), assume that there is a fair tree $\gamma_0 \in \Gamma_G^f$, such that $\gamma_0 \cap H_0$ is not a spanning tree of H_0 and get a contradiction.
Hint: Use the observation that if η^* is the extremal density for FEU, *then*

$$\min_{e \in H_0} \eta^*(e) \leq \frac{1}{|E(H_0)|} \sum_{e \in H_0} \eta^*(e).$$

(iii) Finally, show that parts (b) and (c) of Theorem 18.7 (ii) follow as in [5, Theorem 5.5, Lemma 5.6 and Theorem 5.7].

VI

Algorithms for Modulus

CHAPTER 19

Algorithms for Modulus

SINCE THE MODULUS FOR a finite family of objects, Γ, takes the form of a standard convex optimization problem, there is a variety of algorithms that can be used for its numerical computation. As yet, however, there is no comprehensive theory determining definitively which algorithm is most efficient for computing modulus; in practice, the performance of any algorithm depends on a number of factors. The following is an abbreviated list of some of these factors and their implications.

- **The value of p.** When $p = 2$, the modulus is a quadratic program; the Goldfarb-Idnani algorithm (see Section 19.2) tends to perform well for these problems. When $p \in \{1, \infty\}$, the modulus problem can be cast as a linear program, which can be treated by specialized linear program solvers. For other values of p, the modulus is a general nonlinear convex optimization problem; less is known about these problems, but the *basic algorithm* described in Section 19.1 seems to perform reasonably well in many cases.
- **The size of Γ.** For relatively small families (e.g., $|\Gamma| = O(|E|^p)$ for small p) it may be possible to generate and store the entire usage matrix, $\mathcal{N}$. For example, if Γ is the set of stars (all edges incident on a single vertex) or triangles, then $|\Gamma|$ is of order $O(|E|^{1/2})$ and $O(|E|^{3/2})$ respectively. For these simple families, $\mathcal{N}$ is a relatively small (and sparse) matrix. A good option for computing modulus in these cases is to generate the usage matrix and use a standard convex optimization solver. For other families, though, this approach is not feasible. The family of spanning trees provides a prototypical example. Cayley's formula states that the number of spanning trees on the complete graph, K_n, is n^{n-2}. So K_{60}, for

DOI: 10.1201/9781003024866-25

example, which has 1770 edges, has over 10^{103} spanning trees. It would be impossible to enumerate, let alone store all of the rows of $\mathcal{N}$. For such families, one must use methods that do not require the explicit construction of the usage matrix.

- **The existence of an efficient violated-constraint search.** When the family, Γ, is too large to allow the explicit calculation of the usage matrix, numerical methods often rely on a subroutine (an "oracle") to generate a violated constraint for a given approximation of ρ^*. Both the basic algorithm and the Goldfarb-Idnani implementation for modulus make repeated calls to such an oracle; their efficiency depends strongly on the existence of an efficient oracle, which depends, in turn, on the family, Γ. For *st*-paths or spanning trees, for example, the oracle can be constructed using Dijkstra's or Kruskal's algorithm respectively.

- **The structure of Γ.** Certain families of objects have special structural properties that can be exploited in the development of algorithms. For example, the fact that the spanning trees of a graph form a matroid allows the use of specialized algorithms like Cunningham's algorithm discussed in Section 19.4.

19.1 A BASIC ALGORITHM

The *basic algorithm* for modulus [14] is relatively straightforward to implement. Pseudocode for this method can be found in Algorithm 3. A Python implementation of this algorithm along with a step-by-step demonstration can be found in [12].

Inputs. The inputs to the basic algorithm are the family, Γ, the exponent, $1 \le p \le \infty$, and a small tolerance $0 < \epsilon_{\text{tol}} < 1$. In fact, as described below, Γ is not passed explicitly to the algorithm. Instead, it is given implicitly through a violated constraint oracle.

Initialization. The first two lines of Algorithm 3 initialize two variables. The first, a vector $\rho \in \mathbb{R}^E$, is initialized to zero. Over time, ρ will be repeatedly modified to approximate the optimal density, ρ^*. The second initialized variable is a subset $\Gamma' \subseteq \Gamma$, which is initialized as the empty set. This subset will grow over time as we discover important constraints for the modulus problem.

Iteration loop. Lines 3–10 form the iteration loop of the basic algorithm. The initialization step established some important *loop invariants*

Algorithm 3 Approximates $\text{Mod}_{p,\sigma}(\Gamma)$ with an error tolerance of $0 < \epsilon_{\text{tol}} < 1$.

1: $\rho \leftarrow 0$
2: $\Gamma' \leftarrow \varnothing$
3: **loop**
4: $\quad \gamma \leftarrow \texttt{shortest}_\Gamma(\rho)$ {violated constraint oracle}
5: $\quad$ **if** $\ell_\rho(\gamma) \geq 1 - \epsilon_{\text{tol}}$ **then**
6: $\quad\quad$ `stop`
7: $\quad$ **end if**
8: $\quad \Gamma' \leftarrow \Gamma' \cup \{\gamma\}$
9: $\quad \rho \leftarrow \arg\min\{\mathcal{E}_{p,\sigma}(\rho) : \rho \in \text{Adm}\,\Gamma'\}$ {solution update}
10: **end loop**

that can be proved to remain true throughout all iterations of the loop. Those invariants are

I1. $\Gamma' \subseteq \Gamma$, and

I2. $\rho = \arg\min\{\mathcal{E}_{p,\sigma}(\rho) : \rho \in \text{Adm}\,\Gamma'\}$.

In words, throughout all iterations of the basic algorithm, Γ' remains a subset of the family Γ and ρ remains optimal for the modulus of the subfamily Γ'. (See Exercise 19.2.)

Violated constraint oracle. Line 4 checks the admissibility of the current guess for the density ρ by finding $\gamma \in \Gamma$ that minimizes the ρ-length, $\ell_\rho(\gamma)$. This is where the oracle described above comes in. Algorithm 3 relies on an external subroutine, denoted $\texttt{shortest}_\Gamma$, with the following behavior. Given a density $\rho \in \mathbb{R}^E_{\geq 0}$, $\texttt{shortest}_\Gamma(\rho)$ returns an object $\gamma \in \Gamma$ satisfying

$$\texttt{shortest}_\Gamma(\rho) = \arg\min\left\{\sum_{e\in E} \mathcal{N}(\gamma, e)\rho(e) : \gamma \in \Gamma\right\}.$$

The oracle must be supplied by the user and is the only way in which the basic algorithm "knows" about the family Γ. Ties can be broken arbitrarily; if there are multiple shortest objects, the oracle only needs to return one of them. Implementation specifics for the oracle depend strongly on Γ; examples include Dijkstra's algorithm for the family of *st*-paths and Kruskal's algorithm for the family of spanning trees. Because of this dependence on the oracle, the basic algorithm can only be efficient if there is an efficient oracle for Γ.

Stopping criterion. After the object, γ, with the shortest ρ-length is found, the ρ-length of this object is compared with 1. If $\ell_\rho(\gamma) \geq 1-\epsilon_{\text{tol}}$, the algorithm terminates. Because of the way γ is chosen, this implies that $\ell_\rho(\gamma') \geq 1-\epsilon_{\text{tol}}$ for every $\gamma' \in \Gamma$. Intuitively, if we stop the loop when ρ is "nearly admissible" then ρ should be "nearly optimal." This idea is formalized in Theorem 19.1.

Update. If the loop does not terminate on the current pass, then the variables are updated in Lines 8 and 9. The first line simply adds the object γ to the growing family Γ'. The second performs an update of ρ to ensure that the I2 loop invariant remains true. The exact details of the implementation of this update step are not specified by the algorithm. Essentially, we need to move from an optimal density, ρ, for the previous subfamily, Γ', to an optimal density for the new subfamily with one additional object, $\Gamma' \cup \{\gamma\}$. A simple implementation of this idea is to keep track of the usage matrix, $\mathcal{N}'$, of Γ'. On each pass through the loop, adding γ to Γ' is equivalent to adding a new row to $\mathcal{N}'$; then the optimal ρ can be found using a convex optimization solver such as CVXPY [1,33] or MOSEK [16]. Although this idea works, it is inefficient; on each pass through the loop all prior knowledge about the optimal density is thrown out, and we solve a new (and slightly larger) optimization problem from scratch. A better approach would be to reuse our past work to update ρ. After all, from iteration to iteration all that changes is the addition of a single constraint. Efficient update for general modulus problems is an area of ongoing research. However, for the special case $p = 2$, a method due to Goldfarb and Idnani [44] provides an efficient update by using Lagrangian duality. This method is described in Section 19.2.

Theorem 19.1. *Suppose $G = (V, E, \sigma)$ is a graph and Γ is a finite family of objects on G as in Corollary 10.1 and let $1 \leq p \leq \infty$. Then, given a tolerance $0 < \epsilon_{\text{tol}} < 1$, Algorithm 3 terminates with a density, ρ, and a subfamily, $\Gamma' \subset \Gamma$, with the following properties.*

(a) The density ρ is almost admissible in the sense that

$$(1-\epsilon_{\text{tol}})^{-1}\rho \in \operatorname{Adm}(\Gamma).$$

(b) The relative error is controlled:

$$0 \leq \frac{\operatorname{Mod}_{p,\sigma}(\Gamma) - \operatorname{Mod}_{p,\sigma}(\Gamma')}{\operatorname{Mod}_{p,\sigma}(\Gamma)} \leq \begin{cases} p\epsilon_{tol} & \text{if } 1 \leq p < \infty, \\ \epsilon_{tol} & \text{if } p = \infty. \end{cases}$$

Proof. See Exercise 19.3. □

Remark 19.1. When $1 < p < \infty$ (so that a unique optimal density, ρ^*, exists), Theorem 19.1 can be combined with Lemma 11.3 to show that the basic algorithm terminates with ρ close to ρ^*.

19.2 THE GOLDFARB-IDNANI APPROACH

The Goldfarb-Idnani algorithm [44] is a dual-based method for solving quadratic programs in general. Here, we give an overview of the algorithm in the special case that arises in modulus,

$$\begin{aligned} \text{minimize} \quad & \rho^T \Sigma \rho \\ \text{subject to} \quad & \mathcal{N}\rho \geq \mathbf{1}, \end{aligned}$$

where $\mathcal{N}$ is the nonnegative usage matrix and $\Sigma = \text{diag}(\sigma)$. The Goldfarb-Idnani paper includes a multiplicative factor of $\frac{1}{2}$ in the objective function, which is omitted here to keep the notation uniform.

From Proposition 12.1, the KKT conditions for a primal optimal ρ^* and dual optimal λ^* are

$$\begin{aligned} &\mathcal{N}\rho^* \geq \mathbf{1}, \quad \lambda^* \geq 0, \\ &(\lambda^*)^T(\mathbf{1} - \mathcal{N}\rho^*) = 0, \\ &2\Sigma\rho^* - \mathcal{N}^T\lambda^* = 0. \end{aligned}$$

The iterative scheme

The Goldfarb-Idnani algorithm keeps track of the set of active constraints, $\mathcal{N}_0$ (a subset of rows of $\mathcal{N}$) and the corresponding values of the Lagrangian dual variables, λ_0, for the current set of active constraints. This is essentially equivalent to keeping track of the subfamily Γ' in the basic algorithm. A key difference is that the Goldfarb-Idnani algorithm keeps track of a set of *active* constraints. Unlike with the basic algorithm, in which the subfamily Γ' grows monotonically, the Goldfarb-Idnani must at times remove rows from $\mathcal{N}_0$, when it is advantageous to move away from the constraint.

At the end of each iteration of the algorithm, the current choice of dual variables is optimal for the current active constraint set. Thus, from the KKT conditions, we have that

$$\mathcal{N}_0\rho_0 = \mathbf{1}, \quad \text{where} \quad \rho_0 = \frac{1}{2}\Sigma^{-1}\mathcal{N}_0^T\lambda_0.$$

If this choice of dual variables does not correspond to the minimum, it must be because we can find a row n^T of $\mathcal{N}$ with the property that $n^T\rho_0 < 1$. This can be found, for example, by using the same oracle one would use for the basic algorithm. The Goldfarb-Idnani algorithm proceeds by appending n^T to the bottom of $\mathcal{N}_0$ and modifying the dual variables accordingly.

More generally, as a single iteration progresses, we will track an augmented matrix

$$\mathcal{N}_1 = \begin{bmatrix} \tilde{\mathcal{N}} \\ \\ n^T \end{bmatrix}$$

and a corresponding set of dual variables

$$\lambda_1 = \begin{bmatrix} \tilde{\lambda} \\ \\ \tau \end{bmatrix}$$

maintaining the property that

$$\tilde{\mathcal{N}}\rho_1 = \mathbf{1}, \quad \text{where} \quad \rho_1 = \frac{1}{2}\Sigma^{-1}\mathcal{N}_1^T\lambda_1.$$

When the new constraint is first discovered, $\tilde{\mathcal{N}} = \mathcal{N}_0$, $\tilde{\lambda} = \lambda_0$ and $\tau = 0$. The goal is to update the dual variables (and the active set, if necessary) to ensure that the new constraint is satisfied. Ultimately this involves adding n^T to the constraint set, but it may also require the removal of some of the rows of $\mathcal{N}_0$ along the way.

Consider the parameterized dual vector

$$\lambda_t = \begin{bmatrix} \tilde{\lambda} - tr \\ \\ \tau + t \end{bmatrix},$$

where r is a vector to be determined shortly. The corresponding density vector is

$$\begin{aligned} \rho_t &= \frac{1}{2}\Sigma^{-1}\mathcal{N}_1^T\lambda_t \\ &= \frac{1}{2}\Sigma^{-1}\mathcal{N}_1^T\lambda_1 + \frac{t}{2}(\Sigma^{-1}n - \Sigma^{-1}\tilde{\mathcal{N}}^T r) \\ &= \rho_1 + \frac{t}{2}(\Sigma^{-1}n - \Sigma^{-1}\tilde{\mathcal{N}}^T r). \end{aligned}$$

So,

$$\tilde{\mathcal{N}}\rho_t = \mathbf{1} + \frac{t}{2}(\tilde{\mathcal{N}}\Sigma^{-1}n - \tilde{\mathcal{N}}\Sigma^{-1}\tilde{\mathcal{N}}^T r).$$

If we choose

$$r = \left(\tilde{\mathcal{N}}\Sigma^{-1}\tilde{\mathcal{N}}^T\right)^{-1}\tilde{\mathcal{N}}\Sigma^{-1}n,$$

then changing t won't change the lengths of the objects in $\tilde{\mathcal{N}}$.

Starting from $t = 0$, then, we can increase t, maintaining the current active set of constraints, $\tilde{\mathcal{N}}$, and attempting to satisfy the new constraint n^T. Two things can happen as we do this.

Losing an active constraint. This can happen if one of the entries in $\tilde{\lambda} - tr$ becomes zero. This signals that we must remove the corresponding row from the active set. To do this, we drop the row from $\tilde{\mathcal{N}}$ along with the corresponding dual variable in $\tilde{\lambda}$, and then continue as before.

Satisfying the new constraint. If we avoid losing an active constraint then, necessarily, we will reach a value of t for which the new constraint is satisfied. This value of t can be computed by considering

$$n^T\rho_t = n^T\rho_1 + \frac{t}{2}(n^T\Sigma^{-1}n - n^T\Sigma^{-1}\tilde{\mathcal{N}}^T r).$$

So, the critical value of t is

$$t = \frac{2(1 - n^T\rho_1)}{n^T\Sigma^{-1}n - n^T\Sigma^{-1}\tilde{\mathcal{N}}^T r}.$$

When the new constraint is finally met, a single Goldfarb-Idnani step is complete and we proceed by again searching for an unmet constraint.

An efficient implementation. One of the most expensive parts of the algorithm tends to be solving the linear system

$$\left(\tilde{\mathcal{N}}\Sigma^{-1}\tilde{\mathcal{N}}^T\right) r = \tilde{\mathcal{N}}\Sigma^{-1}n$$

for r. To improve efficiency, one can maintain a Cholesky decomposition

$$\tilde{\mathcal{N}}\Sigma^{-1}\tilde{\mathcal{N}}^T = LL^T.$$

Since L is a triangular matrix, solving the linear system can be performed efficiently using forward and back substitution. Moreover, the two operations performed on the $\tilde{\mathcal{N}}$ matrix (adding or subtracting a row) have efficient update formulas. To simplify notation, we'll drop the tilde and describe how to perform an update on a Cholesky decomposition of the form $\mathcal{N}\Sigma^{-1}\mathcal{N}^T = LL^T$.

Adding a row. First, suppose we wish to augment $\mathcal{N}$ by appending a row, n^T. Note that

$$\begin{bmatrix} \mathcal{N} \\ n^T \end{bmatrix} \Sigma^{-1} \begin{bmatrix} \mathcal{N}^T & n \end{bmatrix} = \begin{bmatrix} \mathcal{N}\Sigma^{-1}\mathcal{N}^T & \mathcal{N}\Sigma^{-1}n \\ n^T\Sigma^{-1}\mathcal{N}^T & n^T\Sigma^{-1}n \end{bmatrix}. \tag{19.1}$$

Consider a factorization of this new matrix of the form

$$\begin{bmatrix} L & 0 \\ \ell^T & t \end{bmatrix} \begin{bmatrix} L^T & \ell \\ 0 & t \end{bmatrix} = \begin{bmatrix} LL^T & L\ell \\ \ell^T L^T & \ell^T\ell + t^2 \end{bmatrix}. \tag{19.2}$$

Comparing the right-hand sides of (19.1) and (19.2) shows that, in order to update the Cholesky factorization, we should choose $\ell = L^{-1}\mathcal{N}\Sigma^{-1}n$ and $t = \sqrt{n^T\Sigma^{-1}n - \ell^T\ell}$. Computing ℓ requires solving a triangular system along with matrix-vector multiplication while computing t requires only vector and scalar operations. Thus, the update operation is much cheaper than computing the entire factorization from scratch.

Removing a row. In order to remove a row from $\mathcal{N}$, consider a block form of the Cholesky factorization

$$\begin{bmatrix} \mathcal{N}_1 \\ n^T \\ \mathcal{N}_2 \end{bmatrix} \Sigma^{-1} \begin{bmatrix} N_1^T & n & N_2^T \end{bmatrix} = \begin{bmatrix} L_{11} & 0 & 0 \\ \ell_{21}^T & t & 0 \\ L_{31} & \ell_{32} & L_{33} \end{bmatrix} \begin{bmatrix} L_{11}^T & \ell_{21} & L_{31}^T \\ 0 & t & \ell_{32}^T \\ 0 & 0 & L_{33}^T \end{bmatrix}.$$

After the row n^T has been removed, we are left with

$$\begin{bmatrix} \mathcal{N}_1 \\ \mathcal{N}_2 \end{bmatrix} \Sigma^{-1} \begin{bmatrix} \mathcal{N}_1^T & \mathcal{N}_2^T \end{bmatrix} = \begin{bmatrix} L_{11} & 0 & 0 \\ L_{31} & \ell_{32} & L_{33} \end{bmatrix} \begin{bmatrix} L_{11}^T & L_{31}^T \\ 0 & \ell_{32}^T \\ 0 & L_{33}^T \end{bmatrix}.$$

The first matrix on the right is almost lower triangular. Each row has at most one nonzero entry to the right of the diagonal. To fix the factorization, we simply need to post-multiply by an orthogonal matrix that removes those excess superdiagonal entries. This orthogonal matrix does not need to be computed explicitly. Instead, we can apply a sequence of Givens rotations that rotate away one superdiagonal entry at a time. When this procedure is completed, the final column will contain only zeros, and can be removed to obtain a square triangular matrix factorization. Since each Givens rotation only affects two columns of L at a time, this procedure of restoring the triangular structure can be done efficiently.

19.3 THE PLUS-ONE ALGORITHM

The Plus-One algorithm was introduced in [29] as a way to compute the spanning tree modulus. In fact, it can be used to find the 2-modulus of any family on an unweighted graph. The algorithm does not converge especially fast; its value lies in its simplicity, as can be seen in Algorithm 4.

Algorithm 4 Approximates $\text{Mod}_2(\Gamma)$ with an error tolerance of $\epsilon_{\text{tol}} > 0$.

1: $w_0 \leftarrow 0$
2: **for** $k = 0, 1, 2, 3, \ldots$ **do**
3:     $\gamma_k \leftarrow \texttt{shortest}_\Gamma(w_k)$ {violated constraint oracle}
4:     **if** $\ell_{w_k}(\gamma_k) \geq (1 - \epsilon_{\text{tol}})k^{-1}\|w_k\|^2$ **then**
5:         `stop`
6:     **end if**
7:     $w_{k+1} \leftarrow w_k + \gamma_k$
8: **end for**

Line 1 initializes an edge weight, $w_0 \in \mathbb{R}^E$. On each pass through the loop, with iteration counter $k = 0, 1, 2, \ldots$, the violated constraint oracle from the basic algorithm is called, using w_k as the edge density. If the loop does not terminate, the weights are updated by adding the object γ (associating the object with its corresponding usage vector). The name "Plus-One" describes what the algorithm does on a family with a $0/1$ usage matrix. In this case, the weights are updated by adding one to each edge used by γ.

The stopping condition on Line 4 is understood to fail when $k = 0$ (the loop repeats at least once). Before proving that the algorithm terminates, let's first understand what the final result tells us, assuming that it does terminate.

Suppose that we reach Line 7 of the algorithm after some number of iterations, and let $\{\gamma_0, \gamma_1, \ldots, \gamma_{k-1}\}$ denote the sequence of spanning trees found thus far by the search on Line 3. Then, at this point,

$$w = w_k := \sum_{r=0}^{k-1} \gamma_r.$$

It is this w that will be used in Line 3 on the next iteration.

Let μ_k be the pmf defined by

$$\mu_k := \frac{1}{k} \sum_{r=0}^{k-1} \delta_{\gamma_r}.$$

That is, μ_k chooses uniformly from among $\{\gamma_0, \gamma_1, \ldots, \gamma_{k-1}\}$, accounting for multiplicity if any objects have been chosen more than once. Using the natural association between the objects and the rows of $\mathcal{N}$, the associated edge usage vector, η_k, satisfies

$$\begin{aligned} \eta_k(e) &:= \sum_{\gamma\in\Gamma} \mu_k(\gamma)\mathcal{N}(\gamma, e) = \frac{1}{k} \sum_{\gamma\in\Gamma} \sum_{r=0}^{k-1} \gamma(e)\delta_{\gamma_r}(\gamma) \\ &= \frac{1}{k} \sum_{r=0}^{k-1} \gamma_r(e) = \frac{w_k(e)}{k}. \end{aligned} \tag{19.3}$$

In other words, the vector w_k/k has the interpretation of being the expected usage vector associated with μ_k.

Theorem 19.2. *Suppose that the Plus-One algorithm terminates (necessarily with $k > 0$). Let $\eta = \eta_k$ be defined as in* (19.3)*. Then*

$$\|\eta\|^{-2} \le \text{Mod}_2(\Gamma) \le (1 - \epsilon_{tol})^{-2} \|\eta\|^{-2}.$$

Proof. See Exercise 19.5 □

All that remains, then, is to verify that the Plus-One algorithm terminates.

Theorem 19.3. *Assuming that Γ is finite, the Plus-One algorithm in Algorithm 4 terminates in a finite number of iterations.*

Proof. See Exercise 19.6. □

19.4 SPECIALIZED ALGORITHMS

Although the general modulus algorithms described above work well in practice, exploiting the specific structure of certain families can yield more efficient specialized approaches. An example is the algorithm for computing 2-modulus of spanning trees [8, 50] based on a matroid algorithm of Cunningham. This algorithm not only computes spanning tree modulus efficiently, but it can also be implemented in exact arithmetic (as opposed to floating point arithmetic), finding the modulus and extremal density without numerical round-off errors. Developing and improving modulus algorithms is an area of ongoing research and it seems likely that other efficient specialized algorithms exist for specific families.

EXERCISES

19.1 *Basic algorithm on the paw.* Consider the Paw graph of Exercise 13.1 and let $\Gamma = \Gamma_{\text{path}}(0,3)$ bet the connecting family of paths from 0 to 3. Let $1 < p < \infty$ be arbitrary and show the steps of the basic algorithm by hand. Remember, in the oracle, ties can be broken arbitrarily. Verify that the solution you obtain does not depend on the choice of the shortest object when there are multiple shortest objects.

19.2 *Loop invariants for the basic algorithm.* Consider the loop invariants I1–I2 on page 361. Show that these invariants are true immediately after the initialization step (the first two lines of Algorithm 3). Then show that if the invariants are true at the beginning of any pass through the loop, they are also true at the end of that pass. This shows that, as long as the algorithm terminates, it will terminate with the loop invariants satisfied.

19.3 *Proof of Theorem 19.1.* Prove the theorem. You may want to use the following proof outline. (Focus on finite $1 \leq p < \infty$ first, then make some small modifications for the $p = \infty$ case.

- Prove that the algorithm terminates after a finite number of steps. (Hint: Remember that Γ is finite. What can you say about $|\Gamma'|$ after each pass through the loop?)
- Use the loop invariants in Exercise 19.2 to prove that

$$\mathcal{E}_{p,\sigma}(\rho) = \text{Mod}_{p,\sigma}(\Gamma') \leq \text{Mod}_{p,\sigma}(\Gamma).$$

(Hint: You might find Proposition 10.4 helpful.)

- Use the stopping criterion to prove part (a) of the theorem.
- Use the above results to show that

$$\operatorname{Mod}_{p,\sigma}(\Gamma) \le (1-\epsilon_{\text{tol}})^{-p}\operatorname{Mod}_{p,\sigma}(\Gamma').$$

- Now you should be able to argue that

$$0 \le \frac{\operatorname{Mod}_{p,\sigma}(\Gamma) - \operatorname{Mod}_{p,\sigma}(\Gamma')}{\operatorname{Mod}_{p,\sigma}(\Gamma)} \le 1-(1-\epsilon_{\text{tol}})^p.$$

To complete the proof from here, it may help to recall that $x \mapsto (1-x)^p$ is a convex function for $x \le 1$.

19.4 *Plus-One algorithm on the paw.* Repeat Exercise 19.1 but using the Plus-One algorithm (with $p = 2$). Show at least ten iterations. Describe how w_k behaves as $k \to \infty$.

19.5 *Proof of Theorem 19.2.* To prove the theorem, define $\mu = \mu_k$, $w = w_k$ and $\eta = \eta_k$ as in Section 19.3. Now define $\rho = \eta/\|\eta\|^2$ and show that the stopping condition of Algorithm 4 implies that $\ell_\rho(\gamma) \ge 1-\epsilon_{\text{tol}}$ for all $\gamma \in \Gamma$. In other words, ρ is almost admissible. This is similar to the argument in Exercise 19.3. Use an analogous argument to complete the proof of Theorem 19.2.

19.6 *Proof of Theorem 19.3.* The theorem can be proved in two steps; first show that $\eta_k \to \eta^*$ and then show that this implies that the algorithm terminates.

For the first part, start with the observation that

$$\eta_k = \frac{k-1}{k}\eta_{k-1} + \frac{1}{k}\gamma_{k-1},$$

which means that

$$\|\eta_k - \eta^*\|^2 = \left\| \frac{k-1}{k}(\eta_{k-1}-\eta^*) + \frac{1}{k}(\gamma_{k-1}-\eta^*)\right\|^2.$$

The key estimate is in showing that the inner product in the cross term,

$$(\eta_{k-1}-\eta^*)^T(\gamma_{k-1}-\eta^*) = \left(\|\eta^*\|^2 - \gamma_{k-1}^T\eta^*\right) + \left(\eta_{k-1}^T\gamma_{k-1} - \eta_{k-1}^T\eta^*\right)$$

is nonpositive. For the first term in parentheses nonpositivity is related to the fact that $\ell_{\rho^*}(\gamma_{k-1}) \geq 1$. For the second term, it is related to the fact that γ_{k-1} minimizes η_{k-1}-length. Nonpositivity of the inner product implies that

$$\|\eta_k - \eta^*\|^2 \leq \left(\frac{k-1}{k}\right)^2 \|\eta_{k-1} - \eta^*\|^2 + \frac{1}{k^2}\|\gamma_{k-1} - \eta^*\|^2.$$

By induction, this can be shown to imply that

$$\|\eta_k - \eta^*\|^2 \leq \frac{1}{k^2}\|\eta_1 - \eta^*\|^2 + \frac{C(k-1)}{k^2},$$

for some constant C depending on the usage matrix $\mathcal{N}$. This shows that $\|\eta_k - \eta^*\| = O(k^{-\frac{1}{2}})$ as $k \to \infty$.

To show that the algorithm terminates, you must show that the stopping condition is eventually satisfied. To start, write

$$\ell_\eta(\gamma) = \eta^T\gamma = (\eta - \eta^*)^T\gamma + (\eta^*)^T\gamma \geq (\eta - \eta^T)\gamma + \|\eta^*\|^2.$$

Apply the Cauchy-Schwarz inequality to the first term to show that the stopping condition is eventually satisfied.

Bibliography

[1] Agrawal, A., Verschueren, R., Diamond, S., and Boyd, S. A rewriting system for convex optimization problems. *Journal of Control and Decision 5*, 1 (2018), 42–60.

[2] Ahlfors, L. V. *Conformal invariants.* AMS Chelsea Publishing, Providence, RI, 2010. Topics in geometric function theory, Reprint of the 1973 original, With a foreword by Peter Duren, F. W. Gehring and Brad Osgood.

[3] Albin, N., Brunner, M., Perez, R., Poggi-Corradini, P., and Wiens, N. Modulus on graphs as a generalization of standard graph theoretic quantities. *Conform. Geom. Dyn. 19* (2015), 298–317. http://dx.doi.org/10.1090/ecgd/287.

[4] Albin, N., Clemens, J., Fernando, N., and Poggi-Corradini, P. Blocking duality for *p*-modulus on networks and applications. *Ann. Mat. Pura Appl. (4) 198*, 3 (2019), 973–999. https://doi-org.er.lib.k-state.edu/10.1007/s10231-018-0806-0.

[5] Albin, N., Clemens, J., Hoare, D., Poggi-Corradini, P., Sit, B., and Tymochko, S. Fairest edge usage and minimum expected overlap for random spanning trees. *Discrete Math. 344*, 5 (2021), 112282.

[6] Albin, N., Fernando, N., and Poggi-Corradini, P. Modulus metrics on networks. *Discrete Contin. Dyn. Syst. Ser. B 24*, 1 (2019), 1–17.

[7] Albin, N., Kottegoda, K., and Poggi-Corradini, P. A polynomial-time algorithm for spanning tree modulus. https://arxiv.org/pdf/2009.03736.pdf.

[8] Albin, N., Kottegoda, K., and Poggi-Corradini, P. An exact-arithmetic algorithm for spanning tree modulus, 2024.

[9] Albin, N., Lind, J., Pankka, P., and Poggi-Corradini, P. Modulus is always one. Preprint.

[10] Albin, N., Lind, J., and Poggi-Corradini, P. Convergence of the probabilistic interpretation of modulus. https://arxiv.org/pdf/2106.11418.pdf.

[11] Albin, N., Lind, J., and Poggi-Corradini, P. Minimizing the determinant of the graph Laplacian.

[12] Albin, N., and Poggi-Corradini, P. The Modulus Book. `https://nathan-albin.github.io/modulus_book`. Accessed: 2024-07-23.

[13] Albin, N., and Poggi-Corradini, P. Minimal subfamilies and the probabilistic interpretation for modulus on graphs. *The Journal of Analysis* (2016), 1–26. http://dx.doi.org/10.1007/s41478-016-0002-9.

[14] Albin, N., Poggi-Corradini, P., Darabi Sahneh, F., and Goering, M. Modulus of families of walks on graphs. In *Proceedings of Complex Analysis and Dynamical Systems VII* (2017), M. L. Agranovsky, M. Ben-Artzi, C. Bénéteau, L. Karp, D. Khavinson, S. Reich, D. Shoikhet, G. Weinstein, and L. Zalcman, Eds., vol. 699 of *Contemporary Mathematics*, pp. 35–56. arXiv:1401.7640.

[15] Alon, N., and Yuster, R. Threshold functions for H-factors. *Combin. Probab. Comput. 2*, 2 (1993), 137–144.

[16] ApS, M. Mosek optimizer api for python 10.2.1. `https://docs.mosek.com/latest/pythonapi/index.html`, 2024.

[17] Barvinok, A. *A course in convexity*, vol. 54 of *Graduate Studies in Mathematics*. American Mathematical Society, Providence, RI, 2002.

[18] Bertsimas, D., and Tsitsiklis, J. N. *Introduction to linear optimization*, vol. 6. Athena scientific Belmont, MA, 1997.

[19] Bollobás, B. *Modern graph theory*, vol. 184 of *Graduate Texts in Mathematics*. Springer-Verlag, New York, 1998.

[20] Borradaile, G., Iglesias, J., Migler, T., Ochoa, A., Wilfong, G., and Zhang, L. Egalitarian graph orientations. *J. Graph Algorithms Appl. 21*, 4 (2017), 687–708.

[21] Borradaile, G., Migler, T., and Wilfong, G. Density decompositions of networks. *J. Graph Algorithms Appl. 23*, 4 (2019), 625–651.

[22] Böttcher, J. Large-scale structures in random graphs. In *Surveys in combinatorics 2017*, vol. 440 of *London Math. Soc. Lecture Note Ser.* Cambridge Univ. Press, Cambridge, 2017, pp. 87–140.

[23] Bourdon, M., and Kleiner, B. Combinatorial modulus, the combinatorial Loewner property, and Coxeter groups. *Groups Geom. Dyn. 7*, 1 (2013), 39–107.

[24] Boyd, S., and Vandenberghe, L. *Convex optimization.* Cambridge University Press, Cambridge, 2004.

[25] Cannon, J. W., Floyd, W. J., and Parry, W. R. Squaring rectangles: the finite Riemann mapping theorem. In *The mathematical legacy of Wilhelm Magnus: groups, geometry and special functions (Brooklyn, NY, 1992)*, vol. 169 of *Contemp. Math.* Amer. Math. Soc., Providence, RI, 1994, pp. 133–212.

[26] Chand, A. PhD thesis, Kansas State University.

[27] Chopra, S. On the spanning tree polyhedron. *Operations Research Letters 8*, 1 (1989), 25–29.

[28] Chung, F. R. K. *Spectral graph theory*, vol. 92 of *CBMS Regional Conference Series in Mathematics.* Published for the Conference Board of the Mathematical Sciences, Washington, DC; by the American Mathematical Society, Providence, RI, 1997.

[29] Clemens, J. *Spanning tree modulus: deflation and hierarchical graph structure.* PhD thesis, Kansas State University, 2018.

[30] Cornuéjols, G. *Combinatorial optimization*, vol. 74 of *CBMS-NSF Regional Conference Series in Applied Mathematics.* Society for Industrial and Applied Mathematics (SIAM), Philadelphia, PA, 2001. Packing and covering.

[31] Demmel, J. W. *Applied numerical linear algebra.* Society for Industrial and Applied Mathematics (SIAM), Philadelphia, PA, 1997.

[32] Devriendt, K., and Lambiotte, R. Discrete curvature on graphs from the effective resistance*. *J. Phys.: Complex. 3*, 2 (jun 2022), 025008.

[33] DIAMOND, S., AND BOYD, S. CVXPY: A Python-embedded modeling language for convex optimization. *J. Mach. Learn. Res. 17*, 83 (2016), 1–5.

[34] DIESTEL, R. *Graph theory*, fifth ed., vol. 173 of *Graduate Texts in Mathematics*. Springer, Berlin, 2018. Paperback edition of [MR3644391].

[35] DRISCOLL, T. Schwarz–Christoffel Toolbox for conformal mapping in MATLAB. `https://github.com/tobydriscoll/sc-toolbox`, Aug. 2021.

[36] DUFFIN, R. The extremal length of a network. *J. Math. Anal. App. 5*, 2 (1962), 200 – 215.

[37] DURRETT, R. *Probability: theory and examples*, fourth ed., vol. 31 of *Cambridge Series in Statistical and Probabilistic Mathematics*. Cambridge University Press, Cambridge, 2010.

[38] ERIKSSON-BIQUE, S., AND POGGI-CORRADINI, P. On the sharp lower bound for duality of modulus. *Proc. Amer. Math. Soc. 150*, 7 (2022), 2955–2968.

[39] FERNANDO, N. *New Metrics on Networks Arising From Modulus and Applications of Fulkerson Duality.* PhD thesis, Kansas State University, 2018.

[40] FOLLAND, G. B. *Real analysis.* Pure and Applied Mathematics (New York). John Wiley & Sons, Inc., New York, 1984. Modern techniques and their applications, A Wiley-Interscience Publication.

[41] FULKERSON, D. Blocking polyhedra. Tech. rep., DTIC Document, 1968.

[42] GHOSH, A., BOYD, S., AND SABERI, A. Minimizing effective resistance of a graph. *SIAM Rev. 50*, 1 (2008), 37–66.

[43] GOERING, M., ALBIN, N., SAHNEH, F., SCOGLIO, C., AND POGGI-CORRADINI, P. Numerical investigation of metrics for epidemic processes on graphs. 1317–1322. 2015 49th Asilomar Conference on Signals, Systems and Computers.

[44] GOLDFARB, D., AND IDNANI, A. A numerically stable dual method for solving strictly convex quadratic programs. *Math. Program. 27*, 1 (Sept. 1983), 1–33.

[45] GRIMMETT, G. *Probability on graphs*, vol. 1 of *Institute of Mathematical Statistics Textbooks.* Cambridge University Press, Cambridge, 2010. Random processes on graphs and lattices.

[46] GRIMMETT, G., AND WELSH, D. *Probability—an introduction*, second ed. Oxford University Press, Oxford, 2014.

[47] GUREL-GUREVICH, O., JERISON, D. C., AND NACHMIAS, A. The Dirichlet problem for orthodiagonal maps. *Adv. Math. 374* (2020), 107379, 53.

[48] HAÏSSINSKY, P. Empilements de cercles et modules combinatoires. *Annales de l'Institut Fourier 59*, no. 6 (2009), 2175–2222. Version revisée et corrigée.

[49] HEINONEN, J. *Lectures on analysis on metric spaces.* Universitext. Springer-Verlag, New York, 2001.

[50] KOTTEGODA, K. *Spanning tree modulus and secure broadcast games.* PhD thesis, Kansas State University, 2020.

[51] KWAPISZ, J. Conformal dimension via p-resistance: Sierpiński carpet. *Ann. Acad. Sci. Fenn. Math. 45*, 1 (2020), 3–51.

[52] LAWLER, G. F. A self-avoiding random walk. *Duke Math. J. 47*, 3 (1980), 655–693.

[53] LAWLER, G. F. The Laplacian-b random walk and the Schramm-Loewner evolution. *Illinois J. Math. 50*, 1-4 (2006), 701–746.

[54] LAWLER, G. F. Topics in loop measures and the loop-erased walk. *Probab. Surv. 15* (2018), 28–101.

[55] LAWLER, G. F. *Random explorations*, vol. 98 of *Student Mathematical Library.* American Mathematical Society, Providence, RI, [2022] ©2022.

[56] LEVIN, D. A., PERES, Y., AND WILMER, E. L. *Markov chains and mixing times.* American Mathematical Society, Providence, RI, 2009. With a chapter by James G. Propp and David B. Wilson.

[57] LINDQVIST, P. *Notes on the stationary p-Laplace equation.* SpringerBriefs in Mathematics. Springer, Cham, 2019.

[58] LOVÁSZ, L. *An algorithmic theory of numbers, graphs and convexity*, vol. 50 of *CBMS-NSF Regional Conference Series in Applied Mathematics*. Society for Industrial and Applied Mathematics (SIAM), Philadelphia, PA, 1986.

[59] LOVÁSZ, L. Energy of convex sets, shortest paths, and resistance. *J. Combin. Theory Ser. A 94*, 2 (2001), 363–382.

[60] LYONS, R., AND OVEIS GHARAN, S. Sharp bounds on random walk eigenvalues via spectral embedding. *Int. Math. Res. Not. IMRN*, 24 (2018), 7555–7605.

[61] LYONS, R., PEMANTLE, R., AND PERES, Y. Resistance bounds for first-passage percolation and maximum flow. *J. Comb. Theory Ser. A 86*, 1 (Apr. 1999), 158–168.

[62] LYONS, R., AND PERES, Y. *Probability on Trees and Networks*. Cambridge University Press, 2016. http://pages.iu.edu/ rdlyons/.

[63] MATOUŠEK, J. *Lectures on discrete geometry*, vol. 212 of *Graduate Texts in Mathematics*. Springer-Verlag, New York, 2002.

[64] MELIKYAN, A. *Random spanning trees on homogeneous graphs*. PhD thesis, Kansas State University, December 2019.

[65] NAKAMURA, T., AND YAMASAKI, M. Generalized extremal length of an infinite network. *Hiroshima Math. J. 6*, 1 (1976), 95–111.

[66] NAKAMURA, T., AND YAMASAKI, M. Extremal length of an infinite network which is not necessarily locally finite. *Hiroshima Math. J. 7*, 3 (1977), 813–826.

[67] ORTIZ-AQUINO, A. *Modulus of Edge Covers, Fractional Edge Covers, and Stars*. PhD thesis, Kansas State University, 2024.

[68] RICKMAN, S. *Quasiregular mappings*, vol. 26 of *Ergebnisse der Mathematik und ihrer Grenzgebiete (3) [Results in Mathematics and Related Areas (3)]*. Springer-Verlag, Berlin, 1993.

[69] ROCKAFELLAR, R. T. *Convex analysis*. Princeton Landmarks in Mathematics. Princeton University Press, Princeton, NJ, 1997. Reprint of the 1970 original, Princeton Paperbacks.

[70] RUDIN, W. *Principles of mathematical analysis*, third ed. McGraw-Hill Book Co., New York-Auckland-Düsseldorf, 1976. International Series in Pure and Applied Mathematics.

[71] SAEKI, S. Handwritten class notes.

[72] SCHRAMM, O. Square tilings with prescribed combinatorics. *Isr. J. Math. 84*, 1-2 (1993), 97–118.

[73] SHEWCHUK, J. R. Triangle: Engineering a 2D Quality Mesh Generator and Delaunay Triangulator. In *Applied Computational Geometry: Towards Geometric Engineering*, M. C. Lin and D. Manocha, Eds., vol. 1148 of *Lecture Notes in Computer Science*. Springer-Verlag, May 1996, pp. 203–222. From the First ACM Workshop on Applied Computational Geometry.

[74] SKOPENKOV, M. The boundary value problem for discrete analytic functions. *Adv. Math. 240* (2013), 61–87.

[75] SPIELMAN, D. http://cs-www.cs.yale.edu/homes/spielman/462/ps4-13.pdf.

[76] SPIELMAN, D. *Spectral and Algebraic Graph Theory.* In progress. http://cs-www.cs.yale.edu/homes/spielman/sagt/sagt.pdf.

[77] SPIELMAN, D. A. Shortest paths. http://www.cs.yale.edu/homes/spielman/365/shortestPaths.pdf.

[78] SPIELMAN, D. A., AND TENG, S.-H. Nearly-linear time algorithms for graph partitioning, graph sparsification, and solving linear systems. In *Proceedings of the 36th Annual ACM Symposium on Theory of Computing* (2004), ACM, New York, pp. 81–90 (electronic).

[79] TETALI, P. Random walks and the effective resistance of networks. *J. Theoret. Probab. 4*, 1 (1991), 101–109.

[80] TRUONG, H. PhD thesis, Kansas State University.

[81] TRUONG, H., AND POGGI-CORRADINI, P. Fulkerson duality for modulus of spanning trees and partitions. https://arxiv.org/pdf/2306.15984.pdf.

[82] VAN MIEGHEM, P. *Graph spectra for complex networks.* Cambridge University Press, Cambridge, 2011.

Index

Made in the USA
Las Vegas, NV
19 September 2026

45952ffc-ecba-40e7-9618-c94179cf62d4R01